Lecture Notes in Mathematics 1944

Alfonso Di Bartolo · Giovanni Falcone
Peter Plaumann · Karl Strambach

Algebraic Groups
and Lie Groups
with Few Factors

 Springer

Authors

Alfonso Di Bartolo
Dipartimento di Matematica e Applicazioni
Università degli Studi di Palermo
via Archirafi 34
90123 Palermo
Italy
alfonso@math.unipa.it

Giovanni Falcone
Dipartimento di Metodi
  e Modelli Matematici
Università degli Studi di Palermo
Viale delle Scienze Ed. 8
90138 Palermo
Italy
gfalcone@unipa.it

Peter Plaumann
Karl Strambach
Mathematisches Institut
Bismarckstrasse 1 1/2
91054 Erlangen
Germany
plaumann@mi.uni-erlangen.de
strambach@mi.uni-erlangen.de

ISBN: 978-3-540-78583-5   e-ISBN: 978-3-540-78584-2
DOI: 10.1007/978-3-540-78584-2

Lecture Notes in Mathematics ISSN print edition: 0075-8434
                             ISSN electronic edition: 1617-9692

Library of Congress Control Number: 2008921922

Mathematics Subject Classification (2000): 20G10, 14L10, 22E25, 17B30, 20E15

Cover design: WMXDesign GmbH

Printed on acid-free paper

9 8 7 6 5 4 3 2 1

springer.com

Preface

In the theory of locally compact topological groups, the aspects and notions from abstract group theory have conquered a meaningful place from the beginning (see New Bibliography in [44] and, e.g. [41–43]). Imposing group-theoretical conditions on the closed connected subgroups of a topological group has always been the way to develop the theory of locally compact groups along the lines of the theory of abstract groups.

Despite the fact that the class of algebraic groups has become a classical object in the mathematics of the last decades, most of the attention was concentrated on reductive algebraic groups. For an affine connected solvable algebraic group G, the theorem of Lie–Kolchin has been considered as definitive for the structure of G, whereas for connected non-affine groups, the attention turns to the analytic and homological aspects of these groups, which are quasi-projective varieties (cf. [79, 80, 89]). Complex Lie groups and algebraic groups as linear groups are an old theme of group theory, but connectedness of subgroups does not play a crucial rôle in this approach, as can be seen in [97]. Non-linear complex commutative Lie groups are a main subject of complex analysis (cf. [1, 7]).

In these notes we want to include systematically algebraic groups, as well as real and complex Lie groups, in the frame of our investigation. Although affine algebraic groups over fields of characteristic zero are related to linear Lie groups (cf. [11–13]), the theorems depending on the group topology differ (cf. e.g. Remark 5.3.6). For algebraic groups we want to stress the differences between algebraic groups over a field of characteristic $p > 0$ and over fields of characteristic zero.

One essential task of group theory is the description of a given group by its composition of more elementary groups. There are two kinds of most elementary Lie groups and algebraic groups. One class is formed by such groups that have a dense cyclic subgroup; such groups are commutative. In the class of locally compact groups these groups are determined in [15], in the class of algebraic groups (over a not necessarily algebraically closed field) they are classified in [24]. The other kind of elementary groups are those which have a chain as lattice of their subgroups. In the class of finite groups they are cyclic groups of prime power order. In Lie groups and in algebraic groups the lattice of closed connected subgroups is a chain precisely in the following cases. If such groups do not have dimension one then in the class of Lie groups they are Shafarevich extensions of simple complex tori (cf. [7], Chapter 1, Section 6). In the class of algebraic groups over fields of characteristic zero they are either simple abelian varieties or extensions of a one-dimensional affine group by a simple abelian variety (cf. Theorem 4.1.3). In the class of algebraic groups over fields of positive characteristic the situation is much more complicated. Besides simple abelian varieties and extensions of a one-dimensional torus by a simple abelian variety there are also affine algebraic groups having a chain as lattice of connected closed subgroups. If they are commutative then they are Witt groups (cf. [89], Chapter 7, Sections 8 and 10), if they are not commutative then they form a very rich family of unipotent groups as our work shows.

Already J. Dieudonné was interested in groups having a chain as their lattice of closed connected subgroups; for such groups we introduce in this book the term *chain*. Namely, in [19], Section 7, he deals with non-commutative two-dimensional groups of this type and remarks that they are counter-examples to conjectures derived from the case of characteristic zero. In general, one reason for the importance of chains is the fact that a precise knowledge of them is indispensable for group theoretical investigations referring to the lattice of connected subgroups. Since the lattices of connected subgroups of the pre-image and the image of an algebraic or topological epimorphism with finite kernel are isomorphic (see [71], Lemma 1.3, p. 256), we consider groups related in this way as equivalent and use for them the term isogenous. More precisely, we use the notion of isogeny for algebraic and topological groups in the sense of [77], p. 417 (see also Section 2.1). This aspect motivated us to open, up to isogeny, the door to the exotic world of algebraic non-commutative chains which consists of unipotent chains since a reduction of algebraic chains to unipotent groups can be easily achieved. Even though over fields of positive characteristic not every connected algebraic group is generated by chains (see Remark 4.1.5), our work documents that they are the fundamental ingredients of unipotent groups over fields of positive characteristic.

Already the unipotent chains of nilpotency class two are difficult to treat. Namely, for a complete classification of them one needs a classification of all non-commutative extensions G of n-dimensional Witt groups by m-dimensional Witt groups such that G has an n-dimensional commutator subgroup. Despite these great obstacles, using the classification of two-dimensional unipotent groups given in [18], II, § 3, 4.6, p. 197, we could concretely determine (through hard and involved computations) all unipotent chains of dimension three over perfect fields of characteristic greater than two. In particular we obtain that any three-dimensional unipotent non-commutative chain over a perfect field of characteristic greater than two has nilpotency class two, its commutator subgroup has dimension one and the center is two-dimensional. An induction yields that for any unipotent chain over a perfect field of characteristic greater than two, the commutator sub-group has co-dimension at least two and the center has dimension at least two (cf. Corollary 4.2.10). Moreover, the knowledge of three-dimensional chains allows us to classify, up to isogenies, all three-dimensional unipotent groups over perfect fields of characteristic greater than two (cf. Section 6). These results demonstrate the richness of examples in dimension three compared with dimension two. The plethora of unipotent k-groups over non-perfect fields and of dimension less or equal two (see [51]) justifies our restriction to algebraic groups over perfect fields.

Using the regular factor systems determining Witt groups as extensions of Witt groups by Witt groups (cf. [18], V, § 1, 1.4, p. 542 or [102]) and [18], II, § 3, 4.6, p. 197, we obtain a classification of unipotent chains over perfect fields of positive characteristic having a one-dimensional commutator subgroup (cf. Theorem 4.3.1).

As Remark 4.1.6 and Example 4.3.4 show, there are in any dimension unipotent chains of nilpotency class two having two-dimensional and three-dimensional commutator subgroup. But the involved structure of these extensions gives no hope that a complete classification of unipotent chains of nilpotency class two having a commutator subgroup of dimension greater than one could be achieved.

The classification of chains having a one-dimensional commutator subgroup yields a classification of connected algebraic groups G over perfect fields of characteristic $p > 2$ such that G has a central subgroup of co-dimension one. These groups have a representation as an almost direct product of a commutative group and a group which is a direct group of chains with amalgamated factor group (cf. Theorem 4.3.12). Moreover, we prove that in an algebraic group G having a central maximal connected subgroup the commutator subgroup G' is a (central) vector group. Conversely, if G' is a central vector group and $G/{}_3G$ is isogenous to a Witt group, then the center ${}_3G$ has co-dimension one in G (cf. Theorem 3.2.8). In contrast to this, a non-commutative algebraic group over a field of characteristic zero cannot have its center of co-dimension one.

Our investigations on chains G with one-dimensional commutator subgroup G' yield conditions under which an automorphism of the factor group G/G' can be extended to an automorphism of G. Using these results we can illustrate that the non-commutative chains are much more rigid than Witt groups. Namely, any connected algebraic group of algebraic automorphisms of a non-commutative unipotent chain of dimension greater than two is unipotent (cf. Corollary 4.3.11).

The lattice of normal connected algebraic subgroups of unipotent algebraic groups for which the nilpotency class is equal to their dimension n forms a chain of length n (cf. Proposition 3.1.13). Such unipotent groups occur only over fields of positive characteristic and play an opposite rôle to the one of chains which cannot have maximal nilpotency class (see Corollary 4.2.10). In Section 3.1 we introduce for any n a significant class of unipotent groups

$\mathfrak{J}_n(\alpha)$ of dimension n and nilpotency class n, characterise these groups as linear groups and study their structure. These groups show that a group with maximal nilpotency class can have a trivial adjoint representation (cf. Remark 3.1.6). At various places we use the groups $\mathfrak{J}_n(\alpha)$ as a source of counter-examples to find the limits of our theorems.

The nilpotency class of n-dimensional algebraic groups over fields of characteristic zero as well as n-dimensional real or complex Lie groups is at most $n - 1$. The Lie algebras corresponding to these groups of maximal nilpotency class are called *filiform Lie algebras* and form a class thoroughly studied for thirty years (cf. [36, 37]). The filiform groups, i.e. the groups having filiform Lie algebras, play in our results on groups in characteristic zero the same rôle as the unipotent chains in positive characteristic.

The simple structure of the lattice of connected subgroups of an algebraic or analytic chain motivated us to study to which extent individual properties of chains restrict the structure of algebraic and analytic groups. Most of these properties remain invariant under isogenies.

In Section 5.2 we investigate connected algebraic groups and connected Lie groups having exactly one maximal connected closed subgroup (*uni-maximal* groups) as well as connected algebraic groups and connected Lie groups having exactly one minimal connected closed subgroup (*uni-minimal* groups). The description of non-affine algebraic groups, respectively complex Lie groups, which are uni-minimal or uni-maximal easily reduces to extensions of affine groups of dimension at most one by abelian varieties, respectively to toroidal groups (cf. Theorem 5.2.7 and Proposition 5.2.4). Connected affine algebraic groups which are uni-minimal or uni-maximal and have dimension greater than one are unipotent algebraic groups over fields of positive characteristic (cf. Proposition 5.2.6).

A non-commutative connected unipotent algebraic group G is uni-maximal if and only if the commutator subgroup of every proper connected algebraic subgroup of G is smaller than the commutator subgroup of G (cf. Theorem 5.2.17). Any group in which the commutator subgroup is a maximal connected subgroup is uni-maximal; in particular the unipotent algebraic groups over fields of positive characteristic having maximal nilpotency class are of such type (cf. Section 3). But we construct in Remark 3.1.12 also numerous examples of uni-maximal algebraic groups in which the commutator subgroup is not maximal. Moreover, any non-commutative connected three-dimensional unipotent algebraic group over a field of positive characteristic which is not a product of two non-commutative chains is uni-maximal (see Theorem 6.4.5).

Uni-minimal non-commutative groups G turn out to be products of chains, where at most one factor C has dimension greater than two; if C is not commutative then the commutator subgroup of G coincides with the commutator subgroup of C (cf. Theorem 5.2.34). This result shows that the structure of uni-minimal groups is less complicated than the structure of uni-maximal groups.

In Corollary 5.2.35 we prove that the conditions to be uni-minimal and uni-maximal are strong enough to characterise the chains over fields of characteristic greater than two. Also the condition that in algebraic groups over fields of characteristic greater than two every proper algebraic subgroup is a chain characterises the chains up to two exceptions of small dimension (cf. Theorem 5.2.30). Moreover, a connected affine algebraic group over a field of arbitrary prime characteristic, containing a chain M as a maximal connected algebraic subgroup, is either a chain or a product of M with a chain of dimension at most two (cf. Theorem 5.2.40).

In chains with a one-dimensional commutator subgroup, any connected algebraic subgroup as well as any proper epimorphic image is commutative. In general however, for algebraic groups over fields of positive characteristic none of these two conditions is sufficient for a concrete description (cf. Corollary 5.2.23 and Proposition 5.2.38). In contrast to this, for real or complex Lie groups, for formal groups and for algebraic groups over fields of characteristic zero the assumption of commutativity of all proper connected subgroups as well as the dual condition of commutativity of all proper epimorphic images is strong enough for a classification. A powerful tool to achieve this goal is the classification of Lie algebras with one of these two properties. The Lie algebras in which every subalgebra is commutative have been studied thoroughly for thirty years (cf. [21, 30–32]). If G is a non-commutative connected affine algebraic group over a field of characteristic zero such that any connected algebraic subgroup is commutative, then G is at most three-dimensional (cf. Proposition 5.3.4). For formal groups we find an analogous situation (cf. Proposition 5.3.5). In contrast to this there exist real and complex Lie groups of any dimension having only commutative proper connected subgroups, they are precisely the extra-special real or complex Lie groups (see Remark 5.3.6). A connected non-simple non-commutative affine algebraic group of dimension at least three over a field of characteristic zero such that every epimorphic image of G is commutative is a Heisenberg group (cf. Corollary 5.3.9 and Proposition 5.3.11). A connected real or complex non-simple non-commutative Lie group of dimension greater than three having only commutative proper epimorphic images is an extra-special complex Lie group having as center a simple complex torus of dimension at least two.

An affine chain of dimension n has exactly one connected algebraic subgroup for any dimension $d \leq n$ and any two epimorphic images of the same dimension are isogenous. Investigating these two properties for connected algebraic groups, respectively for real or complex Lie groups, we call any such group *aligned* if any two proper connected closed subgroups of the same dimension are isomorphic, respectively *co-aligned* if all epimorphic images of the same dimension are isogenous. For algebraic groups over fields of positive characteristic the properties to be aligned and co-aligned are too weak to obtain a concrete description for such groups (see Section 5.7 and Theorem 6.4.9). Also for connected algebraic groups over fields k of characteristic zero and for real or complex Lie groups, the condition to be co-aligned alone is not strong enough to obtain a reasonable description for groups having this property. In particular there is a rich family of nilpotent co-aligned algebraic k-groups as well as of nilpotent co-aligned real or complex Lie groups (see Remark 5.5.5). Only if we assume that these affine groups have nilpotency class two and k is algebraically closed we obtain Heisenberg groups (cf. Proposition 5.5.4).

In the class of solvable non-nilpotent connected affine algebraic groups G of dimension greater than three over algebraically closed fields of characteristic zero such that the unipotent radical of G is commutative, the co-aligned groups are precisely those for which the lattice of connected algebraic subgroups forms a projective geometry (see Theorem 5.5.16 and [71], Lemma 4.7, p. 262). If the unipotent radical U has nilpotency class two and G has dimension greater than four, then the property to be co-aligned characterises the semi-direct products of Heisenberg groups H with a one-dimensional torus T acting on H such that any closed connected subgroup of H is normalized but not centralised by T (see Theorem 5.5.21). However, the filiform groups admitting a non-trivial action of a one-dimensional torus show that a classification of co-aligned solvable non-nilpotent affine algebraic groups with unipotent radical of nilpotency class greater than two is not accessible. The same results hold for solvable non-nilpotent connected linear complex Lie groups of dimension grater than three, respectively four (cf. Theorem 5.5.16 and Theorem 5.5.21).

For co-aligned real Lie groups we meet the same difficulties as for algebraic groups and hence we arrive at a classification only for special subclasses. A connected non-commutative solvable real Lie group G of dimension greater than six having commutative commutator subgroup and containing non-trivial compact elements is co-aligned if and only if G is the direct product of a torus of dimension at most one and a semi-direct product of an even-dimensional vector group V with a one-dimensional torus such that any irreducible subspace of V has dimension two (see Proposition 5.5.23). A connected solvable real Lie group G such that the commutator subgroup G' is not commutative and the factor group G/G' has a non-trivial compact subgroup of dimension ≥ 2 is co-aligned if and only if G is a semi-direct product of two Heisenberg groups with amalgamated centre by a two-dimensional torus (cf. Proposition 5.5.27).

In contrast to the condition to be co-aligned, the property to be aligned is strong. This is documented by the fact that a non-commutative connected affine algebraic group over a field of characteristic zero or a linear complex Lie group is aligned if and only if it is unipotent and has dimension three (see Theorem 5.4.4 and Proposition 5.4.9). Moreover, the classification of the three-dimensional unipotent algebraic groups in Chapter 6 yields that a three-dimensional non-commutative connected unipotent algebraic group G over a perfect field of characteristic $p > 2$ which is aligned is uni-maximal (cf. Theorem 6.4.7). Furthermore, if G is an aligned uni-minimal group then G is a chain (cf. Corollary 6.4.8).

For non-linear complex Lie groups the condition to be aligned creates a situation which is more complicated (cf. Example 5.4.10). However, a

non-commutative connected real Lie group of dimension $n \geq 4$ is aligned
if and only if it is locally isomorphic to one of the following compact Lie
groups: $SO_2(\mathbb{R}) \times SO_3(\mathbb{R})$, $SO_3(\mathbb{R}) \times SO_3(\mathbb{R})$, $SU_3(\mathbb{C}, 0)$, $SO_5(\mathbb{R})$ and the
14-dimensional exceptional Lie group G_2 (cf. Theorem 5.4.8).

In Section 5.7 we characterise chains by the fact that they have only few
non-isogenous factors. Namely a connected unipotent algebraic group is a
chain if and only if it has only finitely many connected algebraic subgroups
(cf. Theorem 5.7.1). This result allows far-reaching generalisations. For instance, a non-commutative unipotent algebraic group G is a chain if and only
if every epimorphic image of G is isogenous to a subgroup of G and any
two connected algebraic subgroups of G of the same dimension are isogenous
(cf. Corollary 5.7.4). The dual conditions also give a characterisation of non-
commutative unipotent chains (cf. Corollary 5.7.8).

In the theory of abstract groups, a group is called hamiltonian or sometimes a Dedekind group if all of its subgroups are normal. For algebraic groups
this condition applied to all algebraic subgroups would not be interesting (see
Theorem 7.2.1). Hence we say that a connected algebraic group is *hamiltonian*
if all its connected algebraic subgroups are normal. For connected algebraic
groups over a field of characteristic zero also this definition is too strong.
Namely, we show that any connected algebraic k-group over a field k of characteristic zero such that any connected k-subgroup is normal is commutative (cf. Theorem 7.2.10). But the situation changes drastically if we consider
hamiltonian groups over fields of positive characteristic. Any non-commutative
chain, more generally any uni-minimal connected algebraical group, is hamiltonian (see Theorem 7.2.19). Other examples of connected hamiltonian algebraic groups are the groups in which the centre has co-dimension one. These
groups in addition are *quasi-commutative*, i.e. algebraic groups where every
commutative connected algebraic subgroup is central. We remark that non-
commutative, but quasi-commutative algebraic groups exist only over fields
of positive characteristic; they have nilpotency class two (Proposition 3.2.26).

Quite often replacing the condition of normality for certain subgroups by
the condition of quasi-normality one obtains for abstract groups results of the
same significance (see [93]). If G is an algebraic group and Q is a connected
algebraic subgroup of G, then there are two natural possibilities to say that
Q is quasi-normal. The stronger version is to demand that $QX = XQ$ for any
algebraic subgroup of G. But with respect to this definition we can prove a
sharper version of Theorem 1 in [87] : A connected algebraic k-subgroup P
of a connected affine algebraic group G defined over an infinite perfect field
k such that $PH = HP$ for any k-closed subgroup H of G is normal in G
(see Theorem 7.1.1). Because of this result we call a connected algebraic subgroup Q of an algebraic group G *quasi-normal* if it is permutable with every
connected algebraic subgroup of G. Quasi-normal, but not normal algebraic
subgroups exist only in algebraic groups G over fields of positive characteristic
(see Corollary 7.1.5). Essentially they are contained in the unipotent radical of
G (see Corollary 7.1.8). Moreover, there are non-commutative algebraic groups

over fields of positive characteristic in which every algebraic subgroup is quasi-normal. Among these groups that we call *quasi-hamiltonian* there are groups which are not hamiltonian (e.g. Example 7.1.18 and Remark 7.1.20). A consequence of Corollary 7.1.8 is the fact that every connected quasi-hamiltonian algebraic group is nilpotent (cf. Theorem 7.1.11).

In Sections 7.1 and 7.2 we give many examples of quasi-hamiltonian and hamiltonian algebraic groups which are neither chains nor uni-minimal (see e.g. Example 7.2.18 and Remark 7.2.20). A big class of such groups is formed by the algebraic groups such that the factor group over their centers is a chain (see Theorem 7.1.12). Moreover, in these sections we describe some product constructions to obtain hamiltonian groups from given ones, e.g. from chains, and discuss which limitations occur.

A subclass of the class of hamiltonian algebraic groups are those groups in which every connected algebraic subgroup is characteristic; we call these groups *super-hamiltonian*. A classification of super-hamiltonian algebraic groups G is easy if G is a direct product of chains (see Corollaries 7.4.2, 7.4.3 and Proposition 7.4.4). However, the decision whether a product of chains is super-hamiltonian is difficult if the factors do not intersect trivially.

The experience from algebraic groups over fields of characteristic zero, from abelian varieties and from finite groups makes it surprising that over fields of prime characteristic there are many examples of three-dimensional non-commutative connected unipotent algebraic groups which are super-hamiltonian. In Section 7.4 we use our classification of three-dimensional connected unipotent groups over perfect fields of characteristic greater than two to decide which of these groups are super-hamiltonian. Although non-hamiltonian three-dimensional unipotent groups G exist if and only if the centre of G is one-dimensional, super-hamiltonian three-dimensional unipotent groups G do not exist only under severe restrictions on the centre, the commutator subgroup G' and the factor group G/G' (see Propositions 7.2.6 and 7.4.6).

Connected quasi-normal subgroups of connected affine algebraic groups over fields of characteristic zero, respectively of connected Lie groups, are treated in [87], respectively [86]. A connected closed subgroup Q of a topological group G is defined to be quasi-normal if it is *topologically permutable* with any closed subgroup of G, i.e. if the sets QP and PQ have the same closure for any closed subgroup P of G (see [52]). This fact motivated us to seek in Section 7.3 a unified method for the study of quasi-normal subgroups in topological and algebraic groups G. It turns out that a unified treatment is possible using a suitable closure operator on the set of subgroups of G (cf. Definition 7.3.4). This procedure allows to prove that in a connected real or complex Lie group G, any connected closed subgroup which is topologically permutable with every closed connected subgroup of G must be normal in G (see Theorem 7.3.5). In the case of real or complex Lie groups this is a positive answer to a conjecture in [53]. Moreover, as a consequence we obtain that every connected real or complex Lie group in which every connected closed subgroup is topologically permutable with any other connected closed sub-group must be commutative (see Corollary 7.3.7). Our point of view has also the advantage that p-adic Lie groups are included in the considerations, pro-vided we modify topological permutability to locally topological permutability (cf. Definition 7.3.10). Using this we can for instance show that any p-adic Lie group G (over an ultrametric field of characteristic zero) contains an open commutative subgroup if any family of subgroups which corresponds to some subalgebra in the Lie algebra of G is locally permutable with every other such family (cf. Theorem 7.3.14).

In our work we extend all results about affine algebraic groups, respectively about linear complex Lie groups, to non-affine algebraic groups, respectively to non-linear complex Lie groups. To do this we use for algebraic groups a series of well-known results of M. Rosenlicht (cf. Section 1.3) and for complex Lie groups some theorems on complex tori and toroidal groups (cf. Section 1.1). Moreover, for algebraic groups we discuss rationality questions and try to generalize our results to algebraic k-groups.

In Chapter 1 we collect known results on real and complex Lie groups, formal groups, p-adic groups and algebraic groups as far as they are needed for our investigations.

The main part of Section 2.1 is devoted to the theory of extensions of alge-braic groups, in view of our use of regular factor systems. In particular, we need to know sufficient conditions for an algebraic k-group G, which is an extension of an algebraic k-group A by an algebraic k-group B, to be k-isomorphic to a k-group defined on the k-variety $B \times A$. Although such a representation of G is not always possible (see e.g. Remarks 2.1.4 and 2.1.5), it exists if the normal subgroup A is k-split and unipotent and B is affine (cf. [82], Theorem 1, p. 99). Another aim of Section 2.1 is to discuss relations between factor systems and isogenies. If there exists an isogeny from a connected commutative unipotent group G_1 onto G_2, then an isogeny from G_2 onto G_1 exists as well (see [89], Proposition 10, p. 176). One could ask if this holds for *unipotent* connected non-commutative algebraic groups. Example 2.1.14 answers this question neg-atively. We establish in Proposition 2.1.12 necessary and sufficient conditions for the existence of an isogeny between central extensions of A_1 by B_1 and A_2 by B_2 which extends two given isogenies between A_1 and A_2 and between B_1 and B_2. In Proposition 2.1.13 we give sufficient conditions for central exten-sions of a Witt group or of a vector group by an algebraic group to be isogenous in sense of [77], p. 417 (see also Section 2.1). Moreover, for two-dimensional non-commutative unipotent groups over perfect fields of characteristic $p > 2$ and of exponent p we give another procedure to obtain, up to a coboundary, all other factor systems from a suitable one (see Proposition 2.1.9).

In Section 2.2 we deal with commutative extensions of commutative con-nected complex Lie groups. Since any such group is (holomorphically) iso-morphic to the direct product of a linear torus $(\mathbb{C}^*)^m$, a vector group $\mathbb{C}^l$ and a toroidal group X, the theory of commutative extensions of commuta-tive connected complex Lie groups reduces to the case of extensions which are toroidal groups. The complex n-dimensional toroidal groups X are completely described by means of period matrices, the columns of which are vectors of the lattice Λ determining X. If the $\mathbb{R}$-span of the lattice Λ has dimension $2n$, then X is a compact group, namely a complex torus. The description of complex tori by period matrices given in [7], Chapter 1, suggests to us a generalization to toroidal groups. Using period matrices, necessary and sufficient conditions for two toroidal groups to be isogenous, respectively isomorphic, are obtained. Propositions 2.2.2 and 2.2.3 allow us to recognize suitable closed subgroups and factor groups of a toroidal group within the corresponding period matrix. Moreover, a non-compact toroidal group can contain more than one maximal closed complex linear torus (cf. Example 2.2.4). Using these results we can concretely decide under which circumstances commutative holomorphic extensions of toroidal groups by toroidal groups split, and we show that, analogously to the case of complex tori, Shafarevich extensions of toroidal groups by toroidal groups exist.

If one considers, as for connected algebraic groups, connected real and complex Lie groups in which every connected closed subgroup is characteristic, then the Shafarevich extensions are prominent examples of super-hamiltonian Lie groups. If a super-hamiltonian Lie group has dimension greater than two, then it is a commutative complex Lie group which is a closed subgroup of the direct product having as factors a one-dimensional vector group, a one-dimensional linear torus and a non-trivial toroidal group which is super-hamiltonian. To settle the question when in a toroidal group X any connected closed subgroup is characteristic is easy if X is the direct product of Shafarevich extensions of a simple torus by a simple torus. In this case X is super-hamiltonian if and only if there is no non-trivial homomorphism between two distinct direct factors of X. In general, however, only a thorough analysis of the period matrix determining X allows to decide whether X is super-hamiltonian or not (see Proposition 7.4.11, Example 7.4.12 and Proposition 7.4.13).

Acknowledgement

Supported by the Deutsche Forschungsgemeinschaft and Università di Palermo (Co.R.I.).

Contents

1

Prerequisites

1.1 Real and Complex Lie Groups

For a general reference about Lie groups see e.g. [11–13, 27, 40, 95]. We collect now some needed results about Lie groups.

Let G be a connected real or complex Lie group and let $\mathfrak{g}$ be its Lie algebra. The group G is *quasi-simple* if its Lie algebra $\mathfrak{g}$ is simple. The group G is *semi-simple* if $\mathfrak{g}$ is a product of simple Lie algebras.

The group G is an *almost direct product* of Lie groups G_i if $G = G_1 \cdots G_n$, where G_i is a normal closed subgroup of G and $G_i \cap G_1 \cdots G_{j-1}G_{j+1} \cdots G_n$ is a closed discrete subgroup of G. In particular G_i is centralised by any G_j, $i \neq j$. If G is semi-simple, then G is an almost direct product of quasi-simple Lie groups G_i, each corresponding to a simple factor $\mathfrak{g}_i$ of $\mathfrak{g}$.

The commutator subgroup G' of a connected Lie group G is the closure of the abstract commutator subgroup, generated by $[x, y] = x^{-1}y^{-1}xy$, with $x, y \in G$. Clearly G' is connected. For an example of a real Lie group where the abstract commutator subgroup is not closed see Remark 5.3.6.

A connected real or complex Lie group G is *solvable* if the series of commutator subgroups $G^{(i)}$ ends with the trivial subgroup.

Denote by $Z^\circ(G)$ the connected component of the centre of a connected Lie group G and by $Z_i^\circ = Z_i^\circ(G)$ the connected component of the group Z_i, where Z_i/Z_{i-1}° is the centre of G/Z_{i-1}°.

A connected real or complex Lie group G is *nilpotent* if the series $Z_i(G)$ ends with G.

For any real or complex Lie group G there exists a maximal connected normal solvable subgroup R of G, called the *radical* of G, such that G/R is a semi-simple Lie group. In any real or complex Lie group G there is a maximal connected nilpotent Lie subgroup $N \leq R$, called the *nilradical*.

If G is a connected real Lie group and it is solvable, then any maximal compact subgroup T of G is a torus and the subgroup $N \cap T$ of the nilradical N of G is central in G. In this case there exists a commutative connected Lie subgroup S such that $G = NS$ and $N \cap S$ is a central compact subgroup of G.

Moreover if the factor group G/G' is compact, then G is a product $G = G'S$, where S is a torus such that $S \cap G'$ is finite (cf. [42]).

As usual we call a real or complex Lie group linear if it has a faithful real respectively complex analytic linear representation. For group-theoretical question the linearity of real Lie groups plays no particular rôle since two real Lie groups which are isomorphic as topological groups are isomorphic as real analytic groups (see [58], 4.10.1. Theorem, p. 185). This is not the case for complex Lie groups, e.g. any elliptic curve is isomorphic to $SO_2(\mathbb{R}) \times SO_2(\mathbb{R})$ as topological group (cf. [7], p. 2). Hence it is not surprising that in group-theoretical investigations usually only complex linear Lie groups are treated (see [55]).

It is a remarkable feature of complex linear Lie groups that they do not contain compact complex analytic subgroups of positive dimension (see [55], Corollary 4.5, p. 103). The main results in the theory of complex linear Lie groups are analogous to those in the theory of Lie groups; the rôle of compact subgroups in real groups is played by groups isomorphic to $(\mathbb{C}^*)^m$. As an example for this phenomenon we mention the fact that every complex linear Lie group is a semi-direct product of a simply connected normal solvable complex Lie subgroup and an almost direct product of groups which are either simple or isomorphic to $\mathbb{C}^*$ (see [55], Theorem 4.43, p. 130). For this reason we shall call subgroups of a complex Lie group which are isomorphic to $(\mathbb{C}^*)^m$ *linear complex tori*.

A connected complex Lie group which is not linear always has an epimorphic image which contains a compact connected complex Lie subgroup (cf. [59], Lemma 6, p. 259 and [1], Proposition 1.1.17, p. 15). Every compact connected complex Lie group is commutative and homeomorphic to the space $(SO_2(\mathbb{R}) \times SO_2(\mathbb{R}))^m$ (see [64], p. 1-2). In this paper we call a compact connected complex Lie group a *complex torus*.

For a complex torus X one considers the *endomorphism algebra* $\mathsf{End}_{\mathbb{Q}}(X) = \mathsf{End}(X) \otimes_{\mathbb{Z}} \mathbb{Q}$. The algebra $\mathsf{End}_{\mathbb{Q}}(X)$ is a finite-dimensional $\mathbb{Q}$-algebra, and conversely, every finite-dimensional $\mathbb{Q}$-algebra can be represented this way (cf. [7], Theorem 9.1, p. 28). The indecomposable complex tori correspond to the local $\mathbb{Q}$-algebras, and the simple complex tori to the skew fields (cf. [7], Proposition 7.3, p. 24). For an abelian variety X the structure of the algebra $\mathsf{End}_{\mathbb{Q}}(X)$ is seriously restricted. Necessary conditions for a skew field of finite dimension over $\mathbb{Q}$ to be the endomorphism algebra of a simple abelian variety can be found in [64], p. 202, whereas sufficient conditions were given in [90].

In contrast to the class of abelian varieties, there are uncountably many non-split extensions of a complex torus by a complex torus ([7], Corollary 6.3, p. 23). In particular there are complex tori containing only one proper subtorus ([7], Example 7.2, p. 23).

Whereas any real commutative connected Lie group is isomorphic to the direct product of the vector group $\mathbb{R}^n$ by a torus SO_2^n, the structure of complex commutative connected Lie groups is more delicate. In particular there exist analoga to non-affine algebraic groups. Indeed any complex commutative connected Lie group G is holomorphically isomorphic to a group having a unique decomposition as direct product $(\mathbb{C}^+)^l \times (\mathbb{C}^*)^m \times X$ (see 1.1.5 in [1]), where $\mathbb{C}^+$ and $\mathbb{C}^*$ are the additive and multiplicative groups of the field of complex numbers, and X is a *toroidal* group in the following sense:

1.1.1 Remark. Let $\mathbb{C}^n$ be the n-dimensional complex vector space and Λ a discrete subgroup of $\mathbb{C}^n$ of complex rank n. Let $\mathbb{R}_\Lambda$ be the real span of Λ and $M\mathbb{C}_\Lambda$ be the maximal complex subspace of $\mathbb{R}_\Lambda$. An n-dimensional complex Lie group G is called toroidal if it is isomorphic to $\mathbb{C}^n/\Lambda$ and the maximal complex subgroup $M\mathbb{C}_\Lambda/(M\mathbb{C}_\Lambda \cap \Lambda)$ is dense in the maximal real torus $\mathbb{R}_\Lambda/\Lambda$ of $\mathbb{C}^n/\Lambda$ (cf. [1], 1.1.4 Theorem, p. 5).

A complex Lie group isomorphic to $(\mathbb{C}^+)^l \times (\mathbb{C}^*)^m$ is an affine group. For $n = 1$ a toroidal group is a complex torus. In a connected complex Lie group G any toroidal subgroup is contained in center of G (cf. [59], Lemma 6, p. 259), and any non-trivial epimorphic image of a toroidal group is again toroidal (cf. [59], Lemma 3, p. 259).

A complex toroidal group $G = \mathbb{C}^n/\Lambda$ is of type q if the maximal complex subspace $M\mathbb{C}_\Lambda$ of the real span $\mathbb{R}_\Lambda$ has the complex dimension q. The group $G = \mathbb{C}^n/\Lambda$ has rank $n + q$ which is equal to the dimension of the maximal topological torus in G. Moreover G is a $(\mathbb{C}^*)^{n-q}$-fiber bundle over a complex torus T as base space (cf. [1], Section 1.14, p. 12).

If G is a toroidal group, then every closed connected affine subgroup N of G is isomorphic to $(\mathbb{C}^+)^l \times (\mathbb{C}^*)^m$ where $2l + m \leq n - q$ (in contrast to the case of algebraic groups, it is possible that a toroidal group contains closed subgroups isomorphic to $\mathbb{C}^+$). If N is maximal in G, then G/N is a complex torus (cf. [1], Proposition 1.1.17, p. 15). If G/N is even an abelian variety, then $2l + m = n - q$ and G/N has dimension $q + l$ (cf. [1], 3.1.16 Fibration Theorem, p. 70 and 3.2.21 Main Theorem, p. 91).

Since an n-dimensional toroidal group G of type q always contains a closed subgroup S isomorphic to $(\mathbb{C}^*)^{n-q}$, the product of S with a closed subgroup $L \leq G$ isomorphic to $\mathbb{C}^+$ is not closed.   □

In 1965 Morimoto proved that each connected complex Lie group G has a unique maximal toroidal subgroup X which is central and which is the smallest closed connected subgroup such that the factor group G/X is linear (see [1], Theorem 1.2.2, p. 6 and [59]).

1.2 Formal Groups and p-Adic Lie Groups

A Lie group over a (non-discrete) complete ultrametric field F of characteristic zero is a group in the variety of analytic manifolds over F ([9], Chapter III, § 1, p. 95). To every Lie group G over F there belongs a Lie algebra $\mathfrak{g}$ ([9], Chapter III, § 3, p. 125). The correspondence between a finite-dimensional Lie group G over F and its Lie algebra $\mathfrak{g}$ is described in [9], Chapter III, § 7,

p. 214ff. Given a subalgebra $\mathfrak{h}$ of $\mathfrak{g}$, the closed subgroups of G having $\mathfrak{h}$ as its Lie algebra form a family $\mathcal{E}_{\mathfrak{h}}$ of closed subgroups of G such that any two elements of $\mathcal{E}_{\mathfrak{h}}$ are locally isomorphic.

As in [39] for us a formal group is given by a formal group law over a $\mathbb{Q}$-algebra. Also for the category of formal groups over a field K of characteristic zero there is a correspondence between Lie algebras and groups given by the exponential map and the Campbell–Hausdorff formula (see [39], II.14, pp. 79-87).

1.3 Algebraic Groups

In this paper we consider algebraic groups in the sense of A. Weil (see [98,99]), defined over a field k. We call these groups *groups over* k or k-*groups*. If the field k is algebraically closed, then the groups over k are separated prevarieties in the sense of [8], 6.2, p. 15. For an arbitrary field k, the k-groups in the sense of Weil are just F-varieties over an algebraically closed field F containing k which carry a k-structure (cf. [8,45,91]). There exists a smallest field k over which a variety is defined (see [81]), and if L is a field extension of k then a k-variety is also an L-variety.

An *algebraic subgroup* of an algebraic group G is a closed subgroup with respect to the Zariski topology. An algebraic subgroup of an algebraic k-group G is not necessarily defined over k. If an algebraic subgroup of a k-group G is also defined over k, we call it a k-subgroup. If k is perfect, then a k-closed subgroup of a k-group is a k-subgroup, but in general it is defined only over a finite purely inseparable extension of k.

The connected components of a k-group G are defined over the separable closure k_s of k, and the Galois group of k_s over k acts on them as a group of permutations, leaving stable precisely the ones which are defined over k (cf. [8], 12.3, p. 25; 14.3-4, p. 31-2; 1.2, p. 46). In particular, the connected component G°, containing the neutral element of G, is defined over k (cf. [8], 1.2, p. 46-47).

If H is an algebraic k-closed subgroup of the k-group G, then it is in general not true that the connected components of H are defined over k_s. For instance, a connected algebraic k-closed subgroup H of the k-group G which is not defined over k cannot be defined over k_s (e.g. see Remark 1.3.1).

The product HN of two k-subgroups of the k-group G, with N normalizing H, is a k-subgroup. The normal closure of a k-subgroup is a k-subgroup, whereas centralisers of k-subgroups are k-closed (cf. [8] 2.3 p. 59, 1.7 p. 52). For instance, the centre of a k-group is a k-closed algebraic subgroup, not necessarily defined over k (e.g. see Remark 1.3.1), whereas the group generated by all the commutators is always a k-subgroup.

A homomorphism ϕ from a k-group G_1 into a k-group G_2 is a k-*homomorphism* if ϕ is an everywhere defined rational map and k is a field of definition for ϕ. The image of a k-homomorphism is a k-subgroup of

G_2 (cf. [77] p. 409). The kernel of a k-homomorphism is a k-closed subgroup of G. If ϕ is *separable*, that means that $k(x)$ is a separable extension of $k(\phi(x))$ for a generic point $x \in G_1$, the connected components of the kernel K are defined over k_s and in particular K° is defined over k (cf. [77], Corollary p. 412). Conversely, if H is an algebraic k-closed normal subgroup of the k-group G such that the connected components of H are defined over k_s, then there exists a k-group G/H and a surjective separable k-homomorphism $\phi : G \longrightarrow G/H$, having H as kernel (cf. [77], Th. 4 p. 413).

1.3.1 Remark. In [82], pp. 115-6, a series of examples is given. Here we recall the following: let k be a non-perfect field of characteristic $p > 0$, and let G be the connected 4-dimensional affine algebraic k-group G defined on the affine space by

$$(x_0, x_1, x_2, x_3) \cdot (y_0, y_1, y_2, y_3) = (x_0 + y_0, x_1 + y_1, x_2 + y_2, x_3 + y_3 + (x_0^p + a x_1^p) y_2).$$

The centre $_3G$ of G is the connected algebraic subgroup with $x_0^p + a x_1^p = x_2 = 0$. If $a^{1/p} \notin k$ then $_3G$ is not defined over k. $\qquad\square$

The general structure of k-groups has been clarified by a series of theorems of Rosenlicht in [77], which are collected in the following

1.3.2 Theorem (Rosenlicht [77]). *Let G be a connected algebraic k-group. Then there exists a unique maximal connected affine algebraic subgroup $L = L(G)$ and a unique minimal connected normal algebraic subgroup $D = D(G)$ such that the factor group G/D is affine. The following statements hold:*

i) $G = LD$ and $(L \cap D)^\circ$ is defined over k.
ii) L is a k-closed characteristic subgroup of G, and G/L is an abelian variety.
iii) D is defined over k, central and characteristic in G. It has no non-trivial affine epimorphic image, and it has only a finite number of elements of any given finite order.
iv) Any k-closed abelian subvariety A of G is defined over k and is contained in D. Moreover, for each abelian subvariety A there exists a connected k-closed algebraic subgroup G_1 of G such that $G = G_1 A$ and $G_1 \cap A$ is finite. $\qquad\square$

For a connected algebraic subgroup $H \leq G$ we clearly have that $L(H) \leq L(G)$. Moreover we have $D(H) \leq D(G)$, because the affine factor group $HD(G)/D(G)$, isomorphic to $H/H \cap D(G)$, is affine, which forces $D(H) \leq H \cap D(G)$.

A theorem in [4] says that, if G is defined over a finite field, then $G = AL$, where A is its maximal abelian subvariety, $L = L(G)$, $A \cap L$ is finite and both A and L are defined over the (perfect) finite field of definition of G.

It follows from *iii)* that in the case of positive characteristic the connected component of the maximal affine algebraic subgroup of $D(G)$ is a torus. For suitable infinite fields of definition there exist connected algebraic groups G for which $G = D(G)$ and $L(G)$ is a torus. Rosenlicht called these groups *toroidal* (see [80], p. 986) in analogy to the complex toroidal groups. The mentioned theorem in [4] yields however the following

1.3.3 Proposition. *Let G be a connected algebraic group such that $G = D(G)$. Then either G is an abelian variety or G is not defined over any finite field.* $\qquad\square$

Algebraic groups $G = D(G)$ are investigated in [79] and [80] as non-splitting extensions of affine algebraic groups by abelian varieties. The main results in [79] yield the following

1.3.4 Proposition. *Let G be a connected algebraic k-group such that $G = D(G)$ and $L(G)$ is a non-trivial vector group. Then the characteristic of k is zero and $G/L(G)$ is an abelian variety of dimension greater than or equal to the dimension of $L(G)$.* $\qquad\square$

In contrast to this proposition, for commutative complex Lie group any extension of a complex vector group by a complex torus splits (cf. [78], p. 153).

1.3.5 Proposition. *Let G be a connected algebraic group over a field of positive characteristic such that the maximal affine connected subgroup $L(G)$ is solvable. Then there exists a connected algebraic subgroup H of $L(G)$ such that $G = HD(G)$ and $H \cap D$ is finite. Moreover H is the semi-direct product of the unipotent radical U of $L(G)$ with a torus T which intersects $D(G)$ trivially.*

Proof. The group $L(G)$ is a semi-direct product of U with a maximal torus Q. Since the group $D(G)$ is central in G, also the group $D(G) \cap L(G)$ is central in G. As the ground field has positive characteristic, the group $(D(G) \cap L(G))^\circ$ is a torus S and there exists a torus T such that $Q = T \times S$. $\qquad\square$

Algebraic affine k-groups have the fundamental property to be linear groups, that is any affine k-group is k-isomorphic to an algebraic k-subgroup of GL_n, for a suitable natural integer n (see [8], Proposition 1.10, p. 54). Let G be an affine k-group and let H and K be algebraic connected k-subgroups of G. Then the commutator subgroup $[H, K] = \langle h^{-1}k^{-1}hk : h \in H, k \in K \rangle$ is a connected algebraic k-group (cf. [8], p. 58). A connected algebraic k-group is solvable if there exists a natural number n such that for the derived series $\{G_i\}$ we have

$$G = G_0 \geq G_1 = [G, G] \geq G_2 = [G_1, G_1] \geq \cdots \geq G_n = [G_{n-1}, G_{n-1}] = \{e\}.$$

A connected algebraic k-group is called nilpotent if there exists a natural number n such that for the descending central series $\mathcal{C}^{i+1}(G) = [G, \mathcal{C}^i(G)]$ (where $\mathcal{C}^0(G) = G$), we have $\mathcal{C}^n(G) = e$.

When one has to show that an algebraic group G is solvable (resp. nilpotent, commutative) and no question of rationality is involved, then no assumption on the field k of definition is needed. But rationality questions arise naturally when one deals with homomorphisms.

Mostly the class of solvable and the class of nilpotent algebraic groups are decisive for our theorems. To deal with these classes, for our needs it is often sufficient to take a perfect, not necessarily algebraically closed, field k as field of definition. Then the notion of split solvable and split nilpotent affine k-group becomes crucial (see [8], Chapter V):

A connected solvable algebraic affine k-group G is called k-split if it has a composition series $G = G_0 > G_1 > \cdots > G_s = \{1\}$ consisting of connected k-subgroups such that its factors G_i/G_{i+1} are k-isomorphic either to $\mathbf{G}_a$ or to $\mathbf{G}_m$. In particular, a torus defined over k is k-split if it is k-isomorphic to the group $(\mathbf{G}_m)^n$.

We collect the main results about k-split solvable affine groups in the following

1.3.6 Theorem. *For a connected solvable affine* k-*group* G *the following holds:*

(1) *G is trigonalizable over* k *if and only if the unipotent radical G_u of G is defined over* k *and the factor group G/G_u is a* k-*split torus. G is* k-*split if and only if the unipotent radical G_u of G is defined over* k *and* k-*split and the factor group G/G_u is a* k-*split torus.*

(2) *G is* k-*split if and only if G as a variety is* k-*isomorphic to the variety $(\mathbf{G}_a)^r \times (\mathbf{G}_m)^s$, where r is the dimension of G_u and s is the dimension of G/G_u.*

(3) *Let G be unipotent. If* k *is perfect, then G is* k-*split. Moreover, G is* k_s-*split if and only if G is* k-*split.*

(4) *Let G be a torus. If* k *is separably algebraically closed, then G is* k-*split. Moreover, G is* $k^{p^{-\infty}}$-*split if and only if G is* k-*split.*   □

The above results (1), (2), (3) can be found in [91], Proposition 14.1.2 p. 237, Exercise 14.3.12 p. 248, Corollary 14.2.7 p. 243, Theorem 14.3.8 p. 245, whilst (4) can be found in [8], Proposition 8.11, p. 117. For a general description of connected solvable affine groups we will always refer to [8], Theorem 10.6, p. 137. A maximal connected solvable algebraic subgroup of a connected affine group G is called a *Borel subgroup* of G. It turns out that the only case where B is nilpotent is when $G = B$ (see [8], Corollary 11.5, p. 148).

1.3.7 Remark. Let G be a k-split one-dimensional connected algebraic group. Then G is k-isomorphic either to the additive group $\mathbf{G}_a$ or to the multiplicative group $\mathbf{G}_m$; in the first case, G is unipotent, and in the second case G is diagonalizable. If the characteristic of k is zero, then the ring of k-endomorphisms of $\mathbf{G}_a$ is isomorphic to k, where the homothety $x \mapsto \alpha x$ corresponds to a given $\alpha \in$ k. In contrast to this, if the characteristic of k is $p > 0$, then the ring of k-endomorphisms of $\mathbf{G}_a$ is isomorphic to the non-commutative ring of polynomials $k[\mathbf{F}]$, where $\mathbf{F}a = a^p\mathbf{F}$. Any element $f(\mathbf{F}) = \sum_{i=0}^n a_i\mathbf{F}^i \in k[\mathbf{F}]$ can be identified with the p-polynomial $f(x) = \sum_{i=0}^n a_i x^{p^i}$ (with $a_i \in$ k) and the homomorphism $x \mapsto f(x)$ corresponds to any such p-polynomial. In particular, the Frobenius homomorphism

$x \mapsto x^p$ corresponds to **F**. Up to such an identification, we will always denote by $k[\mathbf{F}]$ the ring of k-endomorphisms of the additive group.

On the other hand, we have in addition that the ring of k-endomorphisms of $\mathbf{G}_m$ is isomorphic to $\mathbb{Z}$, where the homomorphism $x \mapsto x^k$ corresponds to $k \in \mathbb{Z}$. $\qquad\qquad\square$

1.3.8 Remark. The Lie algebra $\mathfrak{g}$ of a connected affine algebraic group G can be defined as the Lie algebra of derivations of the affine algebra $\bar{k}[G]$ of classes of polynomials modulo the ideal vanishing on G. If the characteristic of k is $p > 0$, then we have also that $\mathfrak{g}$ is a *restricted* Lie algebra (see [8], 3.1, p. 62).

For algebraic affine groups over algebraically closed fields of characteristic zero, the Lie algebra $\mathfrak{g}$ of G is an important tool, since there is a one-to-one correspondence between subalgebras of $\mathfrak{g}$ and connected algebraic subgroups of G and between ideals of $\mathfrak{g}$ and normal connected algebraic subgroups of G (see [45], Chapter V). We remark that we use this correspondence also for affine algebraic groups over an arbitrary field of characteristic zero, if we disregard the k-structure of G.

It can be useful to see G as an algebraic k-subgroup of the general linear group GL_n: in this case, $\mathfrak{g}$ is a Lie subalgebra of the general linear algebra $\mathfrak{gl}_n$ of all $n \times n$ matrices, with the usual Lie product $[X, Y] = XY - YX$.

Since G acts on itself by conjugation, the inner automorphism $\mathsf{Int}_x(a) = xax^{-1}$ induces an automorphism $\mathsf{Ad}(x)$ of $\mathfrak{g}$. If we see G as an algebraic k-subgroup of GL_n and $\mathfrak{g}$ as the corresponding Lie subalgebra of $\mathfrak{gl}_n$, then $\mathsf{Ad}(x)(Y) = xYx^{-1}$ (see [8], 3.13, p. 69). The homomorphism of abstract groups

$$\mathsf{Ad} : G \longrightarrow GL(\mathfrak{g})$$

is indeed a k-homomorphism of algebraic groups, called the *adjoint representation* of G (see [8], 3.13, p. 69).

Let G be a k-split unipotent affine group. If no linear representation of the group G is given, a useful tool for the calculation of the adjoint representation of G is the *tangent bundle* $T(G)$ of G.

We recall that the ring of *dual numbers* of $\bar{k}$, the algebraic closure of k, is the factor ring of the polynomial ring $\bar{k}[x]$ by the principal ideal generated by x^2; therefore we identify the ring of dual numbers of $\bar{k}$ with $\bar{k}[\varepsilon]$, where $\varepsilon^2 = 0$.

In the case where G is a k-split n-dimensional unipotent affine group, G is defined on the n-dimensional affine space by a multiplication $\mu : G \times G \longrightarrow G$

$$\mu(x_1, \cdots, x_n; y_1, \cdots, y_n) = (x_1, \cdots, x_n) \cdot (y_1, \cdots, y_n).$$

Since μ is a k-morphism, it is given by polynomial maps with coefficients in k. Clearly, μ induces also a group multiplication on the set $T(G)$ of n-tuples with entries in $\bar{k}[\epsilon]$.

The subgroup consisting of those elements $(x_1, \cdots, x_n)$ having coefficients x_i in $\bar{\mathsf{k}}$ is clearly isomorphic to G, whereas the subgroup consisting of those elements $(\varepsilon x_1, \cdots, \varepsilon x_n) := \varepsilon(x_1, \cdots, x_n)$ having coefficients in $\varepsilon\bar{\mathsf{k}}$ is isomorphic to the additive group of the restricted Lie algebra $\mathfrak{g}$ of G. Identifying G and $\mathfrak{g}$ with these subgroups of $T(G)$ we have

$$T(G) = \mathfrak{g} \rtimes G$$

and the action of G on $\mathfrak{g}$ by conjugation in $T(G)$ is precisely the action of G on $\mathfrak{g}$ by the adjoint representation (see [8] 3.20, p. 76).   □

1.3.9 Remark. In this paper we consider algebraic groups G defined over a perfect field k; sometimes, if we use Lie algebras, k will be algebraically closed. If k is perfect and infinite and G is a connected affine group, then already the subgroup $G(\mathsf{k})$ of k-rational points of G determines to a great extent the structure of G, since in this case $G(\mathsf{k})$ is dense in G (see [8] Corollary 18.3, p. 220 or [91], 13.3.10 Corollary, p. 232).

Algebraic groups over non-perfect fields are thoroughly treated in [67]. There in Chapitre V a theory of unipotent algebraic groups is developed which shows the complications for non-perfect fields.

Connected algebraic groups G in which the lattice of connected algebraic subgroups is a chain will play a fundamental rôle. This means that for any proper connected algebraic subgroup H of G there is exactly one connected algebraic subgroup of G containing H as the unique connected maximal algebraic subgroup. We call such a group itself a *chain*. A well known class of commutative chains are the Witt groups $\mathfrak{W}_n$ (cf. [18], 6.11 p. 595).

The following examples illustrate concretely why we define affine chains only in the context of algebraic groups defined over a field k such that G is k-split. One more reason lies in the fact that if k is perfect, one is allowed to represent elements of a unipotent algebraic group G as n-tuples of field elements and to use the language of factor systems, an extremely useful tool in our computations.

1) Let E be a finite extension of k and let G be an affine algebraic k-group. As shown in Sections 11.4 and 12.4 of [91], it is possible to associate to the (connected, respectively commutative) k-group G a (connected, respectively commutative) k-group ΠG and a surjective E-homomorphism $\pi : \Pi G \longrightarrow G$ such that, for any k-group H and E-homomorphism $\phi : H \longrightarrow G$, there is a unique k-homomorphism $\psi : H \longrightarrow \Pi G$ with $\phi = \pi \circ \psi$. In particular, for $H = G$ we find a k-homomorphism $\sigma : G \longrightarrow \Pi G$ such that $\pi \circ \sigma = 1$. The kernel of π contains no algebraic normal k-subgroups of ΠG and, if E/k is purely inseparable, it is a connected unipotent subgroup of ΠG. Therefore, if E/k is purely inseparable and G is reductive, $\ker \pi$ is the unipotent radical of the k-group ΠG and it is not defined over k.

For instance, let k be a non-perfect field of characteristic two and let E be a purely inseparable extension of k of degree $[E : k] = 2$. For $G = \mathbf{G}_m$ the connected commutative k-group ΠG has dimension two and $\ker \pi$ is the unipotent radical of ΠG. It has dimension one and is not defined over k, so ΠG is a commutative two-dimensional k-group with a unique connected k-subgroup of dimension one.

2) A k-torus T is said to be k-*irreducible* if it does not contain proper non-trivial k-subtori. Since a k-torus is the product of its largest k-anisotropic k-subtorus and its largest k-split k-subtorus (see [8], Proposition 8.15, p. 121), any k-irreducible k-torus of dimension greater than one is clearly k-anisotropic. But, $SO_2 \times SO_2$ is a two-dimensional $\mathbb{R}$-anisotropic torus which is not $\mathbb{R}$-irreducible. Examples of irreducible k-tori are given in [72].   □

2

Extensions

2.1 Extensions of Unipotent Groups and Isogenies

Let G_1 and G_2 be connected algebraic k-groups. A k-homomorphism $\eta : G_1 \longrightarrow G_2$ is a k-*isogeny* if it is an epimorphism with finite kernel. By [77], Corollary p. 412, any element of the kernel of a separable k-isogeny is defined over the separable closure k_s of k. The groups SL_2 and PSL_2 give an example for the fact that the existence of an isogeny is not a symmetric relation. But if G_1 and G_2 are connected commutative unipotent algebraic group and $\eta : G_1 \longrightarrow G_2$ is an isogeny, then there always exists an isogeny $\theta : G_2 \longrightarrow G_1$ (see [89], Proposition 10, p. 176). The equivalence relation generated by isogenies can be defined as follows: G_1 and G_2 are *isogenous* if there exists a group G_3 and two isogenies $\eta_i : G_3 \longrightarrow G_i$ $(i = 1, 2)$, see [77], section 3, p. 417.

A main result in the context of commutative unipotent algebraic groups, defined over a perfect field k, states that any such connected k-group is k-isogenous to a direct product of Witt groups $\mathfrak{W}_n$ (cf. [18], 6.11 p. 595). The class of Witt groups $\mathfrak{W}_n$ is thoroughly described in [18], Chapitre V, or in [89], VII, 8. The class of unipotent algebraic groups isogenous to a Witt group is therefore the class of commutative unipotent chains in positive characteristic.

In this section we treat the theory of extensions of algebraic groups, as developed in a series of papers by Weil [98], Rosenlicht [77, 79, 82, 83], Serre [89]. The purpose is to describe an algebraic group G if a normal connected algebraic subgroup A and the factor group G/A are given.

A bit more general than the direct product are the *direct product with amalgamated central subgroup* $G = G_1 \curlyvee G_2$ and the *direct product with amalgamated factor group* $G = G_1 \curlywedge G_2$, defined as follows:

2.1.1 Definition. *Let* G_1, G_2 *be connected algebraic groups, let* $Z_i \in G_i$ *be isogenous connected central algebraic subgroups and let* $\mu_i : H \longrightarrow Z_i$ *be two isogenies. The factor group* $(G_1 \times G_2)/\Delta$, *where*

$$\Delta = \{(g_1, g_2^{-1}) \in G_1 \times G_2 : g_i = \mu_i(x) \text{ for } i = 1, 2 \text{ with } x \in H\}$$

is called the direct product with amalgamated central subgroup *and is denoted by* $(G_1 \curlyvee G_2)_H$ *or simply by* $G_1 \curlyvee G_2$, *if H is understood.*

2.1.2 Definition. *Let G_1, G_2 be connected algebraic groups, let $N_i \in G_i$ be connected normal algebraic subgroups such that G_1/N_1 is isogenous to G_2/N_2 and let $\mu_i : H \longrightarrow G_i/N_i$ be two isogenies. The subgroup*

$$(G_1 \curlywedge G_2)_H = \{(g_1, g_2) \in G_1 \times G_2 : g_i N_i = \mu_i(x) \text{ for } i = 1, 2 \text{ with } x \in H\}$$

of $G_1 \times G_2$ is called the direct product with amalgamated factor group *and is denoted by* $(G_1 \curlywedge G_2)_H$ *or simply by* $G_1 \curlywedge G_2$, *if H is understood.*

The following characterisation of the above products follows by standard arguments.

2.1.3 Proposition. *The connected algebraic group G contains two connected algebraic subgroups G_1, G_2 such that $G = G_1 G_2$, $[G_1, G_2] = 1$ and $(G_1 \cap G_2)^\circ = H$ if and only if G is isogenous to the direct product with amalgamated central subgroup $(\overline{G}_1 \curlyvee \overline{G}_2)_{\overline{H}}$, where $\overline{G}_i$ is isogenous to G_i and $\overline{H}$ is isogenous to H.*

The connected algebraic group G contains two connected normal algebraic subgroups N_1, N_2 such that the homomorphism $\pi : G \longrightarrow G/N_1 \times G/N_2$, $\pi(x) = (xN_1, xN_2)$ is an isogeny if and only if G is isogenous to the direct product with amalgamated factor group $(\overline{G}_1 \curlywedge \overline{G}_2)_H$, where $\overline{G}_i$ is isogenous to G/N_i and H is isogenous to $G/(N_1 N_2)$. □

Now we summarize some results concerning the theory of extensions of algebraic groups. These are slightly different from the ones in the general case of abstract groups, especially in the matter of the field of definition and in the non-affine case. One of the purposes is to show that the description of an algebraic group by means of coordinate functions cries for questions of separability of the field of definition. At the end of this section we will therefore abandon the attempt of describing algebraic chains in the context of algebraic k-groups for a general field k and we will assume (mainly) that k is perfect. The principal reference are the papers [82] and [77] of Rosenlicht.

Let A and B be two connected algebraic k-groups, with A commutative and *affine*, and let $\phi : B \times B \longrightarrow A$ be a k-rational regular *factor system*, i.e. an everywhere defined k-rational map satisfying the equation $\delta^2 \phi = 0$, where one defines

$$\delta^2 \phi : (x, y, z) \mapsto \phi(x, y) + \phi(xy, z) - \phi(y, z) - \phi(x, yz). \qquad (2.1)$$

The following multiplication

$$(b_1, a_1)(b_2, a_2) = (b_1 b_2, a_1 + a_2 + \phi(b_1, b_2)) \qquad (2.2)$$

makes $B \times A$ an algebraic k-group G_ϕ, where $1 \times A$ is a central algebraic k-subgroup of G_ϕ, the factor group $G_\phi/(1 \times A)$ is k-isomorphic to B and it is possible to establish an exact sequence

$$1 \longrightarrow A \overset{\imath}{\longrightarrow} G_\phi \overset{\pi}{\longrightarrow} B \longrightarrow 1 \qquad (2.3)$$

of separable k-homomorphisms (cf. [77], Theorem 4, p. 413), where $\imath(a) = (1, a - \phi(1,1))$ and $\pi(b, a) = b$. Since $\imath$ and π are separable, it is possible to identify A with the k-subgroup $1 \times A$ of G_ϕ and B with the factor group $G_\phi/(1 \times A)$. We say that G_ϕ is an *explicit central extension* of A by B, emphasizing that A is, up to a separable k-isomorphism, a *central* k-subgroup of G_ϕ. The set $C^2_k(B, A)$ of all k-rational regular factor systems from $B \times B$ to A is a commutative group with respect to the addition of maps. For any k-rational regular map $\psi : B \longrightarrow A$, the map

$$\delta^1 \psi : (x, y) \mapsto -\psi(y) + \psi(xy) - \psi(x)$$

is a k-rational regular factor system, usually called trivial. The trivial k-rational regular factor systems form a subgroup $B^2_k(B, A)$ of $C^2_k(B, A)$ and the factor group $C^2_k(B, A)/B^2_k(B, A)$ is usually denoted by $H^2_k(B, A)$.

Two explicit central extensions G_{ϕ_1}, G_{ϕ_2} of A by B, given by ϕ_1, ϕ_2, respectively, are k-*equivalent* if there exists a rational k-isomorphism $\gamma : G_{\phi_1} \longrightarrow G_{\phi_2}$ such that the diagram

$$
\begin{array}{ccccccccc}
1 & \longrightarrow & A & \longrightarrow & G_{\phi_1} & \longrightarrow & B & \longrightarrow & 1 \\
 & & id_A\downarrow & & \gamma\downarrow & & id_B\downarrow & & \\
1 & \longrightarrow & A & \longrightarrow & G_{\phi_2} & \longrightarrow & B & \longrightarrow & 1
\end{array}
$$

is commutative (see [50], 6.10, p. 363ff). This happens if and only if $(\phi_1 - \phi_2) \in B^2_k(B, A)$, hence two extensions G_{ϕ_1} and G_{ϕ_2} are equivalent if and only if ϕ_1 and ϕ_2 differ by a trivial factor system. In particular, the extension defined by a trivial factor system $\phi \in B^2_k(B, A)$ is equivalent to the direct product $G_\varphi = A \times B$, which corresponds to the factor system $\varphi(x, y) = 0$ for all $x, y \in B$. In this case we say that the extension *splits*.

Thus we have a bijection between the classes of equivalent explicit central extensions and the classes of $H^2_k(B, A)$. (This bijection is indeed an isomorphism of groups, if one defines a group multiplication on the set of extensions G_ϕ, as described by the well-known methods of Baer [5]).

Now we turn to the problem of describing a connected algebraic k-group G by means of factor systems, once we know a connected central affine k-subgroup A and the corresponding factor group $B = G/A$.

Let G be a connected algebraic k-group and let A be a connected central *affine* algebraic k-subgroup of G. Then the embedding of A in G and the canonical projection π of G onto $B = G/A$ give an exact sequence

$$1 \longrightarrow A \longrightarrow G \overset{\pi}{\longrightarrow} B \longrightarrow 1$$

of separable rational k-homomorphisms (see [77], Theorem 4, p. 413). The question whether the group G is birationally k-isomorphic to an explicit central extension G_ϕ for a suitable k-rational regular factor system ϕ is answered by the existence of a k-rational regular cross section. A rational *cross section* is a rational map from B into G such that $\pi\sigma = id$. Note that a rational cross section is not necessarily a regular one, indeed it could be defined only on an open dense subset of B.

Let σ be a k-rational regular cross section. Since $\pi\sigma = id$, one has that $\delta^1\sigma$ is a k-rational regular function from $B \times B$ to A. As A is a central subgroup of G, writing the multiplication in G additively we have

$$[\sigma(y) - \sigma(xy) + \sigma(x)] + [\sigma(z) - \sigma(xyz) + \sigma(xy)] +$$

$$-[\sigma(z) - \sigma(yz) + \sigma(y)] - [\sigma(yz) - \sigma(xyz) + \sigma(x)] =$$

$$[\sigma(y) - \sigma(xy) + \sigma(x)] + [\sigma(z) - \sigma(xyz) + \sigma(xy)] +$$

$$[-\sigma(y) + \sigma(yz) - \sigma(z)] + [-\sigma(x) + \sigma(xyz) - \sigma(yz)] =$$

$$[\sigma(y) - \sigma(xy) + \sigma(x)] + [-\sigma(x) + \sigma(xyz) - \sigma(yz)] +$$

$$[\sigma(z) - \sigma(xyz) + \sigma(xy)] + [-\sigma(y) + \sigma(yz) - \sigma(z)] =$$

$$\sigma(y) - \sigma(xy) + \sigma(xyz) - \sigma(yz) +$$

$$[\sigma(z) - \sigma(xyz) + \sigma(xy)] + [-\sigma(y) + \sigma(yz) - \sigma(z)] =$$

$$\sigma(y) + [\sigma(z) - \sigma(xyz) + \sigma(xy)] - \sigma(xy) + \sigma(xyz) - \sigma(yz) + [-\sigma(y) + \sigma(yz) - \sigma(z)] =$$

$$\sigma(y) + \sigma(z) - \sigma(yz) + [-\sigma(y) + \sigma(yz) - \sigma(z)] =$$

$$\sigma(y) + [-\sigma(y) + \sigma(yz) - \sigma(z)] + \sigma(z) - \sigma(yz) = 0.$$

Therefore $\phi = -\delta^1\sigma$ is a k-rational regular factor system which makes G birationally k-isomorphic to the explicit central extension G_ϕ defined by (2.3). In fact, define a k-rational regular map $\rho : G_\phi \longrightarrow G$ by $\rho(b, a) = \sigma(b)a$. This turns out to be a birational isomorphism, the inverse of which is $\rho^{-1}(g) = (\pi(g), g(\sigma\pi(g))^{-1})$.

For any other k-rational regular cross section τ we have, by definition, $\pi\sigma = \pi\tau = id$. This forces $(\sigma - \tau)(B) \subseteq \ker \pi$, hence $\sigma - \tau : B \longrightarrow A$ and $\delta^1(\sigma - \tau)$ is a trivial factor system. Thus the factor systems $\delta^1\sigma$ and $\delta^1\tau$ defined by two cross sections give equivalent extensions, and, up to exchanging σ with $\sigma' : x \mapsto \sigma(x)\sigma(1)^{-1}$, we can always assume that $\sigma(1) = 1$. Moreover, a group G having a k-rational regular cross section σ characterises a unique class of equivalent factor systems of $H^2_k(B, A)$, and G is birationally k-isomorphic to the direct product $A \times B$ if and only if there exists a k-rational regular section $\sigma : B \longrightarrow G$ which is a homomorphism (injective since $\pi\sigma = id$).

2.1.4 Remark. As shown, the possibility that G is birationally k-isomorphic to a suitable explicit central extension G_ϕ depends on the existence of a k-rational regular cross section. In [77], Theorem 10, p. 426 (see also [83]), it is proved that, if A is k-split, then a k-rational cross section σ exists. In general, however, σ is not a regular map. In this case $\phi = -\delta^1\sigma$ is a k-rational, but not regular, factor system, defining, according to Weil's construction in [98], a *pre-group* $G(\phi)$ which is not a group, the law of composition being not defined everywhere. However, a birational map exists which transforms the law of composition of $G(\phi)$ into the one of G. (cf. [98], Théorème p. 375 or [101], Théorème 15, p. 136, [89], Lemme 8, p. 89). The group G is therefore not explicitly described, since the factor system giving $G(\phi)$ is not defined everywhere. A natural candidate to describe this situation is a *toroidal group* in the sense of Rosenlicht [80], i.e. a connected algebraic group containing no unipotent element. Tori, abelian varieties and algebraic groups with a torus as the maximal connected affine subgroup are all toroidal. By [80], Theorem 2, p. 986, any regular map $\phi : V \times W \longrightarrow A$, where V and W are varieties and A is a toroidal group, has the shape $\phi(v, w) = \phi_1(v) + \phi_2(w)$ for suitable regular mappings $\phi_1 : V \longrightarrow A$ and $\phi_2 : W \longrightarrow A$. If ϕ is a k-rational regular factor system from $B \times B$ to A, where B is an algebraic group, then

$$0 = \phi(v, w) + \phi(vw, z) - \phi(w, z) - \phi(v, wz) =$$

$$\phi_1(v) + \phi_2(w) + \phi_1(vw) + \phi_2(z) - \phi_1(w) - \phi_2(z) - \phi_1(v) - \phi_2(wz)$$

for all $v, w, z \in B$, and from this it follows that ϕ_1 and ϕ_2 are constant maps. This shows that there exist no regular non-trivial factor systems into a toroidal group.

Let for instance E be a smooth elliptic curve defined over a finite field and let $J_\mathfrak{m}$ be the generalized Jacobian of E defined, according to [76], Theorem 7, p. 518 in § 3, by the modulus $\mathfrak{m} = (M) + (N)$, where M, N are two distinct non-zero points of E. This means that $J_\mathfrak{m}$ is a connected commutative algebraic group having no non-trivial affine image and containing a one-dimensional torus T as maximal affine subgroup such that $J_\mathfrak{m}/T$ is isomorphic to E. By Proposition 1.3.3 the group $J_\mathfrak{m}$ cannot be defined over a finite field.

According to [17], Theorem 5, one can define on the set $T \times E$ a pre-group operation by

$$(k_1, P_1) + (k_2, P_2) = (k_1 \cdot k_2 \cdot \phi(P_1, P_2), P_1 + P_2)$$

putting

$$\phi(P_1, P_2) = \frac{\ell_{P_1, P_2}(M)}{\ell_{P_1 + P_2, O}(M)} \cdot \frac{\ell_{P_1 + P_2, O}(N)}{\ell_{P_1, P_2}(N)}$$

where $\ell_{P,Q}(X) = 0$ is the equation of the line through P and Q (tangent at E if $P = Q$) and O is the zero of E. A birational map exists which transforms the law of composition of the pre-group $T \times E$ into the one of J_{m}, but we observe that ϕ is defined only if $P_1, P_2, \pm(P_1 + P_2) \notin \{M, N\}$. Therefore this is an example of an extension of the one-dimensional torus T by the elliptic curve E defined by a rational factor system, which cannot be defined by a regular factor system. $\square$

2.1.5 Remark. In [82], Theorem 1, p. 99 or Corollary 1, p. 100, Rosenlicht shows that sufficient conditions for the existence of k-rational regular cross sections are that A is k-split and unipotent and that B is affine. In contrast to this, let G be a connected nilpotent linear algebraic group defined over a separably algebraically closed non-perfect field k such that its unipotent part G_u is not defined over k. The maximal torus T of G is defined over k and G is a T-principal fiber space over G/T. Then there is no regular cross section $\sigma : G/T \to G$ that is defined over k (see [82], p. 100).

Now we give a concrete example for the situation taking for $T = \mathbf{G}_m$ a one-dimensional torus defined over a non-perfect field k of characteristic $p > 0$. Let E be a purely inseparable extension of k of degree $[\mathsf{E} : \mathsf{k}] = n = p^t$. We consider the connected commutative k-group ΠT of [91], Section 12.4, that we recalled in Remark 1.3.9. The group ΠT has dimension n and contains T up to a birational k-isomorphism. There exists a surjective E-homomorphism $\rho : \Pi T \longrightarrow T$, the kernel $\ker \rho$ of which is the unipotent radical of ΠT, is connected, has dimension $n - 1$ and does not contain non-trivial algebraic k-subgroups of ΠT. In particular, $\ker \pi$ is not defined over k. As a connected commutative algebraic group, over E the group ΠT is isomorphic to the direct sum of its unipotent radical ΠT_u with its maximal torus T (see [8] Theorem 10.6, p. 137). Assume by contradiction that the exact sequence

$$1 \longrightarrow T \longrightarrow \Pi T \xrightarrow{\ \pi\ } \Pi T/T \longrightarrow 1$$

has a k-rational regular cross section. Then there exists also a k-regular cross section σ with $0 = \sigma\pi(0)$, and the map $\psi : g \mapsto g(\sigma\pi(g))^{-1}$ is a morphism $\Pi T \to T$ sending 0 into 0. Hence ψ is a rational homomorphism, which is separable since ψ is the identity on T (cf. [45], Theorem, p. 44) and defined over k. This implies that its kernel ΠT_u is defined over k, which is a contradiction. $\square$

Before leaving the questions of rationality and turning to the connections between factor systems and isogenies, we want to remark once more that the existence of a k-rational regular factor system is guaranteed if A is a unipotent group defined over a perfect field k and B is affine.

For any factor system $\phi \in \mathsf{C}^2(B, A)$ there is precisely one factor system ϕ_0, equivalent to ϕ, satisfying $\phi_0(1, 1) = 0$. In the group G_{ϕ_0}, equivalent to G_ϕ, we have the useful identity $(b, a) = (b, 0)(1, a)$. Given a factor system $\phi \in \mathsf{C}^2(B, A)$, one can construct others in the following way. If $f : A \longrightarrow A$ and

$g : B \longrightarrow B$ are rational epimorphisms we define $f\phi$ by $f\phi(x, y) = f(\phi(x, y))$ and we define ϕg by $\phi g(x, y) = \phi(g(x), g(y))$. The maps $f\phi$ and ϕg are factor systems and we have:

$$f(\phi + \psi) = f\phi + f\psi, \quad (\phi + \psi)g = \phi g + \psi g, \quad (f\phi)g = f(\phi g).$$

Moreover, we get an induced epimorphism $\hat{f}$ from G_ϕ onto $G_{f\phi}$ and an induced epimorphism $\hat{g}$ from $G_{\phi g}$ onto G_ϕ by:

$$\hat{f}(b, a) = (b, f(a)) \quad \text{and} \quad \hat{g}(b, a) = (g(b), a).$$

We note that f (respectively g) is an isogeny if and only if $\hat{f}$ (respectively $\hat{g}$) is one.

One has $fB^2(B, A) \subseteq B^2(B, A)$ and $B^2(B, A)g \subseteq B^2(B, A)$. Thus we obtain actions of the rational endomorphisms of A, respectively B, on $H^2(B, A)$. We denote by $[\phi]$ the coset $\phi + B^2(B, A)$ and by $G_{[\phi]}$ the set of extensions equivalent to G_ϕ. For rational endomorphisms $f : A \longrightarrow A$, $g : B \longrightarrow B$ and $[\phi] \in H^2(A, B)$, one has the actions given by $f \cdot [\phi] = [f\phi]$, $[\phi] \cdot g = [\phi g]$.

As the group A is commutative, the set $\mathsf{End}(A)$ of rational endomorphisms of A is a ring and the action of $\mathsf{End}(A)$ on $H^2(B, A)$ just defined makes $H^2(B, A)$ an $\mathsf{End}(A)$-module. It must be observed, however, that in general $H^2(B, A)$ is not an $\mathsf{End}(B)$-module, even if B is commutative, because in general the element $\phi(g_1 + g_2) - \phi g_1 - \phi g_2$ does not belong to $B^2(B, A)$, as the following Remark shows.

2.1.6 Remark. To see concretely that the right action of $\mathsf{End}(B)$ on $H^2(B, A)$ does not define a module structure, let $A = B = \mathbf{G}_a$ be the connected unipotent one-dimensional additive group, over a perfect field of characteristic p. It is shown in [18], II, § 3, 4.6, that $H^2(\mathbf{G}_a, \mathbf{G}_a)$ is a free left $\mathsf{End}(\mathbf{G}_a)$-module, having the following family of polynomials as a basis (modulo $B^2(\mathbf{G}_a, \mathbf{G}_a)$):

$$\Phi_1(x, y) = \sum_i \frac{(p-1)!}{i!(p-i)!} x^i y^{p-i};$$
$$\eta_j(x, y) = xy^{p^j} \quad (j = 1, 2, \cdots).$$

Put $g_1(t) = t$ and $g_2(t) = t^p$ and consider the factor system $\theta = \eta_1(g_1 + g_2) - (\eta_1 g_1 + \eta_1 g_2)$. Then we have

$$\theta(x, y) = (x + x^p)(y + y^p)^p - xy^p - x^p y^{p^2} = x^p y^p + xy^{p^2}.$$

Since $\theta(x, y) \neq \theta(y, x)$, we infer that $\theta \notin B^2(B, A)$, i.e. $H^2(\mathbf{G}_a, \mathbf{G}_a)$ is not a right $\mathsf{End}(\mathbf{G}_a)$-module. Moreover, if $p > 2$ then we have $x^p y^p = \frac{1}{2}[(x + y)^{2p} - x^{2p} - y^{2p}] \in B^2(B, A)$, thus θ is equivalent to η_2. On the other hand, for $p = 2$ we have $x^2 y^2 = \Phi_1(x, y)^2$, thus θ is equivalent to $\Phi_1^2 + \eta_2$. $\qquad\square$

2.1.7 Remark. The above computation shows that in characteristic 2 it can happen that the factor set $\eta_k g$ does not belong to the *left* $\mathsf{End}(A)$-submodule

M generated by the set $\{\eta_j : j = 1, 2, \cdots\}$. In odd characteristic this is not possible, because any group G_{η_j} has exponent p, whereas for $\phi \notin M$ the group G_ϕ has exponent p^2. More details on the right action of $\mathsf{End}(B)$ on $\mathsf{H}^2(\mathbf{G}_a, \mathbf{G}_a)$ can be found in Proposition 2.1.9.   □

2.1.8 Remark. For the basis element Φ_1 and an arbitrary p-polynomial $g(t) = \sum_i a_i t^{p^i}$ it is easy to check that

$$[\Phi_1 g] = [\tilde{g}\Phi_1],$$

where $\tilde{g}(t) = \sum_i a_i^p t^{p^i}$. Therefore the submodule of $\mathsf{H}^2(\mathbf{G}_a, \mathbf{G}_a)$ consisting of symmetric factor systems is a two-sided module over the ring $\mathsf{End}(\mathbf{G}_a)$, and this is basic for the fact that, given an isogeny $\gamma_1 : G_1 \longrightarrow G_2$ of two-dimensional commutative unipotent algebraic groups, one finds an isogeny $\gamma_2 : G_2 \longrightarrow G_1$ (see Proposition 2.1.12 *(i)* or [89], § VII, n. 10). More generally this is possible by the same reason for n-dimensional commutative unipotent algebraic groups. But it is by no means possible for non-commutative factor systems, as Example 2.1.14 shows.

If however we restrict our attention to the subspace generated a monomial $g(t) = at^{p^k}$ we obtain

$$\eta_j g(x, y) = a x^{p^k} \cdot (a y^{p^k})^{p^j} = a^{1+p^j} (x y^{p^j})^{p^k} = a^{1+p^j} (\eta_j(x, y))^{p^k}.$$

Putting $\tilde{g}_j(t) = a^{1+p^j} t^{p^k}$ we get $\eta_j g = \tilde{g}_j \eta_j$.   □

In the following proposition we give a general formula for the factor systems $\eta_j \in \mathsf{H}^2(\mathbf{G}_a, \mathbf{G}_a)$, a special case of which has been used in the above Remark 2.1.6. We recall that the ring $\mathsf{End}_k(\mathbf{G}_a)$ of k-endomorphisms of the additive group $\mathbf{G}_a$ is isomorphic to the non-commutative ring $k[\mathbf{F}]$ of p-polynomials, where $\mathbf{F}$ is the Frobenius homomorphism and

$$\sum_i \alpha_i \mathbf{F}^i : x \mapsto \sum_i \alpha_i x^{p^i}.$$

The following proposition shows that in odd characteristic any factor system can be derived by Φ_1 and η_1 only.

2.1.9 Proposition. *If the characteristic of the ground field is greater than 2, then in the free left* $\mathsf{End}(\mathbf{G}_a)$*-module* $\mathsf{H}^2(\mathbf{G}_a, \mathbf{G}_a)$ *we have*

$$[\eta_{2k}] = \sum_{i=0}^{k-1} \mathbf{F}^i [\eta_1 (1 + \mathbf{F}^{2(k-i)-1})] - \left(\sum_{i=0}^{2k-1} \mathbf{F}^i \right) [\eta_1]$$

$$[\eta_{2k+1}] = \sum_{i=0}^{k} \mathbf{F}^i [\eta_1 (1 + \mathbf{F}^{2(k-i)})] + \mathbf{F}^k [\eta_1] - \sum_{i=1}^{k-1} \mathbf{F}^i [\eta_1] - \sum_{i=1}^{k} \mathbf{F}^{k+i} [\eta_1].$$

Proof. Put $\alpha(x) = \frac{1}{2}x^{2p}$, hence $\delta^1\alpha(x,y) = \frac{1}{2}((x+y)^{2p} - x^{2p} - y^{2p}) = x^p y^p$. The assertion follows from the fact that

$$\eta_2 = \eta_1(1+\mathbf{F}) - (1+\mathbf{F})\eta_1 + \delta^1\alpha$$

whereas, for any $k > 1$, we have

$$\eta_k(1+\mathbf{F}) = (1+\mathbf{F})\eta_k + \eta_{k+1} + \mathbf{F}\eta_{k-1}.$$

$\square$

2.1.10 Corollary. *If the characteristic of the ground field is greater than 2, for any $\varphi \in \mathsf{H}^2(\mathbf{G}_a, \mathbf{G}_a)$ there exist $f_0, f_1, \cdots f_n, g_1, \cdots, g_n \in \mathsf{k}[\mathbf{F}]$ such that*

$$\varphi = f_0 \Phi_1 + \sum_{k=1}^{n} f_k \eta_1 g_k.$$

$\square$

The following Remark 2.1.11, which makes Propositions 2.1.12 and 2.1.13 particularly meaningful, plays a certain rôle in Section 4.2.

2.1.11 Remark. We illustrate here the fact that the functor $\mathsf{H}^2(B, A)$ is contra-variant in B and co-variant in A in the special case of isogenies. The arguments and the notations are essentially those of [89], VII, 1. p. 164-165.

1) For any explicit central extension G_{ϕ_1}

$$1 \longrightarrow A_1 \longrightarrow G_{\phi_1} \longrightarrow B \longrightarrow 1$$

of A_1 by B, defined by the factor system $\phi_1 : B \times B \longrightarrow A_1$, and any isogeny $\alpha : A_1 \longrightarrow A_2$ there exists a unique (up to equivalence) explicit central extension G_{ϕ_2}

$$1 \longrightarrow A_2 \longrightarrow G_{\phi_2} \longrightarrow B \longrightarrow 1$$

and an isogeny $\alpha_* : G_{\phi_1} \longrightarrow G_{\phi_2}$, such that the following diagram commutes

$$\begin{array}{ccccccccc} 1 & \longrightarrow & A_1 & \longrightarrow & G_{\phi_1} & \longrightarrow & B & \longrightarrow & 1 \\ & & \alpha\downarrow & & \alpha_*\downarrow & & id_B\downarrow & & \\ 1 & \longrightarrow & A_2 & \longrightarrow & G_{\phi_2} & \longrightarrow & B & \longrightarrow & 1. \end{array}$$

Explicitly we have $[\phi_2] = \alpha[\phi_1]$ and α_* is defined by $\alpha_*(x,y) = (x, \alpha(y))$. Moreover, the group G_{ϕ_2} is the factor group of $G_{\phi_1} \times A_2$ modulo the algebraic subgroup $\Delta = \{(-a, \alpha(a)) : a \in A_1\}$.

2) For any explicit central extension G_{ϕ_1}

$$1 \longrightarrow A \longrightarrow G_{\phi_1} \overset{\pi}{\longrightarrow} B_1 \longrightarrow 1$$

of A by B_1, defined by the factor system $\phi_1 : B_1 \times B_1 \longrightarrow A$, and any isogeny $\beta : B_2 \longrightarrow B_1$ there exists a unique explicit central extension G_{ϕ_2}

$$1 \longrightarrow A \longrightarrow G_{\phi_2} \overset{\pi}{\longrightarrow} B_2 \longrightarrow 1$$

and an isogeny $\beta^* : G_{\phi_2} \longrightarrow G_{\phi_1}$, such that the following diagram commutes

$$\begin{array}{ccccccccc} 1 & \longrightarrow & A & \longrightarrow & G_{\phi_2} & \overset{\pi}{\longrightarrow} & B_2 & \longrightarrow & 1 \\ & & id_A \downarrow & & \beta^* \downarrow & & \beta \downarrow & & \\ 1 & \longrightarrow & A & \longrightarrow & G_{\phi_1} & \overset{\pi}{\longrightarrow} & B_1 & \longrightarrow & 1. \end{array}$$

Explicitly we have $[\phi_2] = [\phi_1]\beta$ and the isogeny β^* is defined by $\beta^*(x, y) = (\beta(x), y)$. Moreover, the group G_{ϕ_2} is the algebraic subgroup of the direct product $B_2 \times G_{\phi_1}$ defined by

$$G_{\phi_2} = \{(b, g) \in B_2 \times G_{\phi_1} : \beta(b) = \pi(g)\}.$$

$\square$

In the case where G_{ϕ_1} and G_{ϕ_2} are two central extensions

$$1 \longrightarrow A_1 \longrightarrow G_{\phi_1} \longrightarrow B_1 \longrightarrow 1$$

$$1 \longrightarrow A_2 \longrightarrow G_{\phi_2} \longrightarrow B_2 \longrightarrow 1$$

we have:

2.1.12 Proposition. *Let G_{ϕ_1} (respectively G_{ϕ_2}) be a central extension of the algebraic affine group A_1 (respectively A_2) by the (not necessarily commutative) algebraic group B_1 (respectively B_2).*

(i) There is an isogeny $i : G_{\phi_1} \longrightarrow G_{\phi_2}$ such that $i(A_1) = A_2$ if and only if there exist isogenies $f : A_1 \longrightarrow A_2$ and $g : B_1 \longrightarrow B_2$ with $f[\phi_1] = [\phi_2]g$.
(ii)If there are isogenies $f : A_1 \longrightarrow A_2$ and $g : B_2 \longrightarrow B_1$ such that

$$[\phi_2] = f[\phi_1]g \tag{2.4}$$

then the groups G_{ϕ_1} and G_{ϕ_2} are isogenous.

Proof. (i) Let $i : G_{\phi_1} \longrightarrow G_{\phi_2}$ be an isogeny such that $i(A_1) = A_2$. Then i is given by

$$i(x_0, x_1) = (g(x_0), f(x_1) + h(x_0)),$$

where $f : A_1 \longrightarrow A_2$, $g : B_1 \longrightarrow B_2$ are isogenies and $h : B_1 \longrightarrow A_2$ is a rational regular map satisfying the equation $f\phi_1 = \phi_2 g + \delta^1 h$.

Conversely, given isogenies g, f as above and a rational regular map $h : B_1 \longrightarrow A_2$ satisfying $f\phi_1 = \phi_2 g + \delta^1 h$, the mapping $i : G_{\phi_1} \longrightarrow G_{\phi_2}$

$$(x_0, x_1) \mapsto (g(x_0), f(x_1) + h(x_0))$$

is an isogeny, since

$$\mathsf{ker}(i) = \{(b,a) : b \in \mathsf{ker}(g), f(a) + h(b) = 0\}$$

is finite, and $i(A_1)$ is clearly equal to A_2.

(ii) By *(i)* we find that G_{ϕ_1} is isogenous to $G_{f_1\phi_1}$ which in turn is isogenous to $G_{f_1\phi_1g_1} = G_{\phi_2}$.   □

The next proposition shows the crucial rôle played by the Ore condition in the context of extensions of algebraic groups and isogenies.

2.1.13 Proposition. *Let A (respectively A_1, A_2) be either the Witt group $\mathfrak{W}_m$ or the vector group $(\mathbf{G}_a)^m$. Let G_ψ (respectively G_{ϕ_1}, G_{ϕ_2}) be a central extension of A (respectively A_1, A_2) by the (not necessarily commutative) algebraic group B (respectively B_1, B_2). If $\eta_i : G_\psi \longrightarrow G_{\phi_i}$ are isogenies with $\eta_i(A) = A_i$, then there exist $h_i : A_i \longrightarrow A$ and $g_i : B \longrightarrow B_i$ such that $h_2[\phi_2]g_2 = h_1[\phi_1]g_1$.*

Proof. By *(i)* there exist isogenies $f_i : A \longrightarrow A_i$ and $g_i : B \longrightarrow B_i$, $(i = 1, 2)$, such that $f_1[\psi] = [\phi_1]g_1$ and $f_2[\psi] = [\phi_2]g_2$. It is shown in [18], V, § 3, 6.9, p. 593, that in the semigroup of isogenies of a Witt group the Ore condition holds. For a vector group this follows from [54], § 10, p. 313. Hence can find two isogenies $h_i : A_i \longrightarrow A$ such that $h_1f_1 = h_2f_2$ and we obtain $h_2[\phi_2]g_2 = h_1[\phi_1]g_1$.   □

We have already mentioned that the groups SL_2 and PSL_2 show that the existence of an isogeny is not a symmetric relation. However, if there exists an isogeny from a connected commutative unipotent group G_1 onto G_2 then an isogeny from G_2 onto G_1 exists as well (see [89], Proposition 10, p. 176). Already for unipotent connected non-commutative algebraic groups this is not any more the case as the following example shows.

2.1.14 Example. In Remark 2.1.6 we denoted by $\eta_1 : \mathbf{G}_a \times \mathbf{G}_a \longrightarrow \mathbf{G}_a$ the factor system defined by $\eta_1(x, y) = xy^p$. As soon as the p-polynomial g is not monomial, the factor system η_1g is no longer contained in the left $\mathsf{End}(\mathbf{G}_a)$-submodule generated by η_1. Therefore a necessary condition to have the equality $f[\eta_1] = [\eta_1]g$, for some p-polynomial f, is that g is monomial, that is $g = a\mathbf{F}^k$ (see Remark 2.1.8). But in this case we have $\eta_1g = \tilde{g}\eta_1$, where $\tilde{g} = a^{1+p}\mathbf{F}^k$. This shows that, if f is a p-polynomial which is not a monomial, it cannot happen that $f[\eta_1] = [\eta_1]g$, because the left $\mathsf{End}(\mathbf{G}_a)$-submodule generated by η_1 is free. By Proposition 2.1.12 *(i)*, there cannot exist an isogeny from $G_{f[\eta_1]}$ to $G_{[\eta_1]}$ whereas by the same Proposition we have an isogeny from $G_{[\eta_1]}$ to $G_{f[\eta_1]}$.

     □

2.1.15 Remark. Non-central extensions of a group A by a group B are in general described by an action of B as a non-trivial group of automorphisms of A

$$a \mapsto a^b \qquad (a \in A, b \in B),$$

and a mapping $F : B \times B \longrightarrow A$, satisfying

$$F(b_1b_2, b_3) \cdot F(b_1, b_2)^{b_3} = F(b_1, b_2b_3) \cdot F(b_2, b_3). \tag{2.5}$$

For the trivial action of B on a commutative group A, the equation (2.5) just reduces to the functional equation of a factor system describing a central extension, as in Section 2.1. With a slight abuse, we call a mapping F satisfying (2.5) a factor system. If A and B are algebraic groups, it is necessary to assume that the factor system F and all the automorphisms $a \mapsto a^b$ for all $b \in B$ are rational maps, in order to have the extension of A by B as an algebraic group.

Let B_α be the central extension

$$1 \longrightarrow B_1 \longrightarrow B_\alpha \longrightarrow B_2 \longrightarrow 1$$

defined on $B_2 \times B_1$ by the product

$$(b_0, b_1)(b'_0, b'_1) = (b_0 \cdot b'_0, b_1 + b'_1 + \alpha(b_0, b'_0))$$

and let G_ϕ be the central extension

$$1 \longrightarrow A \longrightarrow G_\phi \longrightarrow B_\alpha \longrightarrow 1$$

defined on $B \times A$ by the product

$$((b_0, b_1), a) \cdot ((b'_0, b'_1), a') = ((b_0 \cdot b'_0, b_1 + b'_1 + \alpha(b_0, b'_0)!), a + a' + \phi(b_0, b_1, b'_0, b'_1)). \tag{2.6}$$

Let $H = \{(b_0, b_1, a) \in G_\phi : b_0 = 1\}$. Under the assumption that $[G, H] \leq A$ we want to find the factor system γ corresponding to the section $\tau : B_2 \longrightarrow G_\phi$, $\tau(b_0) = (b_0, 0, 0)$ of the non-central extension

$$1 \longrightarrow H \longrightarrow G_\phi \longrightarrow B_2 \longrightarrow 1$$

and we want to compare this factor system with ϕ. With the same argument mentioned in Section 2.1 for central extension, one can easily see that such a factor system $\gamma = (\gamma_1, \gamma_2) : B_2 \times B_2 \longrightarrow H$ is $\gamma = -\delta^1\tau$. (It is remarkable that the effects of changing the section for a non-central extension are not those of adding a trivial factor system, because in this case $\delta^1(\tau - \tau') \neq \delta^1\tau - \delta^1\tau'$. For a concrete example see the proof of Theorem 6.4.7.) A direct computation shows now that

$$\gamma(b_0, b'_0) = (b_0 \cdot b'_0, 0, 0)^{-1} \cdot (b_0, 0, 0) \cdot (b'_0, 0, 0) = (1, \alpha(b_0, b'_0), \beta(b_0, b'_0))$$

where α is the map appearing in (2.6) and

$$\beta(b_0, b'_0) = -\phi(b_0b'_0, 0; (b_0b'_0)^{-1}, -\alpha(b_0b'_0, (b_0b'_0)^{-1})) +$$

$$\phi(b_0, 0; b'_0, 0) + \phi((b_0b'_0)^{-1}, -\alpha(b_0b'_0, (b_0b'_0)^{-1}); b_0b'_0, \alpha(b_0, b'_0)).$$

Therefore the group G_ϕ is isomorphic to the group defined on $B_2 \times H$ by the multiplication

$$((b_0, b_1), a) \cdot ((b_0', b_1'), a') = (b_0, 0, 0) \cdot (1, b_1, a) \cdot (b_0', 0, 0) \cdot (1, b_1', a') =$$

$$(b_0, 0, 0) \cdot (b_0', 0, 0) \cdot (1, b_1, a)^{(b_0, 0, 0)} \cdot (1, b_1', a') =$$

$$(b_0 b_0', 0, 0) \cdot (1, \alpha(b_0, b_0'), \beta(b_0, b_0')) \cdot (1, b_1, a + \sigma_{b_0}(b_1)) \cdot (1, b_1', a') =$$

$$(b_0 b_0', b_1 + b_1' + \alpha(b_0, b_0'), a + a' + \rho(b_0, b_1, b_0', b_1'))$$

where

$$\rho(b_0, b_1, b_0', b_1') = \sigma_{b_0}(b_1) + \beta(b_0, b_0') + \phi(1, b_1, 1, b_1') + \phi(1, \alpha(b_0, b_0'), 1, b_1 + b_1')$$

whereas for any $b_0 \in B_2$ and for any $(1, b_1, a) \in H$ the map $\sigma_{b_0} : B_1 \longrightarrow A$ is a homomorphism such that $(1, b_1, a)^{b_0} = (1, b_1, a + \sigma_{b_0}(b_1))$.

Comparing the representation given by γ with the one given by ϕ we find the remarkable fact that $\gamma_1 = \alpha$ whereas ρ is in general different from ϕ. □

The universal covering $\mathbb{C}^n$ of an arbitrary connected commutative complex Lie group G is a decisive tool for the description of homomorphisms and extensions of connected commutative complex Lie groups. It plays a similar rôle as the Witt group $\mathfrak{W}_n$ for connected commutative unipotent groups. For non-commutative unipotent groups unfortunately no similar tool is available.

2.2 Extensions of Commutative Lie Groups

Since any commutative connected complex Lie group is (holomorphically) isomorphic to the direct product of a linear torus $(\mathbb{C}^*)^m$, a vector group $\mathbb{C}^l$ and a toroidal group X, the theory of commutative extensions of such Lie groups reduces to the case of extensions which are toroidal groups. These groups play a similar rôle as the connected algebraic group $G = D(G)$ with no non-trivial affine epimorphic image.

Homomorphisms and extensions of complex tori X are completely described in [7], Ch. 1, Section 5, by means of *period matrices*, the columns of which are the vectors of the lattice Λ of a suitable representation of $X = \mathbb{C}^n / \Lambda$. This method works also for connected commutative complex Lie groups $G = \mathbb{C}^n / \Lambda$ such that the complex rank of Λ is n, which we will treat now.

Let $X = \mathbb{C}^n / \Lambda$ be a connected commutative complex Lie group. If the complex rank of Λ is $m < n$, then Λ is contained in a complex subspace V of dimension m of $\mathbb{C}^n$. Up to a change of basis and a canonical identification of V with $\mathbb{C}^m$, we can see then that X is isomorphic to $\mathbb{C}^{n-m} \oplus \mathbb{C}^m / \Lambda$. From now on we assume therefore that the complex rank of Λ is n, and we say that such groups have *maximal complex rank*. Let the real rank of Λ be $n + q$, where $0 \le q \le n$.

Up to a change of basis we can assume that $\Lambda = \mathbb{Z}^n \oplus \Gamma$. The corresponding column matrix is

$$P = (I_n, G) = \begin{pmatrix} I_q & 0 & \widehat{T} \\ 0 & I_{n-q} & \widetilde{T} \end{pmatrix} \in M_{n,n+q}(\mathbb{C})$$

where the columns of $G = \begin{pmatrix} \widehat{T} \\ \widetilde{T} \end{pmatrix}$ are $\mathbb{R}$-independent generators of Γ.

In accordance to [7], p. 2, we call P the *period matrix* of X. The imaginary part of G has real rank q, because the columns of P are $\mathbb{R}$-independent. Up to a permutation of the vectors of the basis we can assume that the imaginary part of $\widehat{T}$ is invertible.

For $q = 0$ we have $\Lambda = \mathbb{Z}^n$, hence the group $X = \mathbb{C}^n/\mathbb{Z}^n \cong (\mathbb{C}^*)^n$ is a linear torus, whereas for $q = n$ the group X is a complex torus by definition ([7], p. 1). According to [1], 1.1.11, p. 9, if $P = (I_n\ G)$ is the matrix of a $\mathbb{R}$-basis of the lattice Λ, the group $\mathbb{C}^n/\Lambda$ is toroidal if and only the following *irrationality condition* holds:

for any non-zero $\mathbf{v} \in \mathbb{Z}^n$ the vector $\mathbf{v}G$ is never contained in $\mathbb{Z}^q$.   (2.7)

Homomorphisms of connected commutative complex Lie groups of maximal complex rank can be described in terms of period matrices. In fact, a homomorphism $f : X_1 = \mathbb{C}^{n_1}/\Lambda_1 \longrightarrow X_2 = \mathbb{C}^{n_2}/\Lambda_2$ lifts to a unique homomorphism $\hat{f} : \mathbb{C}^{n_1} \longrightarrow \mathbb{C}^{n_2}$ of $\mathbb{C}$-vector spaces such that $\hat{f}(\Lambda_1) \leq \Lambda_2$. This lifting defines therefore two homomorphisms

$$\rho_a : \mathsf{Hom}(X_1, X_2) \longrightarrow \mathsf{Hom}(\mathbb{C}^{n_1}, \mathbb{C}^{n_2}) \cong M_{n_2,n_1}(\mathbb{C})$$

$$\rho_r : \mathsf{Hom}(X_1, X_2) \longrightarrow \mathsf{Hom}(\Lambda_1, \Lambda_2) \cong M_{n_2+q_2,n_1+q_1}(\mathbb{Z})$$

such that

$$\rho_a(f)P_1 = P_2\rho_r(f),   (2.8)$$

where P_i is a period matrix of X_i ($i = 1, 2$) and where we have identified $\rho_a(f)$ and $\rho_r(f)$ with the matrices corresponding to the chosen basis of $\mathbb{C}^{n_i}$. The homomorphisms ρ_a and ρ_r are called the *analytic* and the *rational representation* of $\mathsf{Hom}(X_1, X_2)$ and the equations in (2.8) are called *Hurwitz relations* (cf. [1], p. 8).

2.2.1 Proposition. *A homomorphism $f : X_1 \longrightarrow X_2$ is an isogeny if and only if $\rho_a(f)$ and $\rho_r(f)$ are square matrices with non-zero determinant. In this case there exists an isogeny $g : X_2 \longrightarrow X_1$ with $fg = l\,id_{X_2}$ and $gf = l\,id_{X_1}$ where $l = |\rho_r(f)|$. In particular, the isogeny f is an isomorphism if and only if $|\rho_r(f)| = \pm 1$.*

Proof. If f is an isogeny, then $\hat{f} = \rho_a(f)$ is bijective for dimensional reasons, hence $|\rho_a(f)| \neq 0$. If we put $\Gamma = \hat{f}^{-1}(\Lambda_2)$, then $\Lambda_1 \leq \Gamma$ and Γ/Λ_1 is the kernel of the isogeny f. If the real rank of Λ_2 were greater than the real rank of Λ_1, then Γ/Λ_1 would be infinite. As $\rho_a(f)P_1 = P_2\rho_r(f)$ we find $\Lambda_1 = \Gamma\rho_r(f)$. Hence we have: 1) the real rank of Λ_1 is not greater than the real rank of Λ_2, since Λ_2 has the same real rank as Γ, 2) $\rho_r(f)$ is a square matrix with non-zero determinant.

Conversely, if $\rho_a(f)$ and $\rho_r(f)$ are square matrices with non-zero determinant, then f is surjective, its kernel is discrete and Λ_1 and $\Gamma = \hat{f}^{-1}(\Lambda_2)$ have the same real rank. As the factor group Γ/Λ_1 is the kernel of f, it has to be finite, proving that f is an isogeny.

Finally, let $l = |\rho_r(f)|$ and let $R = l\rho_r(f)^{-1}$, hence R has integral entries. Since $l\rho_a(f)^{-1}P_2 = P_1R$ we can define a homomorphism $g : X_2 \longrightarrow X_1$ such that $\rho_a(g) = l\rho_a(f)^{-1}$ and $\rho_r(g) = R$. Since $\rho_a(g)$ and $\rho_r(g)$ are square matrices with non-zero determinant, the homomorphism g is an isogeny and it is easy to see that $fg = l\,id_{X_2}$ and $gf = l\,id_{X_1}$.

In particular, if $l = \pm 1$ the isogeny f is an isomorphism. Conversely, if f is an isomorphism, then the rational representation $\rho_r(f) : \Lambda_1 \longrightarrow \Lambda_2$ is an isomorphism of lattices having $\rho_r(f^{-1})$ as the inverse, hence $|\rho_r(f)| = \pm 1$. $\square$

Now we want to study closed subgroups and factor groups of toroidal groups as well as holomorphic commutative extensions of toroidal groups by toroidal groups.

2.2.2 Proposition. *Let $X \cong \mathbb{C}^n/\Lambda$ be a connected commutative complex Lie group of maximal rank n. For any k-dimensional connected closed commutative complex subgroup $X_1 = \mathbb{C}^k/\Lambda_1$ of maximal rank k of X there exists a period matrix P such that*

$$P = \begin{pmatrix} P_1 & \Sigma \\ 0 & P_2 \end{pmatrix}$$

where P_1 is a period matrix of X_1 and P_2 is a period matrix of the factor group X/X_1.

Proof. Let P_1 be a period matrix of the closed subgroup $X_1 \cong \mathbb{C}^k/\Lambda_1$ of X. As X_1 is a closed subgroup of X we can construct an exact sequence

$$0 \longrightarrow \mathbb{C}^k/\Lambda_1 \overset{\imath}{\longrightarrow} \mathbb{C}^n/\Lambda \overset{\pi}{\longrightarrow} \mathbb{C}^{n-k}/\Lambda_2 \longrightarrow 0,$$

where Λ_2 is a lattice corresponding to the connected complex commutative Lie group $X/X_1 \cong \mathbb{C}^{n-k}/\Lambda_2$. Consider the linear maps $\hat{\imath} = \rho_a(\imath)$ and $\hat{\pi} = \rho_a(\pi)$. As $\ker \imath = \hat{\imath}^{-1}(\Lambda)/\Lambda_1$ and $\imath$ is injective we have $\hat{\imath}^{-1}(\Lambda) = \Lambda_1$ from which it follows that also $\hat{\imath}$ is injective. Furthermore by the relation $\ker \pi = \hat{\pi}^{-1}(\Lambda_2)/\Lambda = \imath(\mathbb{C}^k/\Lambda_1) = (\hat{\imath}(\mathbb{C}^k) + \Lambda)/\Lambda$ we have $\hat{\pi}^{-1}(\Lambda_2) = \hat{\imath}(\mathbb{C}^k) + \Lambda$, which yields $\ker \hat{\pi} = \hat{\imath}(\mathbb{C}^k)$. Consequently $\hat{\imath}$ and $\hat{\pi}$ define an exact sequence

$$0 \longrightarrow \mathbb{C}^k \overset{\hat{\imath}}{\longrightarrow} \mathbb{C}^n \overset{\hat{\pi}}{\longrightarrow} \mathbb{C}^{n-k} \longrightarrow 0.$$

By the relation $\hat{\pi}^{-1}(\Lambda_2) = \hat{\imath}(\mathbb{C}^k) + \Lambda$ we have that the homomorphism $\hat{\pi}_{|\Lambda} : \Lambda \longrightarrow \Lambda_2$ is surjective and $\ker \hat{\pi}_{|\Lambda} = \hat{\imath}(\Lambda_1)$. This defines an exact sequence

$$0 \longrightarrow \Lambda_1 \xrightarrow{\hat{\imath}} \Lambda \xrightarrow{\hat{\pi}} \Lambda_2 \longrightarrow 0.$$

Since Λ_2 is a free commutative group we get $\Lambda = \imath(\Lambda_1) \oplus \Gamma$ where $\Gamma \cong \Lambda_2$. Up to a change of basis of the spaces $\mathbb{C}^k$, $\mathbb{C}^n$ and $\mathbb{C}^{n-k}$ we can assume that $\hat{\imath} = \rho_a(\imath) = \begin{pmatrix} I_k \\ 0 \end{pmatrix}$, $\hat{\pi} = \rho_a(\pi) = (0 \ I_{n-k})$, and we can choose a period matrix P of X such that $P = \begin{pmatrix} P_1 & \Sigma \\ 0 & A \end{pmatrix}$, where the columns of the matrix $\begin{pmatrix} \Sigma \\ A \end{pmatrix}$ are $\mathbb{R}$-independent $\mathbb{Z}$-generators of Γ. Furthermore, fixing a period matrix P_2 of X/X_1, up to a change of generators of Γ we can assume that $\rho_r(\pi) = (0 \ I_{n+q-k-q_1})$, where $n + q$ (respectively $n_1 + q_1$) is the real rank of X (respectively of X_1). Now, by the Hurwitz relations we have

$$(0 \ I_{n-k}) \begin{pmatrix} P_1 & \Sigma \\ 0 & A \end{pmatrix} = P_2 (0 \ I_{n+q-k-q_1})$$

from which it follows that $A = P_2$. $\qquad\square$

Now we look for closed linear subtori of a connected commutative complex Lie group $X \cong \mathbb{C}^n/\Lambda$ of maximal rank n with period matrix $P = (I_n \ G)$. Denote by $H = H(l_1, \cdots, l_{n-m})$ the m-dimensional subspace of $\mathbb{C}^n$ defined by

$$H = \{(z_1, \cdots, z_n) \in \mathbb{C}^n : z_{l_k} = 0 \text{ for } l_k \in \{1, \cdots, n\} \text{ and } k = 1, \cdots, n-m\}$$

and let $C_H(P)$ be the matrix obtained from P in the following way: we cancel in P any row with exception of those labeled by $l_1, \cdots, l_{n-m}$ as well as any of the first n columns with exception of those labeled by $l_1, \cdots, l_{n-m}$. Clearly $C_H(P) = (I_{n-m} \ G')$, with $G' \in M_{n-m,q}(\mathbb{C})$.

2.2.3 Proposition. *Let $X \cong \mathbb{C}^n/\Lambda$ be a connected commutative complex Lie group of maximal rank n and let $P = (I_n \ G)$ be a period matrix of X. If the columns of $C_H(P)$ are $\mathbb{R}$-independent, then $X_1 = (H + \Lambda)/\Lambda$ is a closed linear subtorus of X.*

Proof. Let $X_2 \cong \mathbb{C}^{n-m}/\Lambda_2$ be the connected commutative complex Lie group of maximal rank $n - m$ having $C_H(P)$ as a period matrix and let $\hat{f} : \mathbb{C}^n \longrightarrow \mathbb{C}^{n-m}$ be the homomorphism defined by $\hat{f}(z_1, \cdots, z_n) = (z_{l_1}, \cdots, z_{l_{n-m}})$. Since $\hat{f}(\Lambda) \leq \Lambda_2$, a homomorphism $f : X \longrightarrow X_2$ is induced such that X_1 is the kernel. This proves that X_1 is a closed subgroup. In order to prove that X_1 is a linear torus we show that $H \cap \Lambda$ has real rank m. This follows from the fact that the columns of $C_H(P)$ are $\mathbb{R}$-independent, hence no non-trivial linear combination of the columns $l_1, \cdots, l_{n-m}$ of the matrix P with integral (or even real) coefficients enters in H. $\qquad\square$

2.2.4 Remark. The above proposition shows that a toroidal group X with period matrix

$$P = (I_n \ G) = \begin{pmatrix} I_q & 0 & \widehat{T} \\ 0 & I_{n-q} & \widetilde{T} \end{pmatrix} \in M_{n,n+q}(\mathbb{C})$$

(such that the imaginary part of $\widehat{T}$ is invertible) contains a closed linear subtorus L of dimension $n - q$ corresponding to the submatrix $P_1 = I_{n-q}$, because the submatrix $C_H(P) = (I_q \ \widehat{T})$ is the period matrix of a complex torus. Hence L is a maximal closed linear subtorus of X. For instance, in the three-dimensional toroidal group X having

$$P = \begin{pmatrix} 1 & 0 & 0 & i & i \\ 0 & 1 & 0 & i\sqrt{2} & 0 \\ 0 & 0 & 1 & 0 & i\sqrt{2} \end{pmatrix}$$

as a period matrix, the three subgroups $H(2,3)$, $H(1,3)$ and $H(1,2)$ are one-dimensional maximal closed linear subtori. Thus X is a $\mathbb{C}^*$-fiber bundle over the complex tori defined by the period matrices

$$C_{H(2,3)} = \begin{pmatrix} 1 & 0 & i\sqrt{2} & 0 \\ 0 & 1 & 0 & i\sqrt{2} \end{pmatrix}, \qquad C_{H(1,3)} = \begin{pmatrix} 1 & 0 & i & i \\ 0 & 1 & 0 & i\sqrt{2} \end{pmatrix},$$

$$C_{H(1,2)} = \begin{pmatrix} 1 & 0 & i & i \\ 0 & 1 & i\sqrt{2} & 0 \end{pmatrix}$$

. $\qquad\qquad\qquad\qquad\qquad\qquad\qquad\qquad\qquad\qquad\qquad\qquad\square$

Let $X_1 = \mathbb{C}^{n_1}/\Lambda_1, X_2 = \mathbb{C}^{n_2}/\Lambda_2$ be connected commutative complex Lie groups of maximal ranks n_1, n_2 and let P_1, P_2 be the corresponding period matrices. Let

$$0 \longrightarrow X_1 \longrightarrow X \longrightarrow X_2 \longrightarrow 0$$

be an exact sequence of connected commutative complex Lie groups. By Proposition 2.2.2 we find a basis such that the corresponding period matrix is

$$P = \begin{pmatrix} P_1 & \Sigma \\ 0 & P_2 \end{pmatrix} \in M_{n,n+q_1+q_2}(\mathbb{C}).$$

Conversely to each matrix of this form there corresponds a toroidal group X containing a closed subgroup X_1 having P_1 as a period matrix and such that X/X_1 is isomorphic to a toroidal group X_2 having P_2 as a period matrix. In fact, X_1 is the kernel of the homomorphism $f : X \longrightarrow X_2$ which lifts to $\hat{f}(z_1, \cdots, z_n) = (z_{n_1+1}, \cdots, z_n)$.

As a consequence of Hurwitz relations we find that $P = \begin{pmatrix} P_1 & \Sigma \\ 0 & P_2 \end{pmatrix}$ and $Q = \begin{pmatrix} P_1 & \Sigma' \\ 0 & P_2 \end{pmatrix}$ define equivalent extensions if and only if a matrix $A \in M_{n_1,n_2}(\mathbb{C})$ and a matrix $M \in M_{n_1+q_1,n_2+q_2}(\mathbb{Z})$ exists such that

$$\begin{pmatrix} I_{n_1} & A \\ 0 & I_{n_2} \end{pmatrix} \begin{pmatrix} P_1 & \Sigma \\ 0 & P_2 \end{pmatrix} = \begin{pmatrix} P_1 & \Sigma' \\ 0 & P_2 \end{pmatrix} \begin{pmatrix} I_{n_1+q_1} & M \\ 0 & I_{n_2+q_2} \end{pmatrix}.$$

The period matrix $P = \begin{pmatrix} P_1 & \Sigma \\ 0 & P_2 \end{pmatrix}$ defines therefore a split extension of X_1 by X_2 if and only if $\Sigma = P_1 M - A P_2$ with $A \in M_{n_1,n_2}(\mathbb{C})$ and $M \in M_{n_1+q_1,n_2+q_2}(\mathbb{Z})$.

Moreover, if the period matrix $P = \begin{pmatrix} P_1 & \Sigma \\ 0 & P_2 \end{pmatrix}$ is such that $\Sigma = P_1 M - A P_2$ with $M \in M_{n_1+q_1,n_2+q_2}(\mathbb{Q})$, then P defines an extension of X_1 by X_2 which is isogenous to a split one. An isogeny $f : X_1 \to X_2$ is given by $\rho_a(f) = \begin{pmatrix} l I_{n_1} & 0 \\ 0 & I_{n_2} \end{pmatrix}$ and $\rho_r(f) = \begin{pmatrix} l I_{n_1+q_1} & 0 \\ 0 & I_{n_2+q_2} \end{pmatrix}$, where $l \in \mathbb{Z}$ is such that lM has integral entries.

Hence we have the following

2.2.5 Proposition. *Let X_1, X_2 be connected commutative complex Lie groups of maximal rank n_1, n_2 and let P_1, P_2 be the corresponding period matrices. The period matrix*

$$P = \begin{pmatrix} P_1 & \Sigma \\ 0 & P_2 \end{pmatrix} \in M_{n,n+q_1+q_2}(\mathbb{C}) \qquad (n = n_1 + n_2)$$

defines an extension of X_1 by X_2 which is isogenous to a split one, via an isogeny f such that $\rho_a(f) = \begin{pmatrix} l I_{n_1} & 0 \\ 0 & I_{n_2} \end{pmatrix}$ and $\rho_r(f) = \begin{pmatrix} l I_{n_1+q_1} & 0 \\ 0 & I_{n_2+q_2} \end{pmatrix}$, if and only if $\Sigma = P_1 M - A P_2$ with $A \in M_{n_1,n_2}(\mathbb{C})$ and $M \in M_{n_1+q_1,n_2+q_2}(\mathbb{Q})$, where $l \in \mathbb{Z}$ is such that lM has integral entries. $\square$

Extensions of complex tori X_1 and X_2 which are not isogenous to a split analytic extension $X_1 \oplus X_2$ are called *Shafarevich extensions* in [7], Ch. 1, § 6, p. 23. Hence it seems for us to be natural to call also non-split analytic extensions of a toroidal group by a toroidal group Shafarevich extensions.

If X_1 and X_2 are abelian varieties, Shafarevich extensions of X_1 by X_2 are not abelian varieties and hence provide a wide class of non-projective complex tori, since an abelian variety is a complex torus admitting a holomorphic embedding into some projective space (cf. [7], p. xiii).

3

Groups of Extreme Nilpotency Class

In this section we study groups of maximal and minimal nilpotency class.

3.1 Maximal Nilpotency Class

3.1.1 Proposition. *Let $\mathfrak{g}$ be a non-commutative Lie algebra and let $\mathfrak{z}$ be the centre of $\mathfrak{g}$. Then $\dim(\mathfrak{g}/\mathfrak{z}) \geq 2$.*

Proof. If the co-dimension of $\mathfrak{z}$ in $\mathfrak{g}$ were one, then $\mathfrak{g}$ would be generated by $\mathfrak{z}$ and one element of $\mathfrak{g}$: and so $\mathfrak{g}$ would be commutative, a contradiction. $\square$

3.1.2 Corollary. *Let G be a non-commutative group in one of the following classes*

(a) *connected real or complex Lie groups,*
(b) *connected algebraic groups over a field of characteristic zero,*
(c) *formal groups.*

Then $\dim(G/\mathfrak{z}G) \geq 2$.

Proof. First we observe that in case (b) it is enough to consider affine algebraic groups since in general the non-affine subgroup $D(G)$ lies in the centre of G (Theorem 1.3.2). Then in all the three cases the Lie algebra of G is not commutative; for algebraic groups see [8], 7.8 Proposition, p. 108. Now the statements follow from Proposition 3.1.1. $\square$

We see that for any non-commutative nilpotent connected group of dimension n in the classes (a)–(c) of Corollary 3.1.2, the nilpotency class is at most $n-1$. The existence of groups of dimension n and nilpotency class $n-1$ in the cases (a)–(c) follows from the existence of Lie algebras with this property (see [37], Chapter 2, § 1, p. 40). Lie algebras over fields of characteristic zero which have dimension n and nilpotency class $n-1$ are called *filiform*. We shall use the same name for groups in the classes (a)–(c). Filiform algebras of low dimension over algebraically closed fields of characteristic zero have been classified ([3, 33]).

In a filiform Lie algebra $\mathfrak{g}$ of dimension n one can find a basis $\{e_1, \ldots, e_n\}$ such that its centre is $\mathfrak{z} = \langle e_n \rangle$ and $[e_1, e_i] = e_{i+1}$ for $2 \le i \le n - 1$. The isomorphism classes of filiform Lie algebras of dimension n depend on other non-trivial relations between e_h and e_k for $2 \le h, k \le n - 1$.

3.1.3 Remark. The n-dimensional Lie algebra $\mathfrak{g}$ is filiform if and only if $\mathfrak{g}$ is a nilpotent Lie algebra containing a unique ideal $\mathfrak{n}_d$ of any given dimension $d \le n - 2$. More precisely, $\mathfrak{n}_d$ is the term $\mathcal{C}^{n-1-d}\mathfrak{g}$ of the descending central series. □

The situation changes radically for algebraic groups over a field of prime characteristic. We give examples of connected unipotent algebraic groups of dimension n which have nilpotency class n. Apart from its own interest, this class of groups serves us as a source of counter-examples for many questions in this paper.

3.1.4 Example. Let k be a field of characteristic $p > 0$. On the n-dimensional affine space we define a multiplication by

$$(x_0, \cdots, x_{n-1})(y_0, \cdots, y_{n-1}) = (z_0, \cdots, z_{n-1})$$

where $z_0 = x_0 + y_0$ and, for $i = 1, \cdots, n - 1$,

$$z_i = x_i + y_i + \sum_{j=0}^{i-1} x_j y_{i-j-1}^{p^{j+1}}. \tag{3.1}$$

This multiplication describes an n-dimensional unipotent algebraic group $\mathfrak{J}_n$, having the following linear representation

$$M_{n-1}(x_0, \cdots, x_{n-1}) = \begin{pmatrix} 1 & x_0 & x_1 & \cdot & \cdot & \cdot & x_{n-1} \\ & 1 & x_0^p & x_1^p & \cdot & \cdot & x_{n-2}^p \\ & & 1 & x_0^{p^2} & x_1^{p^2} & \cdot & x_{n-3}^{p^2} \\ & & & & \vdots & & \\ & & & & & & 1 \end{pmatrix}. \tag{3.2}$$

This representation is faithful of smallest possible degree, because the nilpotency class of this group is n: in fact, the centre $\mathfrak{z}\mathfrak{J}_n$ of $\mathfrak{J}_n$ is the one-dimensional connected subgroup defined by $x_0 = \cdots = x_{n-2} = 0$, and the factor group $\mathfrak{J}_n/\mathfrak{z}\mathfrak{J}_n$ is isomorphic to $\mathfrak{J}_{n-1}$, hence $\mathfrak{J}_n$ is a group of maximal possible nilpotency class. □

3.1.5 Remark. We want to illustrate with an example the situation of Remark 2.1.15, where we showed that whenever one has two different extensions

$$1 \longrightarrow A_1 \longrightarrow G \longrightarrow B_1 \longrightarrow 1$$

$$1 \longrightarrow A_2 \longrightarrow G \longrightarrow B_2 \longrightarrow 1$$

of the same group, it is not possible to compare the factor systems.
The three-dimensional unipotent group $\mathfrak{J}_3$ is a central extension

$$1 \longrightarrow \mathbf{G}_a \longrightarrow \mathfrak{J}_3 \longrightarrow \mathfrak{J}_2 \longrightarrow 1.$$

The factor system corresponding to such an extension is $\phi(x_0, x_1, y_0, y_1) = x_0 y_1^p + x_1 y_0^{p^2}$ and we have

$$(x_0, x_1, 0)(0, 0, x_2) = (x_0, x_1, x_2),$$

corresponding to the fact that $\phi(x_0, x_1, 0, 0) = 0$. In contrast to this we have

$$(x_0, 0, 0)(0, x_1, x_2) = (x_0, x_1, x_2 + x_0 x_1^p).$$

This shows immediately that the non-central extension

$$1 \longrightarrow H \longrightarrow \mathfrak{J}_3 \longrightarrow \mathbf{G}_a \longrightarrow 1, \tag{3.3}$$

where H is the two-dimensional vector subgroup defined by $x_0 = 0$, is a completely different representation of $\mathfrak{J}_3$. The factor system γ corresponding to the section $\tau(x_0) = (x_0, 0, 0) \in \mathfrak{J}_3$ of the extension (3.3) is $\gamma = -\delta^1 \tau : \mathbf{G}_a \times \mathbf{G}_a \longrightarrow H$, defined by $\gamma(x_0, y_0) = (0, \alpha(x_0, y_0), \beta(x_0, y_0))$ with $\alpha(x_0, y_0) = x_0 y_0^p$ and $\beta(x_0, y_0) = -x_0^p y_0^{p^2} (x_0 + y_0)$. Since H is such that $[G, H] \le {}_3 \mathfrak{J}_3 = \{(x_0, x_1, x_2) \in \mathfrak{J}_3 : x_0 = x_1 = 0\}$, the non-central extension $(\mathfrak{J}_3)_\gamma$ has according to Remark 2.1.15 the following multiplication:

$$(x_0, (0, x_1, x_2)) \cdot (y_0, (0, y_1, y_2)) =$$

$$(x_0 + y_0, (0, x_1 + y_1 + \alpha(x_0, y_0), x_2 + y_2 + \beta(x_0, y_0) + \sigma_{y_0}(x_1))),$$

where $\sigma_{y_0}(x_1) = -y_0 x_1^p + y_0^{p^2} x_1$.
The isomorphism between $\mathfrak{J}_3$ and the explicit extension $(\mathfrak{J}_3)_\gamma$ defined by (3.3) is $\rho : (\mathfrak{J}_3)_\gamma \longrightarrow \mathfrak{J}_3$,

$$\rho(x_0, (0, x_1, x_2)) = \tau(x_0)(0, x_1, x_2) = (x_0, 0, 0)(0, x_1, x_2) = (x_0, x_1, x_2 + x_0 x_1^p),$$

whereas $\rho^{-1}(x_0, x_1, x_2) = \left(\pi(x_0, x_1, x_2), (x_0, x_1, x_2)(\sigma\pi(x_0, x_1, x_2))^{-1} \right)$. $\square$

3.1.6 Remark. Denote by $A(x_0, \cdots, x_{n-1})$ the block matrix

$$A(x_0, \cdots, x_{n-1}) = M_0(x_0) \oplus M_1(x_0^p, x_1) \oplus M_2(x_0^{p^2}, x_1^p, x_2) \oplus \cdots$$

$$\cdots \oplus M_{n-1}(x_0^{p^{n-1}}, x_1^{p^{n-2}}, \cdots, , x_{n-1})$$

having the matrices $M_i(x_0^{p^i}, x_1^{p^{i-1}}, \cdots, x_i)$, $0 \le i \le n-1$, as $(i+2) \times (i+2)$-blocks along the main diagonal, and consider the n-dimensional connected unipotent group $G = \{A(x_0, \cdots, x_{n-1}) : x_0, \cdots, x_{n-1} \in \bar{k}\}$. Note that the group G is a diagonal subgroup of the direct product $\mathfrak{J}_1 \times \mathfrak{J}_2 \times \cdots \times \mathfrak{J}_n$. The Lie algebra $\mathfrak{g}$ of G is then

$$\mathfrak{g} = \{a(x_0, \cdots, x_{n-1}) = M_0^*(x_0) \oplus M_1^*(0, x_1) \oplus M_2^*(0, 0, x_2) \oplus \cdots$$

$$\cdots \oplus M_{n-1}^*(0, 0, \cdots, 0, x_{n-1}) : x_0, \cdots, x_{n-1} \in \bar{k}\},$$

where $M_i^*(x_0, \cdots, x_i) = M_i(x_0, \cdots, x_i) - I$. Clearly $\mathfrak{g}$ is commutative and, in spite of the fact that the centre of G is minimal (G being a group of maximal nilpotency class), the adjoint representation of such a group is trivial, as one immediately checks looking at the action of G on $\mathfrak{g}$ by conjugation. $\qquad \square$

3.1.7 Remark. Introducing the i-dimensional subgroups $H_i \leq \mathfrak{J}_n$ given by $x_0 = \cdots = x_{n-i-1} = 0$, we see that, for $i \leq (n+1)/2$, the subgroup H_i is a vector group. Furthermore, for the i-th term $\mathfrak{z}_i \mathfrak{J}_n$ of the ascending central series we have $\mathfrak{z}_i \mathfrak{J}_n = H_i$ and $\mathfrak{J}_n / \mathfrak{z}_i \mathfrak{J}_n \cong \mathfrak{J}_{n-i}$. $\qquad \square$

3.1.8 Proposition. *Let k be the largest natural integer such that $k \leq (n+1)/2$. Then the k-dimensional vector group H_k contains every connected commutative algebraic subgroup of $\mathfrak{J}_n$.*

Proof. Let L be a connected commutative algebraic subgroup of $\mathfrak{J}_n$. We proceed by induction on n, the assertion being trivial for $n \leq 2$. Consider the canonical projection $\pi : \mathfrak{J}_n \longrightarrow \mathfrak{J}_{n-1}$ with kernel $\mathfrak{z} \mathfrak{J}_n$ and the connected commutative algebraic subgroup $\pi(L)$ of $\mathfrak{J}_{n-1}$. Using the induction hypothesis we see that $\pi(L)$ is contained in the subgroup $T \leq \mathfrak{J}_{n-1}$ given by $x_0 = \cdots = x_{n-l-2} = 0$, where l is the largest natural integer such that $l \leq n/2$. Obviously the pre-image under π of T is precisely H_{l+1}, so we are done if n is odd. Let $n = 2l$ be even and let $a = (0, 0, \cdots, 0, a_{l-1}, \cdots, a_{n-1}), b = (0, 0, \cdots, 0, b_{l-1}, \cdots, b_{n-1}) \in L$. Computing $ab = ba$ we get:

$$ab = (0, 0, \cdots, 0, a_{l-1} + b_{l-1}, \cdots, a_{n-2} + b_{n-2}, a_{n-1} + b_{n-1} + a_{l-1} b_{l-1}^{p^l})$$

whereas

$$ba = (0, 0, \cdots, 0, a_{l-1} + b_{l-1}, \cdots, a_{n-2} + b_{n-2}, a_{n-1} + b_{n-1} + b_{l-1} a_{l-1}^{p^l}).$$

Since the coefficients a_{l-1}, b_{l-1} run over an infinite field, the found identity $a_{l-1} b_{l-1}^{p^l} = b_{l-1} a_{l-1}^{p^l}$ yields $a_{l-1} = b_{l-1} = 0$. $\qquad \square$

We get a generalization of the group $\mathfrak{J}_n$ by replacing in the representation (3.2) the Frobenius endomorphism $\mathbf{F} : x \mapsto x^p$ by any surjective endomorphism $\alpha : x \mapsto \sum_{i=0}^{t} b_i x^{p^i}$ of the additive group $\mathbf{G}_a$, with $b_i \neq 0$ for some $i > 0$. This way we obtain a group $\mathfrak{J}_n(\alpha)$. The elements of the group $\mathfrak{J}_n(\alpha)$ have a representation as matrices $M(x_0, \ldots, x_{n-1}) = (a_{i,j})$, with

$a_{i,j} = 0$, for $i > j$,

$a_{i,i} = 1$,

$a_{1,j} = x_{j-2}$ and

$$a_{i+1,i+1+t} = a_{i,i+t}^{\alpha^i},$$

where we write x^{α^i} for $\alpha^i(x)$. With this notation we can introduce the following family of subgroups of $\mathfrak{J}_n$:

$$\begin{aligned}
S_0 &= \mathfrak{J}_n = \mathfrak{J}_n(\mathbf{F}), \\
S_1 &= \{(0, x_1, 0, x_3, \cdots) : x_i \in \bar{\mathsf{k}}\} \cong \mathfrak{J}_{m_1}(\mathbf{F}^2), \\
S_2 &= \{(0, 0, x_2, 0, 0, x_5, \cdots) : x_i \in \bar{\mathsf{k}}\} \cong \mathfrak{J}_{m_2}(\mathbf{F}^3),
\end{aligned}$$
$$\vdots$$

where $m_i = \lfloor \frac{n}{i+1} \rfloor$.

All the groups $\mathfrak{J}_n(\alpha)$ have the same nilpotency class n and can be characterised by the following

3.1.9 Proposition. *For $n \geq 3$, the group $\mathfrak{J}_n(\alpha)$ is the only group G of linear transformations on an $(n+1)$-dimensional vector space V containing a vector $0 \neq \mathbf{v} \in V$, fixed by any element of $\mathfrak{J}_n(\alpha)$, such that:*

i) on the factor space $V/\langle\mathbf{v}\rangle$ the group $\mathfrak{J}_n(\alpha)$ induces the group of linear transformations represented by a matrix $M^{(\alpha)} = (a_{i,j}^\alpha)$, with $M = (a_{i,j})$ representing an element of $\mathfrak{J}_{n-1}(\alpha)$;

ii) $\mathfrak{J}_n(\alpha)$ induces the group of linear transformations $\mathfrak{J}_{n-1}(\alpha)$ on the factor space $V^*/\langle\mathbf{v}^*\rangle$, where $\mathbf{v}^*, V^*$ are the canonical dual objects of $\mathbf{v}, V$.

Proof. By the first condition we can choose a basis of V such that G is represented by matrices

$$\begin{pmatrix}
1 & a_{1,2} & a_{1,3} & \cdot & & \cdot & a_{1,n} \\
 & 1 & x_0^\alpha & x_1^\alpha & \cdot & \cdot & x_{n-2}^\alpha \\
 & & 1 & x_0^{\alpha^2} & x_1^{\alpha^2} & \cdot & x_{n-3}^{\alpha^2} \\
 & & & & \vdots & & \\
 & & & & & & 1
\end{pmatrix}. \tag{3.4}$$

The second condition yields $a_{1,j} = x_{j-1}$ for $j = 1, \ldots, n-1$. It remains to note that $a_{1,n}$ is independent of $x_0, \ldots, x_{n-2}$. In fact the matrices above form a group, the centre of which is the one-dimensional subgroup consisting of those matrices, where only $a_{1,n}$ is different from zero. $\qquad\square$

3.1.10 Proposition. *If $r \neq s$ and $n > 1$, then there is no surjective homomorphism from $\mathfrak{J}_m(\mathbf{F}^r)$ to $\mathfrak{J}_n(\mathbf{F}^s)$.*

Proof. In order to prove the assertion, we first assume that $m = n = 2$. Let $\gamma : \mathfrak{J}_2(\mathbf{F}^r) \longrightarrow \mathfrak{J}_2(\mathbf{F}^s)$ be a homomorphism and put

$$\gamma(x_0, x_1) = \gamma((x_0, 0)(0, x_1)) = \gamma(x_0, 0)\gamma(0, x_1)$$

$$= (\gamma_1(x_0), \gamma_2(x_0))(0, \gamma_3(x_1)) = (\gamma_1(x_0), \gamma_2(x_0) + \gamma_3(x_1)).$$

It is easy to see that γ_1 and γ_3 define endomorphisms of the additive group $\mathbf{G}_a$ and that $\gamma_1 \neq 0$ precisely when γ is surjective. Compute now

$$\gamma((x_0, x_1)(y_0, y_1)) = \gamma(x_0 + y_0, x_1 + y_1 + x_0 y_0^{p^r})$$

$$= (\gamma_1(x_0 + y_0), \gamma_2(x_0 + y_0) + \gamma_3(x_1 + y_1 + x_0 y_0^{p^r})).$$

On the other hand we find

$$\gamma(x_0, x_1)\gamma(y_0, y_1) = (\gamma_1(x_0), \gamma_2(x_0) + \gamma_3(x_1))(\gamma_1(y_0), \gamma_2(y_0) + \gamma_3(y_1)) =$$

$$(\gamma_1(x_0) + \gamma_1(y_0), \gamma_2(x_0) + \gamma_3(x_1) + \gamma_2(y_0) + \gamma_3(y_1) + \gamma_1(x_0)\gamma_1(y_0)^{p^s})$$

forcing $-\delta^1\gamma_2(x_0, y_0) = \gamma_3(x_0 y_0^{p^r}) - \gamma_1(x_0)\gamma_1(y_0)^{p^s}$. The polynomial $\delta^1\gamma_2$ (x_0, y_0) is symmetric, whereas a necessary condition for the polynomial $\gamma_3(x_0 y_0^{p^r}) - \gamma_1(x_0)\gamma_1(y_0)^{p^s}$ to be symmetric is that $r = s$, which is seen by looking at the monomial of maximal degree. For the general case we can reduce easily to $n = 2$ considering the canonical projection $\mathfrak{J}_n(\mathbf{F}^s) \longrightarrow \mathfrak{J}_2(\mathbf{F}^s)$. Let $\gamma : \mathfrak{J}_m(\mathbf{F}^r) \longrightarrow \mathfrak{J}_2(\mathbf{F}^s)$ be an epimorphism and let K be the kernel of γ. Since $\mathfrak{J}_m(\mathbf{F}^r)$ has maximal nilpotency class m and $\mathfrak{J}_2(\mathbf{F}^s)$ has nilpotency class two, the connected component K° of K coincides with $[\mathfrak{J}_m(\mathbf{F}^r), \mathfrak{J}_m'(\mathbf{F}^r)]$, and the group $\mathfrak{J}_m(\mathbf{F}^r)/K^\circ$ is isomorphic to $\mathfrak{J}_2(\mathbf{F}^r)$. Thus we would obtain an epimorphism from $\mathfrak{J}_2(\mathbf{F}^r)$ to $\mathfrak{J}_2(\mathbf{F}^s)$, which we had excluded before.   □

3.1.11 Remark. The exponent of the group $\mathfrak{J}_n$ as well as of $\mathfrak{J}_n(\alpha)$ is p^t, where t is the smallest natural number such that $p^{(t-1)} < n+1 \leq p^t$. In order to prove this, we identify the element $(x_0, \cdots, x_{n-1})$ with the $(n+1) \times (n+1)$ matrix $M = M(x_0, \cdots, x_{n-1})$ given in (3.2). As $n+1 \leq p^t$, we have

$$M^{p^t} - I = (M - I)^{p^t} = 0.$$

On the other hand, for $x_0 \neq 0$ the rank of the matrix $(M - I)$ is n. Thus the dimension of the eigenspace relative to the (unique) eigenvalue 1 is one and this forces the minimal polynomial of M to be equal to the characteristic polynomial, hence $(M - I)^{p^{(t-1)}} \neq 0$.

We note that for $p > n$, the group $\mathfrak{J}_n(\alpha)$ provides an example of an n-dimensional group of nilpotency class n where every element has exponent p. In the class of finite p-groups to these groups there correspond finite groups of exponent p and of nilpotency class $p - 1$. An example of a finite group of exponent 3 and of nilpotency class 2 as well as of a finite group of exponent 5 and of nilpotency class 4 is given in [38]; a finite group of exponent 7 and of nilpotency class 6 can be obtained from the group G of order 7^{20416} of nilpotency class 28 given in [68] by considering the factor group $G/\mathfrak{z}_{22}G$, where $\mathfrak{z}_{22}G$ is the 22th member of the ascending central series.   □

3.1.12 Remark. Since the group operation of $\mathfrak{J}_n(\alpha)$ is given by polynomials and a surjective endomorphism $\alpha : x \mapsto \sum_{i=0}^{t} b_i x^{p^i}$, with $b_i \neq 0$ for some

$i > 0$, it is possible to get a group with the same operation, but with coefficients in a suitable commutative ring containing $\bar{\mathsf{k}}$. A natural candidate in our context is the Witt ring $\mathfrak{W}$ (see [18], V, § 1), where we put $\alpha = \mathbf{F}$, the Frobenius endomorphism of $\mathfrak{W}$. In particular, if we take the coefficients of $\mathfrak{J}_n$ in the Witt ring $\mathfrak{W}_m$, we obtain an nm-dimensional unipotent algebraic group $\Gamma = \mathfrak{J}_n(\mathfrak{W}_m)$ of nilpotency class n such that the factor group $\mathfrak{z}_i\Gamma/\mathfrak{z}_{i-1}\Gamma$ is isomorphic to $\mathfrak{W}_m$. Denoting by $\mathcal{C}^i(G)$ the terms of the descending central series of a nilpotent group G, we find that $\mathcal{C}^i(\Gamma)/\mathcal{C}^{i+1}(\Gamma)$ is again isomorphic to $\mathfrak{W}_m$.

We exhibit the multiplication in the explicit example of the 4-dimensional unipotent algebraic group $G = \mathfrak{J}_2(\mathfrak{W}_2)$, denoting by Φ_1 the factor system corresponding to the two-dimensional Witt group $\mathfrak{W}_2$ (see Remark 2.1.6):

$$(x_0, x_1; x_2, x_3) \cdot (y_0, y_1; y_2, y_3) = (x_0 + y_0, x_1 + y_1 + \Phi_1(x_0, y_0);$$

$$x_2 + y_2 + x_0 y_0^p, x_3 + y_3 + \Phi_1(x_2, y_2) + x_0^p y_1^p + x_1 y_0^{p^2} + \Phi_1(x_2 + y_2, x_0 y_0^p)).$$

This group G has a unique maximal connected subgroup $M = \{(x_0, x_1; x_2, x_3) \in G : x_0 = 0\}$, which is isomorphic to the direct product $\mathfrak{W}_2 \times \mathbf{G}_a$. The uniqueness of M as a maximal subgroup of G follows from the fact that for the commutator subgroup G' one has $G' = \{(x_0, x_1; x_2, x_3) \in G : x_0 = x_1 = 0\}$ and G/G' is a Witt group. Denoting by C the vector subgroup $C = \{(x_0, x_1; x_2, x_3) \in G : x_0 = x_2 = 0\}$, we see that C is normal and that G/C is isomorphic to $\mathfrak{J}_2$. Finally, the unique normal one-dimensional connected algebraic subgroup $K = \{(x_0, x_1; x_2, x_3) \in G : x_0 = x_1 = x_2 = 0\}$ is such that G/K is isomorphic to the subgroup L of the direct product $\mathfrak{W}_2 \times \mathfrak{J}_2$ defined by

$$L = \{(x_0, x_1) \times (y_0, y_1) \in \mathfrak{W}_2 \times \mathfrak{J}_2 : x_0 = y_0\}.$$

The rôle of such examples will be clear in the next sections (see Remarks 3.2.13, 3.2.11 and 5.2.12). □

The groups $\mathfrak{J}_n$ are examples for unipotent groups in which the lattice of normal connected algebraic subgroups is a chain.

3.1.13 Proposition. *Let G be a connected unipotent algebraic k-group over a perfect field k of positive characteristic. If the dimension n of G is equal to its nilpotency class, then the lattice of normal connected algebraic k-subgroups of G is a chain of length n.*

Proof. Let G be a counter-example to the assertion of minimal dimension $n > 1$. Since the group G must have one-dimensional centre $\mathfrak{z}\,G$, in the factor group $G/\mathfrak{z}^\circ\,G$ the lattice of normal connected algebraic k-subgroups is a chain of length $n-1$. The assertion follows from the fact that any normal connected algebraic k-subgroup of G contains $\mathfrak{z}^\circ\,G$. □

Closing this section we show that the descending central series $\mathcal{C}^{i+1}(G) = [G, \mathcal{C}^i(G)]$ (with $\mathcal{C}^0(G) = G$) of an affine algebraic group of dimension n and nilpotency class n has particular properties. Furthermore, we show that the non-commutative two-dimensional unipotent group never occurs as a commutator subgroup of a nilpotent group. To reach our goal we need two propositions, the first of which holds for abstract groups.

3.1.14 Proposition. *Let G be a group with a subgroup H such that $[G, H] \leq {}_3G$. Then $[G', H] = 0$.*

Proof. Take $g, f \in G, x \in H$ and compute

$$x^{gf} = (x[x, g])^f = x[x, f][x, g] = x^{fg},$$

that is $x^{[g^{-1}, f^{-1}]} = x$. Hence, $G' \leq C_G(H)$. $\square$

We formulate the next proposition for unipotent algebraic groups, but it is true also for unipotent Lie groups.

3.1.15 Proposition. *Let G be a unipotent algebraic k-group and let H be a normal algebraic subgroup of dimension l. Then $\mathcal{C}^{l-1}(G) \leq C_G(H)$.*

Proof. Let $H_1 < \cdots < H_i < \cdots < H_l = H$ be a series of normal subgroups of G, with dim $H_i = i$ ([8], III Theorem 10.6(2), p. 138 and V, 15.5 Corollary, p. 205). Since H_l/H_1 is $(l-1)$-dimensional, by induction it is centralised by $\mathcal{C}^{l-2}(G/H_1) = (\mathcal{C}^{l-2}(G) \cdot H_1)/H_1$. Thus we have $[\mathcal{C}^{l-2}(G) \cdot H_1, H_l] \leq H_1$, and since $H_1 \leq {}_3G$ we get $[\mathcal{C}^{l-2}(G), H_l] \leq H_1 \leq {}_3G$. Furthermore, since H_l/H_{l-1} is one-dimensional, we have that $H_l/H_{l-1} \leq {}_3(G/H_{l-1})$, that is $[G, H_l] \leq H_{l-1}$. Take $h \in H_l, x \in G, y \in \mathcal{C}^{l-2}(G)$ and compute
$h^{xy} = (h[h, x])^y = h^y[h, x]^y.$
Since $[h, x] \in [H_l, G] \leq H_{l-1}$, by induction we have that $[h, x]^y = [h, x]$, therefore we get $h^{xy} = h^y[h, x]$.
On the other hand, we have
$h^{yx} = h^y[h^y, x] = h^y[h[h, y], x].$
Since $[h, y] \in [H_l, \mathcal{C}^{l-2}(G)] \leq {}_3G$, we get $h^{yx} = h^y[h, x]$, that is $[x, y] \in C_G(H_l)$.
$\square$

3.1.16 Corollary. *Let G be a unipotent algebraic k-group and let H be a two-dimensional normal algebraic subgroup of G. Then the commutator subgroup G' centralises H.*

3.1.17 Corollary. *Let G be a connected nilpotent algebraic k-group. If the commutator subgroup G' of G has dimension two, then G' is commutative.*

Proof. Since for the maximal affine subgroup L of G one has $L' = G'$ (Theorem 1.3.2), we may assume that G is affine, even that G is unipotent ([8], III, 10.6(3) Theorem, p. 138). Putting $H = G'$ in Proposition 3.1.15 our statement follows. $\square$

3.1.18 Corollary. *Let G be an n-dimensional connected nilpotent algebraic k-group. If $n \geq 2$, then $\mathcal{C}^{n-2}(G)$ commutes with every maximal connected subgroup. In particular, for $n > 2$ either $\mathcal{C}^{n-2}(G)$ is central or G has a unique maximal connected subgroup.*

Proof. According to 1.3.2 we may assume that G is affine. Moreover we can suppose that G is unipotent ([8], III, 10.6(3) Theorem, p. 138). Then by Proposition 3.1.15 any maximal connected algebraic subgroup of G is centralised by $\mathcal{C}^{n-2}(G)$. $\qquad\square$

Since any finite p-group is supersolvable (cf. [46], p. 716), the proof of Proposition 3.1.15 yields the following

3.1.19 Remark. Let G be a finite group of order p^n and let H be a normal subgroup of G of order p^l. Then $\mathcal{C}^{l-1}(G) \leq C_G(H)$. $\qquad\square$

3.1.20 Remark. The group $\mathfrak{J}_5$, which has nilpotency class $n = 5$, gives an example of a unipotent group G in which the subgroup $\mathcal{C}^{n-3}(G)$ does not commute with the unique maximal connected subgroup G' and in which $\mathcal{C}^{n-2}(G)$ is not central. $\qquad\square$

3.2 Groups with a Big Centre

Corollary 3.1.2 shows that in algebraic groups G over fields k of characteristic zero, as well as in connected real or complex Lie groups, the connected component of the centre never is a maximal connected subgroup of G. In contrast to this, in an algebraic group G over a field of positive characteristic the centre of G can have co-dimension one in G. An example for this phenomenon is the group $\mathfrak{J}_2$ of Example 3.1.4.

3.2.1 Proposition. *Let G be a connected nilpotent affine algebraic group and let A_0 be a closed connected subgroup of $G' \cap \mathfrak{z}^\circ G$. If $A_0 \neq G'$, then there exists a closed connected subgroup A_1 in G' which is normal in G, contains A_0 as a maximal closed connected subgroup and is central in G'.*

Proof. As G is the direct product of a torus with a unipotent group (see [8], Theorem 10.6 (3), p. 138), one only has to consider the case where G is unipotent but not commutative.

Let $A_0 \leq A_1 \leq \cdots \leq G'$ be a chain of closed connected normal subgroups of G such that each of them has co-dimension one in the next (see [8], Theorem 10.6(2), p. 138). As A_1/A_0 is one-dimensional, G/A_0 acts trivially by conjugation on A_1/A_0. This means

$$[G/A_0, A_1/A_0] = A_0/A_0,$$

forcing $[G, A_1] \leq A_0$. Therefore, for any $x \in A_1$ and $g \in G$ we have

$$x^g = x[x, g] \text{ with } [x, g] \in A_0 \leq {}_3{}^\circ G.$$

This implies

$$x^{gf} = x^f \cdot [x, g]^f = x \cdot [x, f] \cdot [x, g] = x^{fg},$$

that is $x^{[g,f]} = x$. Hence G' is contained in $C_G(A_1)$. Because A_1 is contained in G', the assertion follows.   □

3.2.2 Corollary. *Let G be a connected nilpotent affine algebraic group. If the connected component ${}_3{}^\circ G'$ of the centre of G' is contained in the connected component ${}_3{}^\circ G$ of the centre of G, then $G' \leq {}_3{}^\circ G$, and thus G has nilpotency class two.*

Proof. The assertion follows putting $A_0 = {}_3{}^\circ G'$ in Proposition 3.2.1.   □

3.2.3 Corollary. *Let G be a connected nilpotent affine algebraic group such that $\dim(G' \cap {}_3{}^\circ G) = \dim G' - 1$. Then G' is commutative.*

Proof. The assertion follows putting $A_0 = G' \cap {}_3{}^\circ G$ in Proposition 3.2.1.   □

3.2.4 Corollary. *Let G be a connected nilpotent affine algebraic group. Then either $\dim G' = 1$ or $\dim {}_3{}^\circ G' \geq 2$.*

Proof. The assertion follows taking A_0 to be a one-dimensional connected central subgroup of G' in Proposition 3.2.1.   □

If G is a finite p-group and if A_0 is a central subgroup of G', then there exists a chain $A_0 \leq A_1 \leq \cdots \leq A_s = G'$ of subgroups normal in G such that A_i/A_{i-1} has prime order p (cf. [46], p. 716). Hence the proof of Proposition 3.2.1 yields

3.2.5 Proposition. *Let G be a finite p-group and let A_0 be a subgroup of $G' \cap {}_3 G$. If $A_0 \neq G'$, then there exists in G' a normal subgroup A_1 of G which is central in G' and contains A_0 as a maximal subgroup.*   □

As consequences of this proposition we obtain

3.2.6 Remark. *Let G be a finite p-group.*

a) If ${}_3 G'$ is central in G, then $G' \leq {}_3 G$, and thus G has nilpotency class two.
b) If $G'/(G' \cap {}_3 G)$ has order p, then G' is commutative.
c) Either G' has order p or the center of G' has order at least p^2.   □

Finally we remark that Proposition 3.2.1 and its corollaries hold for nilpotent real or complex Lie groups, too.

Groups Where $G/_3G$ Is a Chain

An important topic of this paper are chains, i.e. connected algebraic groups G such that the lattice of connected algebraic subgroups of G is a chain. In this section we show some interesting properties of algebraic groups G for which the factor group $G/_3G$ is a chain, where $_3G$ denotes the centre of G. These groups are examples of groups which are treated in Section 7 (cf. Theorem 7.1.12).

3.2.7 Remark. As for an algebraic group G the factor group $G/_3G$ is affine and with an affine chain also its Borel subgroups are chains, it follows from [8], Corollary 11.5, p. 148, that $G/_3G$ is either a one-dimensional torus or a unipotent group. Hence G is nilpotent and $G/_3G$ is a unipotent chain. This yields that $G/_3G$ has a unique d-dimensional connected algebraic subgroup for any $d \leq \dim G/_3G$.   □

In the first theorem of this section we relate the size of the commutator subgroup to the size of centre for these groups.

3.2.8 Theorem. *Let G be a non-commutative connected algebraic k-group.*
If the connected component $_3°G$ of $_3G$ is a maximal connected subgroup of G, then $_3°G$ has co-dimension one in G, the characteristic of k is positive and the commutator subgroup G' is a (central) vector group.
Conversely, if G' is a central vector group and $G/_3G$ is isogenous to a single Witt group, then $_3G$ has co-dimension one.

Proof. Assume that the connected component $_3°G$ of $_3G$ is a maximal connected subgroup of G. Since $G/_3°G$ is affine, the centre $_3G$ has co-dimension one. Hence $G' \leq _3°G$. Moreover G is nilpotent and G' is unipotent. If the characteristic of the field k is zero, then the Lie algebra of G is commutative which is a contradiction. Let the characteristic of k be $p > 0$. For any given $x \in G$ the mapping $\sigma_x : G \longrightarrow G'$, defined by $\sigma_x(y) = [x, y]$, is a homomorphism, because $G' \leq _3G$. Since the maximal connected subgroup $_3°G$ is contained in the kernel of σ_x, the image $\sigma_x(G)$ is one-dimensional, at most. Therefore, any commutator is central and has order p, i.e. the group G' is a vector group.

Conversely, assume that G' is a central vector group and $G/_3G$ is isogenous to a single Witt group. The above mapping $\sigma_x : G \longrightarrow G'$ is a homomorphism, the kernel Q_x of which contains the centre. Therefore G/Q_x is simultaneously a vector group and isogenous to a Witt group, that is G/Q_x is one-dimensional, at most. This means that $_3G$, which is equal to $\bigcap_{x \in G} Q_x$, has co-dimension one in G, because $G/_3G$ has a unique maximal connected subgroup.   □

3.2.9 Remark. The conditions of the second part of Theorem 3.2.8 are indispensable:

– The three-dimensional group $\mathfrak{J}_3$, exhibited in Example 3.1.4, is such that $\mathfrak{J}_3/\mathfrak{z}\mathfrak{J}_3$ is a chain, but the commutator subgroup $\mathfrak{J}_3'$ is a *non-central* two-dimensional vector group. In this case $\mathfrak{z}°\mathfrak{J}_3$ has dimension one.

– The group of 3×3 unipotent matrices has a two-dimensional vector group as factor group $G/\mathfrak{z}G$ and a one-dimensional commutator subgroup, which coincides with the centre.   □

3.2.10 Theorem. *Let G be a connected algebraic* k*-group. If $G/\mathfrak{z}G$ is a chain, then* $\dim \mathfrak{z}G + \dim G' \geq \dim G$.

Proof. By the decomposition $G = L(G)D(G)$ of Theorem 1.3.2 we get $\mathfrak{z}G = \mathfrak{z}L(G) \cdot D(G)$ and $G' = L(G)'$. Thus the factor group $G/\mathfrak{z}G$ is isomorphic to $L(G)/\mathfrak{z}L(G)$. Since

$$\dim G - \dim \mathfrak{z}G - \dim G' = \dim L - \dim \mathfrak{z}L - \dim L'$$

we only have to consider the case where G is affine.

Together with $G/\mathfrak{z}G$, the group G itself is a nilpotent group, so we may assume that G is unipotent ([8], III Theorem 10.6(3), p. 138). Let G be a counter-example of minimal dimension, let K be a one-dimensional central subgroup of G' and $\overline{G} = G/K$. As $\overline{G}/\mathfrak{z}\overline{G}$ is isomorphic to a factor group of $G/\mathfrak{z}G$, the assumption is true by minimality for $\overline{G}$, forcing

$$\dim H - 1 + \dim G' \geq \dim G,$$

where $H/K = \mathfrak{z}(G/K)$. The assertion will be proved if we show that $\mathfrak{z}G$ has co-dimension one in H.

Since $K \leq \mathfrak{z}G$ and $[G, H] \leq K$, for any given $g \in G$ the mapping $\sigma_g : H \longrightarrow K$, defined by $h \mapsto [h, g]$, is a homomorphism with a subgroup $M \geq \mathfrak{z}G$ of co-dimension at most one in H as kernel.

As $G/\mathfrak{z}G$ is a chain, the connected component $M°$ of M can be either the unique maximal connected subgroup of H, or $M° = \mathfrak{z}°G$. In the latter case we are done, in the former it suffices to note that $M°$, as the unique maximal connected subgroup of H, would centralise now *any* element $g \in G$, a contradiction.   □

3.2.11 Remark. The assumption that $G/\mathfrak{z}G$ is a chain in Theorem 3.2.10 is necessary: The group of 3×3 unipotent matrices is an example of a group G where $G/\mathfrak{z}G$ is a two-dimensional vector group and $\dim \mathfrak{z}G + \dim G' = 2 < \dim G$. Also, in Remark 3.1.12 we have seen that the groups $\mathfrak{J}_n(\mathfrak{W}_m)$ form a class for which $\dim \mathfrak{z}G + \dim G' = \dim G$, such that $G/\mathfrak{z}G$ is not a chain.

We also remark that in positive characteristic p, the difference $(\dim \mathfrak{z}G + \dim G') - \dim G$ may be arbitrarily large. Let G be the $n + 1$-dimensional unipotent algebraic group defined on the affine space by the multiplication

$$(x_0, x_1, \cdots, x_n)(y_0, y_1, \cdots, y_n) = (x_0 + y_0, x_1 + y_1 + x_0 y_0^{p^{t_1}}, \cdots, x_n + y_n + x_0 y_0^{p^{t_n}}),$$

where the natural integers t_i are such that $t_i \neq t_j$ for $i \neq j$. It is clear that $_3G$ is the n-dimensional subgroup defined by $x_0 = 0$. Each commutator in G has the form

$$(0, x_0 y_0^{p^{t_1}} - x_0^{p^{t_1}} y_0, \cdots, x_0 y_0^{p^{t_n}} - x_0^{p^{t_n}} y_0).$$

Assume by contradiction that $\dim G' < n - 1$. Then we find p-polynomials $f_i(T_i)$ such that

$$G' = \left\{ (0, a_1, \cdots, a_n) : \sum_{i=1}^{n} f_i(a_i) = 0 \right\},$$

by [45], Proposition 20.3, p. 129. In particular we have for any x_0, y_0 that

$$\sum_{i=1}^{n} f_i(x_0 y_0^{p^{t_i}} - x_0^{p^{t_i}} y_0) = 0.$$

But for $\phi_i(T) = T^{p^i+1}$ we find $x_0 y_0^{p^{t_i}} - x_0^{p^{t_i}} y_0 = 2 x_0 y_0^{p^{t_i}} - \delta^1 \phi_i(x_0, y_0)$, hence

$$\sum_{i=1}^{n} f_i(2 x_0 y_0^{p^{t_i}}) - f_i(\delta^1 \phi_i(x_0, y_0)) = 0,$$

which is a contradiction to the fact that for $i \in \mathbb{N}$ the polynomials $x_0 y_0^{p^{t_i}}$ are in a basis of the free left $\mathsf{End}(\mathbf{G}_a)$-module $\mathsf{H}^2(\mathbf{G}_a, \mathbf{G}_a)$ (see [18], II, § 3, 4.6), as mentioned already in Remark 2.1.6.

In algebraic groups over fields of characteristic zero or Lie groups G the sum of the dimensions of the commutator subgroup and the center can be smaller, equal or larger than $\dim G$.

Let $\mathfrak{g}$ be a nilpotent Lie algebra over a field of characteristic zero such that $\mathfrak{g}$ is a free Lie algebra of nilpotency class two. If $\mathfrak{g}$ has n generators, then the commutator subalgebra $\mathfrak{g}'$ of $\mathfrak{g}$ coincides with its center $_3$ and has dimension $\frac{n(n-1)}{2}$. One has $\dim \mathfrak{g}' + \dim _3 > \dim \mathfrak{g}$ if and only if $n \geq 4$. $\qquad \square$

3.2.12 Proposition. *Let G be a connected algebraic k-group such that $G/_3G$ is a chain. If G' also is a chain, then $\dim _3G + \dim G' = \dim G$.*

Proof. As in Theorem 3.2.10, it is harmless to assume that G is unipotent, because G is a nilpotent group and $D(G)$ is central in G. Let K be a one-dimensional subgroup of $G' \cap _3G$. Since $(G/K)' = G'/K$ and $(G/K)/_3(G/K)$ are chains, by induction it follows that $\dim G'/K + \dim _3(G/K) = \dim G/K$, that is $\dim G' + \dim _3(G/K) = \dim G$.

Let H be the (unique) connected subgroup of G such that $\dim (H/_3G) = 1$. Since $H/_3G \leq _3(G/_3G)$, we have $[G, H] \leq _3G$ and for any $g \in G$ the mapping $\sigma_g : H \longrightarrow [G, H]$, defined by $\sigma_g(h) = [g, h]$, is a homomorphism, the kernel of which contains $_3G$. Thus $\dim [g, H] \leq 1$ for all $g \in G$. Since G' is a chain, it follows that $[g, H] \leq K$ for any $g \in G$, hence $H/K \leq {}_3(G/K)$. By Theorem 3.2.10 we have

$$\dim(H/K) = \dim {}_3 G \geq \dim G - \dim G' = \dim {}_3(G/K),$$

therefore $H/K = {}_3(G/K)$ and we are done. $\square$

3.2.13 Remark. By the above Proposition 3.2.12, if G is a unipotent algebraic group such that G' is a central chain and G/G' is also a commutative chain, then $\dim G/G' = \dim G'$. In order to get such groups, we recall that in Remark 3.1.12 we have constructed for any n an example $G = \mathfrak{J}_2(\mathfrak{W}_n)$ of a $2n$-dimensional unipotent algebraic group G, where $G' = {}_3 G$ is the n-dimensional Witt group.

For an example L where $\dim L/L' > \dim L'$, let $G = \mathfrak{J}_2(\mathfrak{W}_n)$, $m \geq 2n$, and consider the direct product $G \times \mathfrak{W}_m$. Denote by $(x_1, \cdots, x_n; x_{n+1}, \cdots, x_{2n})$ the generic element of G and by $(y_1, \cdots, y_m)$ the generic element of $\mathfrak{W}_m$. The connected algebraic subgroup $L < G \times \mathfrak{W}_m$ defined as

$$L = \{g = (x_1, \cdots, x_n; x_{n+1}, \cdots, x_{2n}) \times (y_1, \cdots, y_m) \in G \times \mathfrak{W}_m :$$

$$x_i = y_i, \text{ for } i = 1, \cdots, n\}$$

gives an example of a connected unipotent algebraic group, where L' is a central chain, L/L' is a commutative chain and $\dim L/L' = m > n = \dim L'$.
     $\square$

For an affine algebraic group G it is in general not possible to obtain sufficient informations on ${}_3(G/K)$, where K is a central one-dimensional subgroup of G. In the following situation, however, we have:

3.2.14 Proposition. *Let G be a connected algebraic* k-*group such that $G/{}_3 G$ and $W = (G' \cap {}_3 G)^\circ$ are chains and such that $\dim W \geq \dim G' - 1$. Let K_t be the (unique) t-dimensional connected subgroup of W. If $W = G'$ or $t < \dim(G/{}_3 G)$, then $\dim {}_3(G/K_t) = \dim {}_3 G$; otherwise ${}_3(G/K_t)$ has co-dimension one in G/K_t.*

Proof. If $W = G'$ then $G' \leq {}_3 G$ and G' is a chain. By Proposition 3.2.12 we have $\dim G = \dim G' + \dim {}_3 G$.

Since for the group $\overline{G} = G/K_t$ both $\overline{G}'$ and $\overline{G}/{}_3\overline{G}$ are also chains, we find as well that $\dim(G/K_t) = \dim(G'/K_t) + \dim {}_3(G/K_t)$, and we are done in this case by Proposition 3.2.12.

Now let G' be not contained in ${}_3 G$. Let T_t be the connected algebraic subgroup of G such that $T_t/K_t = {}_3^\circ(G/K_t)$, hence $[G, T_t] \leq K_t$. If we had $T_t = G$, then $G' \leq K_t \leq {}_3 G$ which is not the case. Hence $T_t \neq G$.

As $[G, T_t] \leq K_t \leq {}_3 G$, we have that for any $g \in G$ the map $\sigma_g : T_t \longrightarrow [G, T_t]$ defined by $x \mapsto [g, x]$ is a homomorphism, the kernel Q_g of which contains ${}_3^\circ G$ and has at most co-dimension t in T_t. If M is the intersection of all the kernels Q_g, we have $M^\circ = {}_3^\circ G$ and $\dim {}_3 G = \dim M \geq \dim T_t - t = \dim {}_3^\circ (G/K_t)$. This follows from the fact that $G/{}_3 G$ is a chain. A further consequence of this is $\dim W \geq \dim G' - 1 \geq \dim G - \dim {}_3 G - 1$.

Let H_t be the (unique) connected subgroup of G such that ${}_3 G$ has co-dimension t in H_t. Since the previous discussion holds for any $1 \leq t \leq \dim W$, one has $T_t \leq H_t$.

As $\dim W = \dim G' - 1$ and $(G/{}_3 G)' = (G'{}_3 G)/{}_3 G \simeq G'/W$ the factor group $G/{}_3 G$ has a one-dimensional commutator subgroup. Then by Theorem 3.2.8 the centre of $\overline{G} = G/{}_3 G$ has co-dimension one in $\overline{G}$, because $\overline{G}/{}_3 \overline{G}$ is a chain as well as $G/{}_3 G$.

Therefore, if $H_t \neq G$, then the factor group $H_t/{}_3 G$ is central in $G/{}_3 G$. Thus $[G, H_t] \leq (G' \cap {}_3 G)^\circ = W$ and the map $\sigma_g : H_t \longrightarrow [G, H_t]$ defined by $h \mapsto [g, h]$ is a homomorphism, the kernel of which contains ${}_3^\circ G$. Therefore the image of σ_g has at most dimension t from which it follows that $[G, H_t] \leq K_t$. In particular, as $T_t/K_t = {}_3^\circ(G/K_t)$ we have $H_t \leq T_t$. But we have seen that $T_t \leq H_t$. Thus $T_t = H_t$.

Let $s = \dim(G/{}_3 G)$ and $t < s$. Then $H_t \neq G$ and the claim follows since $\dim {}_3(G/K_t) = \dim(T_t/K_t) = \dim(H_t/K_t) = \dim H_t - t = \dim {}_3 G$.

Finally, let $t \geq s$. Since $[G, H_{s-1}] \leq K_{s-1} < K_t$ we have $H_{s-1}/K_t \leq {}_3(G/K_t)$. Moreover, $\dim(H_{s-1}/K_t) = \dim(G/K_t) - 1$ and the assertion follows now from the fact that G/K_t is not commutative. $\qquad\square$

3.2.15 Proposition. *Let G be a connected unipotent group. If $G/{}_3 G$ and $W = (G' \cap {}_3 G)^\circ$ are chains, such that $\dim W = \dim G' - 1 < \dim(G/{}_3 G)$, then*

$$\dim {}_3 G + \dim G' = \dim G.$$

Proof. As the commutator subgroup of G/W is one-dimensional, by Theorem 3.2.8 we find that ${}_3(G/W)$ has co-dimension one in G/W. Now we apply Proposition 3.2.14, finding $\dim {}_3(G/W) = \dim {}_3 G$. From this we conclude $\dim G = \dim(G/W) + \dim W = \dim {}_3(G/W) + 1 + \dim W = \dim {}_3 G + \dim G'$. $\qquad\square$

3.2.16 Remark. The condition that $W = (G' \cap {}_3 G)^\circ$ is a chain is not dispensable in Proposition 3.2.15. For instance, the unipotent group $G = G(i, j)$ defined on the 4-dimensional affine space by the multiplication

$$(x_0, x_1, x_2, x_3)(y_0, y_1, y_2, y_3) =$$

$$(x_0 + y_0, x_1 + y_1 + x_0 y_0^{p^i}, x_2 + y_2 + x_0 y_1^{p^i} + x_1 y_0^{p^{2i}}, x_3 + y_3 + x_0 y_0^{p^j})$$

is such that ${}_3 G$ coincides with the two-dimensional vector subgroup of elements with $x_0 = x_1 = 0$. Moreover the factor group $G/{}_3 G$ is isomorphic to the two-dimensional chain $\mathfrak{J}_2(\mathbf{F}^i)$ of Example 3.1.4.

Let $M = \{(x_0, x_1, x_2, x_3) \in G : x_0 = 0\}$ and $H = \{(x_0, x_1, x_2, x_3) \in M : x_0 = x_1 = x_3 = 0\}$. Since $[G, M] = H$ we have that $H \leq G'$. Moreover the factor group G/H is isomorphic to the three-dimensional unipotent group $\overline{G}$ defined on the affine space by the multiplication

$$(x_0, x_1, x_3)(y_0, y_1, y_3) = (x_0 + y_0, x_1 + y_1 + x_0 y_0^{p^i}, x_3 + y_3 + x_0 y_0^{p^j}).$$

As we have seen in Remark 3.2.11, for $i \neq j$ the two-dimensional commutator subgroup $\overline{G}'$ has dimension two (whereas $\overline{G}'$ is clearly one-dimensional for $i = j$). This proves that the group $G = G(i, j)$ has a three-dimensional commutator subgroup for $i \neq j$ and shows that the assumption that $(G' \cap {}_3 G)^\circ$ is a chain is indispensable in Proposition 3.2.15.   □

Quasi-Commutative Groups

If every proper closed subgroup of an algebraic group or a topological group G lies in the centre of G, then G is commutative. The reason for this is the fact that every element of G lies in a closed commutative subgroup of G; for algebraic groups see [8], 1.21(c), p. 57. This gives a justification for the following

3.2.17 Definition. *We call a connected algebraic or topological group G quasi-commutative, if every closed connected commutative subgroup of G lies in the centre of G.*

We show for Lie groups and algebraic groups over fields of characteristic zero that in these classes quasi-commutative groups are commutative; the proof is trivial.

3.2.18 Proposition. *A Lie algebra $\mathfrak{g}$ is commutative if and only if every commutative subalgebra lies in the centre of $\mathfrak{g}$.*

Proof. The assertion follows from the fact that every one-dimensional subspace of $\mathfrak{g}$ is a subalgebra.   □

3.2.19 Proposition. *Let G be a group belonging to one of the following classes:*

(i)   a formal group over some field of characteristic zero,
(ii) a connected affine algebraic group over a field of characteristic zero,
(iii) a connected real or complex Lie group.

If every formal respectively closed connected subgroup of G lies in the centre of G, then G is commutative.

Proof. If G is a formal group the assertion follows from Proposition 3.2.18 together the description of functorial equivalence between formal groups and their Lie algebras in [39], Theorem 14.2.3, p. 81. In case *(ii)* the same argument holds, now using [45], 13.1 Theorem, p. 87.

In a connected real or complex Lie group G there is neighborhood U of the identity covered by local one-parameter subgroups (see [40], Theorem 3.2,

p. 80 or [55], Theorem 1.15, p. 23). If G is quasi-commutative, then the closures of the one-parameter subgroups lie in the centre of G. Since U generates G, the group G is commutative.   □

For p-adic Lie groups one can only expect a local version of Proposition 3.2.19. Given subgroups A, B of a topological group G, we say that A is *locally* contained in B, if $A \cap B$ is open in B.

3.2.20 Proposition. *Let G be finite-dimensional Lie group over a complete ultrametric field of characteristic zero. If every subgroup of G which has an open commutative subgroup is locally contained in the centre of G, then G contains an open commutative subgroup.*

Proof. Let $\mathfrak{h}$ be a commutative subalgebra of the Lie algebra $\mathfrak{g}$ of G. By [9], Theorem 3, Ch. III, § 4, $n°$ 2, p. 170, there is a subgroup H of G having $\mathfrak{h}$ as Lie algebra. Because $\mathfrak{h}$ is commutative, the group H has an open commutative subgroup. So H is locally contained in the centre of G. It follows that $\mathfrak{h}$ is contained in the centre of $\mathfrak{g}$. Thus by Proposition 3.2.18 the Lie algebra $\mathfrak{g}$ is commutative. Our assertion now is a consequence of [9], Corollaire 3, Ch. III, § 4, $n°$ 1, p. 168.   □

3.2.21 Proposition. *An algebraic k-group G is quasi-commutative if and only if its maximal affine subgroup $L(G)$ is quasi-commutative.*

Proof. Let $G = L(G)D(G)$ be the decomposition described in Theorem 1.3.2 and let $L(G)$ be quasi-commutative. Let H be a commutative connected subgroup of G and consider the product $T = HD(G)$, which is a commutative connected subgroup of G, as well. Let $T = L(T)D(T)$ be its decomposition and observe that, as a commutative connected subgroup of $L(G)$, the subgroup $L(T)$ is central in $L(G)$, hence in G as well. On the other hand, the subgroup $D(T)$ is contained in any connected algebraic subgroup S of T such that T/S is affine (see Theorem 1.3.2). Since $T/D(G)$ is affine, we have $D(T) \leq D(G)$, therefore $H \leq T = L(T)D(T) \leq \mathfrak{z}G$.   □

This proposition and Proposition 3.2.19 show the following fact:

3.2.22 Proposition. *Any connected quasi-commutative algebraic group over a field of characteristic zero is commutative.*   □

The groups of type $\mathfrak{J}_2$ show that in contrast to the case of characteristic zero there are non-commutative, quasi-commutative algebraic groups over fields of positive characteristic.

Subgroups and direct products of quasi-commutative groups are quasi-commutative as well, and the following proposition shows that quasi-commutativity is invariant under isogenies:

3.2.23 Proposition. *Let G be a connected affine algebraic k-group and E be a finite normal subgroup of G. Then G/E is quasi-commutative if and only if G is quasi-commutative.*

Proof. Let G/E be quasi-commutative and U be a connected commutative subgroup of G. As the group UE/E is central in G/E, we have $[G, U] \leq E$. But $[G, U]$ is connected and E is finite, so the last condition holds if and only if $[G, U] = 1$.

Conversely, let G be quasi-commutative and let U/E be a connected commutative subgroup of G/E. Since $U/E = U^\circ E/E$ one has $(U^\circ)' \leq E$ and this forces $[G, U^\circ] = 1$. Thus U/E is central in G/E. □

The following example shows that homomorphic images of quasi-commutative groups in general do not have this property.

3.2.24 Example. As in Example 3.1.4, let $\mathfrak{J}_2$ be the two-dimensional group defined on the affine plane by the operation

$$(x_0, x_1) \cdot (y_0, y_1) = (x_0 + y_0, x_1 + y_1 + x_0 y_0^p).$$

In the direct product $\mathfrak{J}_2 \times \mathfrak{J}_2$, which is a quasi-commutative group, we factor out the one-dimensional central subgroup

$$N = \{((x_0, x_1), (y_0, y_1)) : x_0 = y_0 = 0, x_1 = -y_1\}$$

and obtain a three-dimensional group. Via the homomorphism $\mathfrak{J}_2 \times \mathfrak{J}_2 \longrightarrow G$

$$((x_0, x_1), (y_0, y_1)) \mapsto (x_0, y_0, x_1 + y_1)$$

we identify the factor group $(\mathfrak{J}_2 \times \mathfrak{J}_2)/N$ with the group G defined on the three-dimensional affine space by

$$(x_0, x_1, x_2) \cdot (x_0', x_1', x_2') = (x_0 + x_0', x_1 + x_1', x_2 + x_2' + x_0 x_0'^p + x_1 x_1'^p).$$

The group G is not quasi-commutative, because the subgroup $H = \{(x_0, x_1, x_2) \in G : x_0 = x_1\}$ is a two-dimensional vector group, not contained in the one-dimensional centre $_3 G = \{(x_0, x_1, x_2) \in G : x_0 = x_1 = 0\}$. □

The group $G = \mathfrak{J}_2 \curlyvee \mathfrak{J}_2$ in example 3.2.24 is the direct product with amalgamated centre. We shall show now that for $r \neq s$ the group $\mathfrak{J}_2(\mathbf{F}^r) \curlyvee \mathfrak{J}_2(\mathbf{F}^s)$ is quasi-commutative.

3.2.25 Example: Let $r \neq s$ and identify the group $G = \mathfrak{J}_2(\mathbf{F}^r) \curlyvee \mathfrak{J}_2(\mathbf{F}^s)$ with the algebraic group defined on the three-dimensional affine space through the multiplication

$$(x_0, x_1, x_2)(y_0, y_1, y_2) = (x_0 + y_0, x_1 + y_1, x_2 + y_2 + x_0 y_0^{p^r} + x_1 y_1^{p^s}).$$

Then $_3^\circ G = \{(x_0, x_1, x_2) \in G : x_0 = x_1 = 0\}$ and G is nilpotent of class two, since $G/_3^\circ G$ is a two-dimensional vector group.

Let $\phi : \mathbf{G}_a \longrightarrow G$ with $\phi(t) = (\phi_0(t), \phi_1(t), \phi_2(t))$ be a homomorphism. Then we have

$$\phi_2(t_1 + t_2) = \phi_2(t_1) + \phi_2(t_2) + \phi_0(t_1)\phi_0(t_2)^{p^r} + \phi_1(t_1)\phi_1(t_2)^{p^s}.$$

Thus $\phi_0(t_1)\phi_0(t_2)^{p^r} + \phi_1(t_1)\phi_1(t_2)^{p^s} = \phi_2(t_1 + t_2) - \phi_2(t_1) - \phi_2(t_2)$ is a coboundary. In particular we find that

$$\phi_0(t_1)\phi_0(t_2)^{p^r} + \phi_1(t_1)\phi_1(t_2)^{p^s} = \phi_0(t_2)\phi_0(t_1)^{p^r} + \phi_1(t_2)\phi_1(t_1)^{p^s}.$$

Looking at the degree of the variables, we see that this is possible only if $\phi_0 = \phi_1 = 0$ and this proves that the only one-dimensional connected subgroup of G is $\mathfrak{z}°G$. Analogously let H be a two-dimensional subgroup of G. Since $H > \mathfrak{z}°G$ and $H/\mathfrak{z}°G$ is one-dimensional, we can assume that

$$H = \{(x_0, x_1, x_2) : x_0 = \psi_0(t), x_1 = \psi_1(t)\}$$

where $\psi_i : \bar{\mathsf{k}} \longrightarrow \bar{\mathsf{k}}$ is a p-polynomial. Again the equation

$$\psi_0(t_1)\psi_0(t_2)^{p^r} + \psi_1(t_1)\psi_1(t_2)^{p^s} = \psi_0(t_2)\psi_0(t_1)^{p^r} + \psi_1(t_2)\psi_1(t_1)^{p^s}$$

holds only if $\psi_0 = \psi_1 = 0$. Thus we have no commutative two-dimensional subgroup and G is quasi-commutative.   □

3.2.26 Proposition. *Let G be a quasi-commutative algebraic k-group, not commutative itself. Then G is nilpotent of class two.*

Proof. Since any maximal torus of G is central, the Borel subgroups of G are nilpotent, forcing G to be nilpotent as well (see [8], 11.5, p. 148). Since $\mathfrak{z}°G'$ is commutative, it is central, and the assertion on the nilpotency class follows from Corollary 3.2.2.   □

4

Chains

A connected algebraic group G is called a *chain* if the lattice of connected algebraic subgroups of G is a chain, that is: for any two connected algebraic subgroups K, H of G we have either $H \leq K$ or $K \leq H$. If G is affine, then G is a chain if and only if it has a unique connected algebraic subgroup of dimension d, for any $d = 1, \cdots, n$, because G is, together with a Borel subgroup of G, nilpotent.

4.1 Preliminaries

A connected algebraic k-group G is called a k-*chain* if the lattice of connected algebraic k-subgroups of G is a chain. A $\bar{\mathsf{k}}$-chain is also a k-chain. In contrast to this, an irreducible k-torus of dimension $n > 1$ is a k-chain which is not a $\bar{\mathsf{k}}$-chain (see Remark 1.3.9 or [72]). If G is a connected k-trigonalizable affine group, then the unipotent radical G_u of G is a k-subgroup. By [8], Theorem 18.3, p. 218, if G is not unipotent, then there exists a maximal torus T of G defined over k. Thus a connected k-trigonalizable k-chain is either unipotent or an irreducible torus. The same holds for connected k-group if we substitute the hypothesis that G is k-trigonalizable with the one that k is perfect, as the following proposition shows. In contrast, in Remark 1.3.9 we have seen that there exist commutative k-chains G over a non-perfect field k, not k-trigonalizable, which are not unipotent, nor tori.

4.1.1 Proposition. *If G is an affine k-chain over a perfect field k, then G is either a unipotent $\bar{\mathsf{k}}$-chain or an irreducible torus.*

Proof. We assume first that G is not unipotent. If k is infinite, then the (abstract) subgroup $G(\mathsf{k})$ of k-rational points of G is dense in G (see [8], Corollary 18.3, p. 220). Let T be a maximal torus of G defined over k (see [8], Theorem 18.3, p. 218). Any torus which is conjugate to T by an element of $G(\mathsf{k})$ is also defined over k, hence coincides with T, because G is a k-chain.

As the normalizer of T in G is closed (see [8], 1.7 Proposition, p. 52) and contains the dense subgroup $G(\mathsf{k})$, the k-torus T is normal in G. Together with a Borel subgroup, the group G is nilpotent. Since k is perfect, the unipotent radical G_u of G is defined over k. As G is a k-chain, $G = T$ is an irreducible torus.

If k is finite, then G is k-*quasi-split*, that is G has a Borel subgroup B defined over k (see [8], Proposition 16.6, p. 211). Again the unipotent radical B_u of B is defined over k and by [8], Theorem 18.3, p. 218, we can take a maximal torus S of B defined over k. As B is a k-chain, we have that either $B = S$ or $B = B_u$. In both cases $G = B$, thus G is an irreducible torus, because we are in the case where G is not unipotent.

Let now G be unipotent. If G is commutative, then G is k-isogenous to a Witt group, hence it is a $\bar{\mathsf{k}}$-chain. Let now G be a non-commutative unipotent counter-example to the assertion of minimal dimension $n > 1$. As k is perfect, the unipotent group G is k-split, that is G has a composition series $G = G_0 > G_1 > \cdots > G_s = 1$, consisting of connected k-subgroups such that its factors G_i/G_{i+1} are k-isomorphic to $\mathbf{G}_a$. In particular for any $l \geq 1$ the algebraic k-subgroup G_l is by induction a $\bar{\mathsf{k}}$-chain and it is the unique connected algebraic k-subgroup of G of co-dimension l, because G is a k-chain. Now, as any subgroup of the descending central series $\mathcal{C}^i(G)$ is defined over k (see [8], 2.3 Proposition, p. 58), the commutator subgroup G' contains a connected central k-subgroup H. It follows that G contains a unique maximal connected $\bar{\mathsf{k}}$-subgroup $M = G_1$, because by induction the k-group G/H is a $\bar{\mathsf{k}}$-chain, and we are done. $\square$

4.1.2 Corollary. *If G is a connected k-split affine group over a perfect field k, then G is a k-chain if and only if G is a $\bar{\mathsf{k}}$-chain. Moreover, if G is not unipotent, then G is a one-dimensional torus.* $\square$

Since in characteristic zero any central extension of a one-dimensional group by a one-dimensional group splits, the only unipotent chains over fields of characteristic zero are one-dimensional. The following theorem shows that the investigation of chains reduces to the analysis of affine chains which are unipotent groups over a field of positive characteristic p.

4.1.3 Theorem. *A connected algebraic group G is a non-affine chain if and only if it is either a simple abelian variety or an extension of a one-dimensional affine group C by a simple abelian variety A which is not isogenous to the direct sum $C \oplus A$. The subgroup C is the only connected algebraic subgroup of G. If C is unipotent, then the characteristic of the ground field is zero. If C is a one-dimensional torus, then G is not defined over a finite field.*

Proof. From Rosenlicht's Theorem 1.3.2 we conclude that the chain $G = D(G)$ is an algebraic group with no affine image, so it is an extension of an affine chain $C = L(G)$ by a simple abelian variety A which is not isogenous to the direct sum $C \oplus A$. If the characteristic of the ground field k is zero, then the affine chain C can be either a one-dimensional torus or a one-dimensional unipotent group. If the characteristic of k is positive, C can be only a one-dimensional torus, since $G = D(G)$ has only a finite number of elements of any given order. Proposition 1.3.3 yields that G is not defined over a finite field. Assume that H is another non-trivial connected algebraic subgroup of G. Since C has dimension one, we have $C \leq H$. But G/C is a simple abelian variety, hence $H = G$, a contradiction.

Conversely, assume that G is not a simple abelian variety and that G is not isogenous to the direct sum $C \oplus A$. Since $C = L(G)$ has dimension one, we have that either $L(G) \cap D(G)$ is finite, or $G = D(G)$. The first possibility can be excluded since otherwise $D(G)$ would be an abelian variety and G would be isogenous to $L(G) \oplus D(G)$. Hence $G = D(G)$. Let now $H \neq L(G)$ be any connected algebraic subgroup of G. Since $G/L(G)$ is a simple abelian variety, we have $HL(G) = G$. Once again, as $\dim L(G) = 1$, we have that either $H \cap L(G)$ is finite or $L(G) < H$. Since the first possibility can be excluded we have that G is isogenous to $L(G) \oplus H$, which contradicts our assumptions. Thus we have $H = G$. $\qquad\square$

4.1.4 Remark. Over an algebraically closed field k of arbitrary characteristic, there exist extensions of a one-dimensional torus T by an abelian variety A which are not isogenous to the direct product $A \oplus T$ ([89], § VII n° 16, Théorème 6, p. 184). For instance, take the generalized Jacobian $J_\mathfrak{m}$ of a smooth elliptic curve E, defined over a finite field with respect to the modulus $\mathfrak{m} = (M) + (N)$, where M, N are two distinct non-zero points of E mentioned in Remark 2.1.4.

If the characteristic of k is zero and C is a one-dimensional vector group, such extensions exist, as well, ([89], § VII n° 17, Théorème 7 and Remarque I°, p. 186). In [79] it is proved that for any abelian variety A over k, the group $\mathrm{Ext}(A, \mathbf{G}_a)$ is a k-vector space such that $\dim_k \mathrm{Ext}(A, \mathbf{G}_a) = \dim A$. $\qquad\square$

4.1.5 Remark. The elementary structure of chains may suggest that any connected algebraic group is generated by chains, but the group $\mathfrak{J}_3$ defined in 3.1.4 gives an example of an algebraic group which is not generated by chains, because its unique two-dimensional subgroup is a vector group. $\qquad\square$

We recall that the smallest dimension of unipotent non-commutative chains over fields of characteristic $p > 0$ is two. Such groups are necessarily chains. For instance, the group $\mathfrak{J}_2(\mathbf{F}^r)$ of matrices of the form

$$\begin{pmatrix} 1 & x & y \\ 0 & 1 & x^{p^r} \\ 0 & 0 & 1 \end{pmatrix}$$

is a chain.

Now we describe a class of non-commutative unipotent n-dimensional chains over a field k of characteristic $p > 0$ which generalize this example. They are non-splitting extensions of the $(n-1)$-dimensional Witt group by $\mathbf{G}_a$.

According to Remark 1.3.7, we identify the ring of k-endomorphisms of the additive group with the non-commutative ring of p-polynomials $k[\mathbf{F}]$.

If $\mathcal{S} \subseteq k[\mathbf{F}]$ is the finite set of the p-polynomials $f, g_1, \ldots, g_t$, with $f \neq 0$, we define a central extension $\mathfrak{C}_n = \mathfrak{C}_n(\mathcal{S})$

$$1 \longrightarrow \mathfrak{W}_1 \longrightarrow \mathfrak{C}_n \longrightarrow \mathfrak{W}_{n-1} \longrightarrow 1.$$

For this purpose we denote the elements of the group $\mathfrak{W}_{n-1}$, which is defined on an $(n-1)$-dimensional affine space, by $\mathbf{x}_{n-1} = (x_0, x_1, \cdots, x_{n-2})$ and define a factor system $\Theta : \mathfrak{W}_{n-1} \times \mathfrak{W}_{n-1} \longrightarrow \mathfrak{W}_1$ by

$$\Theta(\mathbf{x}_{n-1}, \mathbf{y}_{n-1}) = f \Phi_{n-1}(\mathbf{x}_{n-1}, \mathbf{y}_{n-1}) + \sum_{i=1}^{t} g_i(x_0 y_0^{p^i}),$$

where Φ_{n-1} is the factor system of the Witt extension

$$1 \longrightarrow \mathfrak{W}_1 \longrightarrow \mathfrak{W}_n \longrightarrow \mathfrak{W}_{n-1} \longrightarrow 1.$$

If every g_i is zero, then $\mathfrak{C}_n$ is isogenous to the Witt group $\mathfrak{W}_n$, whereas if $\mathfrak{C}_n$ is not commutative, then $\mathfrak{C}'_n$ is one-dimensional. As $\mathfrak{C}_n/\mathfrak{C}'_n$ is isomorphic to a Witt group, the group $\mathfrak{C}_n$ has a unique maximal connected algebraic subgroup, which is a chain. Thus $\mathfrak{C}_n$ is a chain, as well. Note that for $f = id$ and $g_i = 0$, we have $\mathfrak{C}_n = \mathfrak{W}_n$.

We will show in Theorem 4.3.1 that every unipotent chain over a perfect field k of characteristic greater than two, having a one-dimensional commutator subgroup, is k-isogenous to a chain $\mathfrak{C}_n(\mathcal{S})$ introduced above. But there exist n-dimensional unipotent chains where the commutator subgroup has dimension greater than one. The simplest examples are the following ones.

4.1.6 Remark. Let $n \geq 4$. We denote by $\mathbf{x}_i$ the i-tuple $(x_0, \cdots x_i)$ and by $\Phi_{i+1}(\mathbf{x}_i, \mathbf{y}_i)$ the polynomials defining the Witt group $\mathfrak{W}_n$ by the operation

$$(x_0, \cdots, x_{n-1}) + (y_0, \cdots, y_{n-1}) = (z_0, \cdots, z_{n-1}),$$

where $z_0 = x_0 + y_0$ and $z_i = x_i + y_i + \Phi_i(\mathbf{x}_{i-1}, \mathbf{y}_{i-1})$.
Since for the non-symmetric map $\phi : \mathfrak{W}_{n-2} \times \mathfrak{W}_{n-2} \longrightarrow \mathfrak{W}_2$ given by

$$\phi(\mathbf{x}_{n-3}, \mathbf{y}_{n-3}) = (x_0 y_0^p, y_0^{p^2} x_1 - y_0^p x_1^p - (x_0 + y_0)^p \Phi_1^p(x_0, y_0))$$

one has $\delta^2 \phi = 0$, the multiplication

$$(x_0, \cdots, x_{n-1}) + (y_0, \cdots, y_{n-1}) = (z_0, \cdots, z_{n-1})$$

where $z_0 = x_0 + y_0$ and

$$\begin{cases} z_i &= x_i + y_i + \Phi_i(\mathbf{x}_{i-1}; \mathbf{y}_{i-1}) \quad \text{for } i = 1, \cdots, n-3; \\ z_{n-2} &= x_{n-2} + y_{n-2} + \Phi_{n-2}(\mathbf{x}_{n-3}; \mathbf{y}_{n-3}) + x_0 y_0^p \\ z_{n-1} &= x_{n-1} + y_{n-1} + \Phi_{n-1}(\mathbf{x}_{n-2}; \mathbf{y}_{n-2}) + y_0^{p^2} x_1 - y_0^p x_1^p + \\ &\quad + \Phi_1(x_{n-2} + y_{n-2} + \Phi_{n-2}(\mathbf{x}_{n-3}; \mathbf{y}_{n-3}), x_0 y_0^p) - (x_0 + y_0)^p \Phi_1^p(x_0, y_0) \end{cases}$$

defines on the n-dimensional affine space a connected unipotent k-group $\mathfrak{K}_n$. The centre of $\mathfrak{K}_n$ is the $(n-2)$-dimensional subgroup

$$_3\mathfrak{K}_n = \{(x_0, \cdots, x_{n-1}) \in \mathfrak{K}_n : x_0 = x_1 = 0\}$$

and the commutator subgroup of $\mathfrak{K}_n$ is the two-dimensional subgroup

$$\mathfrak{K}'_n = \{(x_0, \cdots, x_{n-1}) \in \mathfrak{K}_n : x_0 = x_1 = \cdots = x_{n-3} = 0\}.$$

Since $\mathfrak{K}_n/\mathfrak{K}'_n$ is isomorphic to a Witt group, $\mathfrak{K}_n$ has a unique maximal connected subgroup M, which is defined by $x_0 = 0$. As M is isomorphic to a Witt group, $\mathfrak{K}_n$ is a chain. $\qquad\square$

4.1.7 Remark. In Remark 3.1.6 we have shown that in positive characteristic the kernel of the adjoint representation of a connected unipotent algebraic group G can be much larger than $_3G$. For a chain of type $\mathfrak{C}_n$ the kernel of Ad is precisely the centre of $\mathfrak{C}_n(\mathfrak{S})$. In fact, representing the Lie algebra $\mathfrak{g}$ of $\mathfrak{C}_n$ within the tangent bundle $T(\mathfrak{C}_n)$ by elements $\varepsilon(x_0, \cdots, x_{n-1})$ with $\varepsilon^2 = 0$ (cf. Remark 1.3.8), the adjoint representation Ad is given by

$$(x_0, \cdots, x_{n-1})\,\varepsilon(a_0, \cdots, a_{n-1})(x_0, \cdots, x_{n-1})^{-1} =$$

$$\varepsilon\left(a_0, \cdots, a_{n-1} - \sum_{i=1}^{t} g_i(a_0 x_0^{p^i})\right).$$

But for the chain $\mathfrak{K}_n$ given in Remark 4.1.6 a direct computation shows that

$$(0, x_1, \cdots, x_{n-1})\,\varepsilon(a_0, \cdots, a_{n-1})(0, x_1, \cdots, x_{n-1})^{-1} = \varepsilon(a_0, \cdots, a_{n-1}).$$

Thus the maximal connected subgroup $M = \{(x_0, \cdots, x_{n-1}) \in \mathfrak{K}_n : x_0 = 0\}$, which is not central, is contained in the connected component of the kernel of the adjoint representation of $\mathfrak{K}_n$. The group M is indeed the kernel of Ad, because for the connected one-dimensional subgroup $H = \{(x_0, \cdots, x_{n-1}) \in \mathfrak{K}_n : x_0 = \cdots = x_{n-2} = 0\}$ we have that $\mathfrak{K}_n/H$ is isomorphic to $\mathfrak{C}_{n-1}$, a group which has M/H as the kernel of the adjoint representation. $\qquad\square$

4.2 Three-Dimensional Chains

The aim of the present section is to show that any three-dimensional chain has a one-dimensional commutator subgroup. In Theorem 4.3.1 we will obtain a classification, up to k-isogenies, of the group $G(k)$ of k-rational points of any connected unipotent chain G, defined over an infinite perfect field k, having a one-dimensional commutator subgroup. As k is perfect and G is unipotent, G is k-split, hence we can realise the chain G on the affine space by a normal series of algebraic k-subgroups such that each algebraic factor group is biregularly k-isomorphic to $\mathbf{G}_a$.

Aiming for a contradiction, we begin our discussion by assuming that G is a three-dimensional chain of nilpotency class three, defined over a perfect field k. Hence G' is the two-dimensional connected k-subgroup, $\mathfrak{z}^{\circ}G = [G, G']$ is the one-dimensional connected k-subgroup and by [18] II, § 3, n. 4.6, p. 197, there exists a finite set $J \subset \mathbb{N}$ of indices and a corresponding finite set of p-polynomials $\mathcal{S} = \{f, g_j : j \in J\} \subset k[\mathbf{F}]$, where some g_j is different from zero, such that $G/\mathfrak{z}^{\circ}G$ is a k-group k-isomorphic to $\mathfrak{C}_2(\mathcal{S})$. Therefore G is a central extension

$$1 \longrightarrow \mathfrak{W}_1 \longrightarrow G \longrightarrow \mathfrak{C}_2(\mathcal{S}) \longrightarrow 1.$$

As we have seen in Section 2, the connected unipotent group G is biregularly k-isomorphic to the affine algebraic group defined on the three-dimensional affine space by the following multiplication

$$(x_0, x_1, x_2)(y_0, y_1, y_2) = \tag{4.1}$$
$$\big(x_0 + y_0, x_1 + y_1 + \alpha(x_0, y_0), x_2 + y_2 + \Theta(x_0, x_1; y_0, y_1)\big),$$

where, by [18], II, § 3, n. 4.6, p. 197,

$$\alpha(x_0, y_0) = f\Phi_1(x_0, y_0) + \sum_{j \in J} g_j\left(x_0 y_0^{p^j}\right) \tag{4.2}$$

and $\Theta \in \mathsf{C}^2(\mathfrak{C}_2(\mathcal{S}), \mathfrak{W}_1) \setminus \mathsf{B}^2(\mathfrak{C}_2(\mathcal{S}), \mathfrak{W}_1)$ is a non-trivial regular factor system. Note that $\mathfrak{z}^{\circ}G = \{(x_0, x_1, x_2) : x_0 = x_1 = 0\}$, whereas $G' = \{(x_0, x_1, x_2) : x_0 = 0\}$. In particular, as $\dim G' = 2$, by Corollary 3.2.3 the commutator subgroup G' is a commutative chain. Therefore there exists a non-trivial p-polynomial $h \in k[\mathbf{F}]$, and $\varphi \in \mathsf{B}^2_k(\mathfrak{W}_1, \mathfrak{W}_1)$, such that the multiplication in G' is given by

$$(0, x_1, x_2)(0, y_1, y_2) = (0, x_1 + y_1, x_2 + y_2 + h\Phi_1(x_1, y_1) + \varphi(x_1, y_1)).$$

Now we determine Θ looking at the *non-central* extension

$$1 \longrightarrow G' \xrightarrow{\ i\ } G \xrightarrow{\ \pi\ } \mathfrak{W}_1 \longrightarrow 1. \tag{4.3}$$

Non-central extensions have been discussed in Remark 2.1.15. The non-central extension (4.3) determines the factor system $F = (\theta_1, \theta_2) : \mathfrak{W}_1 \times \mathfrak{W}_1 \longrightarrow G'$, mapping the pair (x, y) to $(0, \theta_1(x, y), \theta_2(x, y))$, and an action of $\mathfrak{W}_1$ on G' as a group of automorphisms. More precisely, since $[G, G'] = \mathfrak{z}^{\circ}G$, computing the commutator $[(y_0, 0, 0), (0, x_1, x_2)]$ for any $y_0 \in \mathfrak{W}_1$ and for any $(0, x_1, x_2) \in G'$ one obtains

$$[(y_0, 0, 0), (0, x_1, x_2)] = \big(0, 0, \sigma_{y_0}(x_1)\big)$$

for a suitable $\sigma_{y_0} \in k[\mathbf{F}]$ such that $\sigma_{y_0+z_0} = \sigma_{y_0} + \sigma_{z_0}$ for any y_0, z_0. The multiplication in G is now given by

$$(x_0, x_1, x_2) \cdot (y_0, y_1, y_2) = \Big(x_0 + y_0, x_1 + y_1 + \theta_1(x_0, y_0),$$
$$x_2 + y_2 + h\Phi_1(x_1, y_1) + \varphi(x_1, y_1)+$$
$$+h\Phi_1\big(x_1 + y_1, \alpha(x_0, y_0)\big) + \sigma_{y_0}(x_1)+$$
$$+\varphi\big(x_1 + y_1, \alpha(x_0, y_0)\big) + \theta_2(x_0, y_0)\Big).$$

According to Remark 2.1.15, we have that θ_1 is equal to the polynomial α appearing in (4.1) and (4.2).

4.2.1 Remark. We remark that, up to k-isomorphisms, it is possible to achieve $\varphi = 0$, without altering the factor system α. Since $\varphi \in B_k^2(\mathfrak{W}_1, \mathfrak{W}_1)$ there exists a regular k-function $c : \mathfrak{W}_1 \longrightarrow \mathfrak{W}_1$ such that $\varphi = \delta^1 c$. Consider the two-dimensional connected commutative unipotent group H defined on the affine plane by the multiplication

$$(0, x_1, x_2)(0, y_1, y_2) = (0, x_1 + y_1, x_2 + y_2 + h\Phi_1(x_1, y_1)).$$

The group H is a central extension of $\mathfrak{W}_1$ by $\mathfrak{W}_1$ which is equivalent to G'. Consider now the non-central extension

$$1 \longrightarrow H \xrightarrow{i} G_1 \xrightarrow{\pi} \mathfrak{W}_1 \longrightarrow 1, \tag{4.4}$$

defined by the factor system $(\alpha, \theta_2 - c \circ \alpha) : \mathfrak{W}_1 \times \mathfrak{W}_1 \longrightarrow H$

$$(x, y) \mapsto \big(\alpha(x, y), \theta_2(x, y) - c(\alpha(x, y))\big)$$

and the following action of $\mathfrak{W}_1$ as a group of k-rational automorphisms of H

$$(x_1, x_2)^{y_0} = (x_1, x_2 + \sigma_{y_0}(x_1)).$$

The multiplication in G_1 is now given by

$$(x_0; x_1, x_2) \cdot (y_0; y_1, y_2) = \Big(x_0 + y_0;$$
$$x_1 + y_1 + \alpha(x_0, y_0),$$
$$x_2 + y_2 + h\Phi_1(x_1, y_1) + \sigma_{y_0}(x_1) +$$
$$+ h\Phi_1\big(x_1 + y_1, \alpha(x_0, y_0)\big) + \theta_2(x_0, y_0)\Big).$$

Therefore we get a k-isomorphism $\imath : G \longrightarrow G_1$ putting

$$(x_0, x_1, x_2) \mapsto (x_0; x_1, x_2 - c(x_1)), \tag{4.5}$$

which in general is not an equivalence of the extensions (4.3) and (4.4), even if it is an equivalence if we consider G and G_1 as central extensions of $\mathfrak{W}_1$ by $\mathfrak{C}_2(8)$. $\qquad\square$

Up to the above isomorphism (4.5), from now on we assume that the multiplication in G is given by

$$(x_0, x_1, x_2) \cdot (y_0, y_1, y_2) = \Big(x_0 + y_0,$$
$$x_1 + y_1 + \alpha(x_0, y_0),$$
$$x_2 + y_2 + h\Phi_1(x_1, y_1) + \sigma_{y_0}(x_1) +$$
$$+ h\Phi_1\big(x_1 + y_1, \alpha(x_0, y_0)\big) + \beta(x_0, y_0)\Big) \tag{4.6}$$

where we put for short $\beta = \theta_2 - c \circ \alpha$. As usual, up to adding a constant to β, we assume $\beta(0,0) = 0$.

Because of the associativity in G, the map β satisfies the identity

$$\delta^2 \beta(x, y, z) = \beta(x, y) + \beta(x + y, z) - \beta(y, z) - \beta(x, y + z) = \\ h\Phi_1(\alpha(y, z), \alpha(x, y + z)) - h\Phi_1(\alpha(x, y), \alpha(x, y + z)) - \sigma_z(\alpha(x, y)). \tag{4.7}$$

In particular, we find $\beta(x, 0) = \beta(0, y) = 0$ and we can write

$$(x_0, x_1, x_2) = (x_0, 0, 0)(0, x_1, x_2)$$

for any $(x_0, x_1, x_2) \in G$.

What we want to show is that no polynomial β satisfies the identity (4.7).

4.2.2 Remark. Since the group G defined by the multiplication in (4.6) is three-dimensional and has a one-dimensional centre, one could think that the adjoint representation of G easily excludes the existence of such a group G.

In fact, if the characteristic of k is $p > 2$, the adjoint representation $\mathsf{Ad}(G)$ has exponent p, being a subgroup of the group of 3×3 unipotent matrices, whereas $G/_3G$ has exponent p^2, as the following shows: if $G/_3G$ had exponent p, then we would have

$$x = y^{-p} x y^p \quad \text{for any } x, y \in G.$$

As $[G', G] \leq {}_3G$, putting $x^y = y^{-1} x y$ and $[x, y] = x^{-1} y^{-1} x y$ we find now

$$x = x^{y^p} = y^{1-p} x [x, y] y^{p-1} = y^{1-p} x y [x, y] \big[[x, y], y \big] y^{p-2} =$$

$$y^{2-p} x [x, y] \, [x, y] y^{p-2} \big[[x, y], y \big] = \cdots =$$

$$y^{k-p} x y^{p-k} [x, y]^k \big[[x, y], y \big]^{(p-1)+(p-2)+\cdots(p-k)} = x [x, y]^p \big[[x, y], y \big]^{p\frac{p-1}{2}}.$$

Since $[G', G]$ has dimension one, we find that $\big[[x, y], y \big]^{p\frac{p-1}{2}} = 1$ and this yields $[x, y]^p = 1$, a contradiction to the fact that G' is isogenous to a Witt group.

This forces G' to be contained in the kernel of Ad. Unfortunately this is not an immediate contradiction, because for a unipotent chain the kernel of Ad can be larger than ${}_3G$, as shown in Remark 4.1.7.

In order to give the adjoint representation of G, we look at the tangent bundle $T(G)$ (see Remark 1.3.8).

Now we have

$$(x_0, x_1, x_2) \, \varepsilon(a_0, a_1, a_2) \, (x_0, x_1, x_2)^{-1} =$$

$$\varepsilon(a_0, a_1 + l_1(x_0, a_0), a_2 + l_2(x_0, x_1, a_0, a_1))$$

with $\varepsilon l_1(x_0, a_0) = \alpha(x_0, \varepsilon a_0) - \alpha(\varepsilon a_0, x_0)$ and

$$\varepsilon l_2(x_0, x_1, a_0, a_1) = h\Phi_1(x_1, \varepsilon a_1) + \sigma_{\varepsilon a_0}(x_1) +$$
$$+ h\Phi_1(x_1 + \varepsilon a_1, \alpha(x_0, \varepsilon a_0)) + \beta(x_0, \varepsilon a_0) -$$
$$- h\Phi_1(x_1, -x_1 - \alpha(x_0, -x_0)) - \beta(x_0, -x_0) +$$
$$+ h\Phi_1(x_1 + \varepsilon a_1 + \alpha(x_0, \varepsilon a_0), -x_1 - \alpha(x_0, -x_0)) +$$
$$+ \sigma_{-x_0}(\varepsilon a_1 + \alpha(x_0, \varepsilon a_0)) +$$
$$+ h\Phi_1(\varepsilon a_1 + \alpha(x_0, \varepsilon a_0) - \alpha(x_0, -x_0), \alpha(x_0 + \varepsilon a_0, -x_0)) +$$
$$+ \beta(x_0 + \varepsilon a_0, -x_0)).$$

For a given $(x_0, x_1, x_2) \in G$ we find

$$\mathsf{Ad}(x_0, x_1, x_2) = \begin{pmatrix} 1 & l_1(x_0, 1) & l_2(x_0, x_1, 1, 0) \\ & 1 & 0 \\ & & 1 \end{pmatrix}.$$

Since $l_1(x_0, 1) = \sum_{j \in J} g_j(x_0) - g_j(x_0^{p^j})$, the group $\mathsf{Ad}(G)$ is one-dimensional. A direct computation shows in fact that $l_2(x_0, x_1, 1, 0)$ actually depends on x_0 only, so we still do not have a contradiction. In order to obtain a contradiction it remains to show that $l_2(x_0, x_1, 1, 0)$ cannot be a p-polynomial in x_0, but to prove it is at least as hard as to prove the non-existence of β by direct computation on equation (4.7).   □

For the rest of the section, we will denote the right-hand side of the above identity (4.7) by $P(x, y, z)$, and with $P_d(x, y, z)$ for any $d \geq 0$ the homogeneous component of degree d of $P(x, y, z)$. The cohomology operator δ^2 is homogeneous, i.e. for any $n \geq 0$ the homogeneous component of degree n of the polynomial $\delta^2 \beta(x, y, z)$ is equal to $\delta^2 \beta_n(x, y, z)$, where $\beta_n(x, y)$ is the homogeneous component of degree n of $\beta(x, y)$. Then it is clear that $P(x, y, z) \in \delta^2(\mathsf{k}[x, y])$ if and only if $P_d(x, y, z) \in \delta^2(\mathsf{k}[x, y])$ for all d.

Moreover, since $P(x, y, z) = Q(x, y, z) - R(x, y, z)$, where

$$Q(x, y, z) = h\Phi_1(\alpha(y, z), \alpha(x, y + z)) - h\Phi_1(\alpha(x, y), \alpha(x, y + z)),$$
$$R(x, y, z) = \sigma_z(\alpha(x, y)),$$

for any d we have $P_d(x, y, z) = Q_d(x, y, z) - R_d(x, y, z)$.

In particular, if $h(t) = \sum_{i=0}^{s} c_i t^{p^i}$, then $Q(x, y, z) = \sum_{i=0}^{s} c_i T(x, y, z)^{p^i}$, where

$$T(x, y, z) = \Phi_1(\alpha(y, z), \alpha(x, y + z)) - \Phi_1(\alpha(x, y), \alpha(x + y, z)).$$

In order to find the homogeneous components of $P(x, y, z)$ we can restrict to the homogeneous components of $T(x, y, z)$ and $R(x, y, z)$.

4.2.3 Proposition. *For any $d = p^m + p^n$ such that $R_d \neq 0$, there is no polynomial $\gamma(x, y)$ such that $\delta^2 \gamma(x, y, z) = R_d(x, y, z)$.*

Proof. As any homogeneous component of the polynomial $\sigma_z(\sum_{j=1}^{t} g_j(xy^{p^j}))$ has degree equal to the sum of three p-powers, R_d coincides with the homogeneous component of the polynomial $\sigma_z(f\Phi_1(x, y))$, hence

$$R_d = a \cdot z^{p^m} \Phi_1^{p^{n-1}} + b \cdot z^{p^n} \Phi_1^{p^{m-1}}$$

with $a, b \in \mathsf{k}$, $(a, b) \neq (0, 0)$.

If there were $\gamma(x,y) \in k[x,y]$ such that $\delta^2\gamma(x,y,z) = R_d(x,y,z)$, then with $\lambda_{y_0}(x_1) = a \cdot y_0^{p^m} x_1^{p^{n-1}} + b \cdot y_0^{p^n} x_1^{p^{m-1}}$, the multiplication

$$(x_0, x_1, x_2)(y_0, y_1, y_2) =$$

$$(x_0 + y_0, x_1 + y_1 + \Phi_1(x_0, y_0), x_2 + y_2 + \lambda_{y_0}(x_1) + \gamma(x_0, y_0))$$

should define a three-dimensional unipotent group G such that $\dim G' = 1$. Moreover, $\dim {}_3G = 1$, as well, and $G/{}_3{}^\circ G$ is a chain. This yields a contradiction to Theorem 3.2.10 since $\dim G' + \dim {}_3G < \dim G$. $\quad\square$

In order to find the homogeneous component $T_d(x,y,z)$, with $d = p^m + p^{m+n}$ and $n > 0$, we need to recall that

$$\Phi_1(x,y) = \sum_{i=1}^{p-1} \left\{ \begin{matrix} p \\ i \end{matrix} \right\} x^i y^{p-i} - x^i y^{p-i}.$$

Hence $T(x,y,z)$ is equal to

$$\sum_{i=1}^{p-1} \left\{ \begin{matrix} p \\ i \end{matrix} \right\} \left\{ \alpha(y,z)^i \alpha(x, y+z)^{p-i} - \alpha(x,y)^i \alpha(x+y,z)^{p-i} \right\}.$$

Moreover, putting

$$f\Phi_1(x,y) = \sum_{i=0}^{t} a_i \Phi_1^{p^i}(x,y),$$
$$\sum_{j \in J} g_j(xy^{p^j}) = \sum_{j \in J} \sum_{k \geq 1} b_{jk} x^{p^j} y^{p^{j+k}},$$
$$\deg \sum_{j \in J} g_j(xy^{p^j}) = p^u + p^{u+v},$$

the polynomial $\alpha(y,z)^h \alpha(x, y+z)^{p-h}$ with α as in equation (4.2) is equal to

$$\left\{ \sum_{r=0}^{h} \binom{h}{r} \left[\sum_{(r_0,\ldots,r_t)} \omega_{(r_0,\ldots,r_t)} \prod_{i=0}^{t} \left(a_i \Phi_1^{p^i}(y,z) \right)^{r_i} \right] \right.$$

$$\left. \left[\sum_{(n_{01},\ldots,n_{uv})} \mu_{(n_{01},\ldots,n_{uv})} \prod_{j,k} \left(b_{jk} y^{p^j} z^{p^{j+k}} \right)^{n_{jk}} \right] \right\} \cdot$$

$$\left\{ \sum_{s=0}^{p-h} \binom{p-h}{s} \left[\sum_{(s_0,\ldots,s_t)} \rho_{(s_0,\ldots,s_t)} \prod_{i=0}^{t} \left(a_i \Phi_1^{p^i}(x, y+z) \right)^{s_i} \right] \right.$$

$$\left. \left[\sum_{(m_{01},\ldots,m_{uv})} \tau_{(m_{01},\ldots,m_{uv})} \prod_{j,k} \left(b_{jk} x^{p^j}(y+z)^{p^{j+k}} \right)^{m_{jk}} \right] \right\},$$

where

i) $\sum_{i=0}^{t} r_i = r$,

ii) $\sum_{j,k} n_{jk} = h - r$,

iii) the coefficients $\omega_{(r_0,\ldots,r_t)}$ are those coming from the computation of the r-th power and the coefficients $\mu_{(n_{01},\ldots,n_{uv})}$ are those coming from the computation of the $(h - r)$-th power,

iv) $\sum_{i=0}^{t} s_i = s$,

v) $\sum_{j,k} m_{jk} = p - h - s$,

vi) the coefficients $\rho_{(s_0,\ldots,s_t)}$ arise from the computation of the s-th power and the coefficients $\tau_{(m_{01},\ldots,m_{uv})}$ are those coming from the computation of the $(p-h-s)$-th power.

The polynomial $\alpha(x,y)^h \alpha(x+y,z)^{p-h}$ is equal to

$$\left\{ \sum_{r=0}^{h} \binom{h}{r} \left[\sum_{(r_0,\ldots,r_t)} \omega_{(r_0,\ldots,r_t)} \prod_{i=0}^{t} \left(a_i \Phi_1^{p^i}(x,y) \right)^{r_i} \right] \right.$$

$$\left. \left[\sum_{(n_{01},\ldots,n_{uv})} \mu_{(n_{01},\ldots,n_{uv})} \prod_{j,k} \left(b_{jk} x^{p^j} y^{p^{j+k}} \right)^{n_{jk}} \right] \right\}.$$

$$\left\{ \sum_{s=0}^{p-h} \binom{p-h}{s} \left[\sum_{(s_0,\ldots,s_t)} \rho_{(s_0,\ldots,s_t)} \prod_{i=0}^{t} \left(a_i \Phi_1^{p^i}(x+y,z) \right)^{s_i} \right] \right.$$

$$\left. \left[\sum_{(m_{01},\ldots,m_{uv})} \tau_{(m_{01},\ldots,m_{uv})} \prod_{j,k} \left(b_{jk}(x+y)^{p^j} z^{p^{j+k}} \right)^{m_{jk}} \right] \right\},$$

where r_i, n_{jk}, s_i, m_{jk} fulfil the respective conditions i), ii), iv), v), the coefficients $\omega_{(r_0,\ldots,r_t)}, \mu_{(n_{01},\ldots,n_{uv})}$ fulfil the condition iii), and $\rho_{(s_0,\ldots,s_t)}$, $\tau_{(m_{01},\ldots,m_{uv})}$ satisfy the condition vi).

Therefore, in order to compute the homogeneous component $T_d(x,y,z)$ it is enough to put, for all $h=1,\ldots,p-1$,

$$r_i + s_i = x_i \text{ for any } i = 0,\ldots,t,$$
$$n_{jk} + m_{jk} = y_{jk} \text{ for any } j,k,$$

and to solve the equation

$$\sum_{i=0}^{t} x_i p^{i+1} + \sum_{j,k} y_{jk} \left(p^j + p^{j+k} \right) = p^m + p^{m+n} \tag{4.8}$$

in the variables x_i, y_{jk} taking values in $\{0,1,\ldots,p\}$ and fulfilling the condition

$$\sum_{i=0}^{t} x_i + \sum_{j,k} y_{jk} = p. \tag{4.9}$$

For each solution of (4.8), (4.9) we will obtain a possible summand $h_i(x,y,z)$ in $T_d(x,y,z)$.

4.2.4 Lemma. *Let $p > 2$ be a prime integer, and let $x_0, x_1, \ldots, x_t, y_{01},$ $y_{02}, \ldots, y_{uv}$ be non-negative integers such that $\sum_i x_i + \sum_{j,k} y_{jk} = p$. For every pair (m,n) of non-negative integers, the solutions of the equations (4.8), (4.9) are the following:*

1) $y_{m-1,n} = p$, $x_i = 0$ for any index i and $y_{jk} = 0$ for all $(j,k) \neq (m-1,n)$;
2) $y_{m,n-1} = 1$, $x_{m+n-2} = p-1$,
    $y_{jk} = 0$ for all $(j,k) \neq (m,n-1)$ and $x_i = 0$ for any index $i \neq m+n-2$;
3) $y_{m-1,n+1} = 1$, $x_{m-2} = p-1$,
    $y_{jk} = 0$, for all $(j,k) \neq (m-1,n+1)$ and $x_i = 0$, for any index $i \neq m-2$;

4) $y_{m+n-2,1} = p - 1$, $y_{m,n-2} = 1$,
$y_{jk} = 0$ for all $(j,k) \notin \{(m+n-2,1),(m,n-2)\}$ e $x_i = 0$ for any index i;

5) $y_{m-2,1} = p - 1$, $y_{m-2,n+2} = 1$,
$y_{jk} = 0$, for all $(j,k) \notin \{(m-2,1),(m-2,n+2)\}$ and $x_i = 0$ for any index i .

Proof. Let r be the largest number among p^t and p^{u+v}. Write the equation (4.8) as

$$a_0 + a_1 p + \cdots + a_r p^r = p^m + p^{m+n}, \qquad (4.10)$$

where

$$a_0 = (y_{01} + \cdots + y_{0n_0}),$$
$$a_s = (x_{s-1} + y_{0s} + y_{1,s-1} + \cdots + y_{s-1,1} + y_{s,1} + \cdots + y_{sn_s}) \text{ for } s = 1, \ldots, r.$$

Note that, by condition (4.9), a_0 as well as any other a_s is not greater than p. If none of the a_s's is equal to p, then the identity (4.10) has a unique solution, given by $a_m = 1 = a_{m+n}$, otherwise we could write $p^m + p^{m+n}$ in two different ways in base p representation. This forces $p \leq \sum_s a_s = 2$, a contradiction.

Moreover if two of the a_s's are equal to p, then all the others are zero: Assume to the contrary that $a_h = a_l = p$ and $a_q \neq 0$, with $h < l$ and $h \neq q \neq l$. As $a_h = p$, every x_i and y_{jk} which is not a term of a_h is equal to zero. Since $y_{h,l-h}$ is the only common summand in a_h and a_l, it follows that $y_{h,l-h} = p$, whereas for all indices i we have $x_i = 0$ and for all pairs $(j,k) \neq (h,l-h)$ we have $y_{jk} = 0$. Because $y_{h,l-h}$ is not among the summands of a_q, we find the contradiction $a_q = 0$.

We now assume that there are two elements among the a_s's equal to p. Since all the others are zero, looking again at the base p representation we find that the only possibility is $a_{m-1} = a_{m+n-1} = p$. Hence we find, with the above argument, that $y_{m-1,n} = p$, and for all indices i we have $x_i = 0$ and for all pairs $(j,k) \neq (m-1,n)$ we have $y_{jk} = 0$, that is solution 1).

Assume now that only one among the a_s's is equal to p.

(i) If $a_{m+n-1} = p$, then $a_m = 1$ and all the other $a_s = 0$. Therefore for the same reason as above $y_{m,n-1} = 1$, hence $x_{m+n-2} = p - 1$, whilst $y_{jk} = 0$ for all $(j,k) \neq (m,n-1)$ and $x_i = 0$ for all $i \neq m+n-1$, that is solution 2).

(ii) If $a_{m-1} = p$, then $a_{m+n} = 1$ and all the other $a_s = 0$. Therefore $y_{m-1,n+1} = 1$ and $x_{m-2} = p - 1$, whilst $y_{jk} = 0$ for all $(j,k) \neq (m-1,n+1)$ and $x_i = 0$ for all $i \neq m-2$, that is solution 3).

(iii) If $a_{m+n-2} = p$, then the equation (4.10) is equivalent to

$$\sum_{s \neq m+n-2} a_s p^s = p^m + (p-1)p^{m+n-1}.$$

Since $a_s < p$ for any $s \neq m+n-2$, the only solution is $a_m = 1$, $a_{m+n-1} = p - 1$ and $a_s = 0$ for all $s \notin \{m, m+n-1, m+n-2\}$.

It follows that $y_{m+n-2,1} = p - 1$ and $y_{m,n-2} = 1$, whilst $y_{j,k} = 0$ for all $(j,k) \notin \{(m+n-2,1),(m,n-2)\}$ and $x_i = 0$ for all indices i, that is solution 4).

(iv) If $a_{m-2} = p$, then the equation (4.10) is equivalent to

$$\sum_{s \neq m-2} a_s p^s = (p-1)p^{m-1} + p^{m+n}.$$

Since $a_s < p$ for any $s \neq m-2$, the only solution is $a_{m-1} = p-1$, $a_{m+n} = 1$ and $a_s = 0$ for all $s \notin \{m-2, m-1, m+n\}$.

It follows that $y_{m-2,1} = p - 1$ and $y_{m-2,n+2} = 1$, whilst $y_{jk} = 0$ for all $(j,k) \notin \{(m-2,1),(m-2,n+2)\}$ and $x_i = 0$ for all indices i, that is solution 5).

(v) To end the proof, we show now that, with the exception of the above coefficients $a_{m-2}, a_{m-1}, a_{m+n-2}, a_{m+n-1}$, one has $a_s \neq p$.

Let $a_{m-1-k} = p$ for a given $k \geq 2$. Then the equation (4.10) is equivalent to

$$\sum_{s \neq m-1-k} a_s p^s = p^m + p^{m+n} - p^{m-k} =$$
$$= (p-1)(p^{m-1} + p^{m-2} + \cdots + p^{m-k}) + p^{m+n}.$$

Since $a_s < p$ for any $s \neq m-1-k$, the only solution is $a_{m+n} = 1$, $a_{m-1} = a_{m-2} = \cdots = a_{m-k} = p-1$ and $a_s = 0$ for any other index s. It follows that $y_{m-1-k,1} = \cdots = y_{m-1-k,k} = p-1$ and $y_{m-1-k,n+k+1} = 1$. By (4.9) we get $p = 1 + k(p-1)$, a contradiction.

Similarly, we obtain a contradiction if we assume $a_{m+n-1-k} = p$ for a given $2 \leq k \leq n$. Moreover, if $s \geq m+n$, no coefficient a_s can be equal to p. Otherwise $\sum_s a_s p^s > p^m + p^{m+n}$.   □

Denoting by $h_i(x,y,z)$ ($i = 1, \cdots, 5$) the polynomial corresponding in Lemma 4.2.4 to solution i) of the equations (4.8), (4.9), we find $T_d(x,y,z) = \sum_{i=1}^{5} h_i(x,y,z)$ for $d = p^m + p^{m+n}$.

Note that $h_1(x,y,z)$ is zero for $m = 0$, whilst for $m \geq 1$ it is equal to

$$b_{m-1,n}^p \cdot \Phi_1\left(y^{p^{m-1}} z^{p^{m+n-1}}, x^{p^{m-1}}(y+z)^{p^{m+n-1}}\right) -$$
$$\Phi_1\left(x^{p^{m-1}} y^{p^{m+n-1}}, (x+y)^{p^{m-1}} z^{p^{m+n-1}}\right).$$

Since

$$\Phi_1\left(y^{p^{m-1}} z^{p^{m+n-1}}, x^{p^{m-1}}(y+z)^{p^{m+n-1}}\right) =$$
$$\Phi_1\left(y^{p^{m-1}} z^{p^{m+n-1}}, x^{p^{m-1}} z^{p^{m+n-1}}\right) +$$
$$\Phi_1\left((x+y)^{p^{m-1}} z^{p^{m+n-1}}, x^{p^{m-1}} y^{p^{m+n-1}}\right) -$$
$$\Phi_1\left(x^{p^{m-1}} z^{p^{m+n-1}}, x^{p^{m-1}} y^{p^{m+n-1}}\right),$$

we have

$$h_1(x, y, z) = b_{m-1,n}^p \cdot \left(z^{p^{m+n}} \Phi_1^{p^{m-1}}(x, y) - x^{p^m} \Phi_1^{p^{m+n-1}}(y, z) \right).$$

Because of the identity

$$\delta^2\left(-(x+y)^{p^m} \Phi_1^{p^{m+n-1}}(x, y) \right)(x, y, z) =$$

$$z^{p^m} \Phi_1^{p^{m+n-1}}(x, y) - x^{p^m} \Phi_1^{p^{m+n-1}}(y, z),$$

it follows that $h_1(x, y, z)$ is equal to the polynomial

$$b_{m-1,n}^p \cdot \left(\delta^2\left(-(x+y)^{p^m} \Phi_1^{p^{m+n-1}}(x, y) \right)(x, y, z) + \right.$$

$$\left. z^{p^{m+n}} \Phi_1^{p^{m-1}}(x, y) - z^{p^m} \Phi_1^{p^{m+n-1}}(x, y) \right).$$

Note that for $i = 2, 3, 4, 5$ the polynomial $h_i(x, y, z)$ is equal to $d_i \cdot f_i(x, y, z)$, with $d_i \in \mathsf{k}$ depending on a_i and b_{jk}, moreover $f_i(x, y, z)$ is equal to

$$l_i(y, z) \left\{ \sum_{h=1}^{p-1} \frac{(p-1)!}{(h-1)!(p-h)!} \, g_i^{h-1}(y, z) \, g_i^{p-h}(x, y+z) \right\} +$$

$$l_i(x, y+z) \left\{ \sum_{h=1}^{p-1} \frac{(p-1)!}{h!(p-h-1)!} \, g_i^{h}(y, z) \, g_i^{p-h-1}(x, y+z) \right\} -$$

$$l_i(x, y) \left\{ \sum_{h=1}^{p-1} \frac{(p-1)!}{(h-1)!(p-h)!} \, g_i^{h-1}(x, y) \, g_i^{p-h}(x+y, z) \right\} -$$

$$l_i(x+y, z) \left\{ \sum_{h=1}^{p-1} \frac{(p-1)!}{h!(p-h-1)!} \, g_i^{h}(x, y) \, g_i^{p-h-1}(x+y, z) \right\},$$

where

a) $l_2(x, y) = x^{p^m} y^{p^{m+n-1}}$, $g_2(x, y) = \Phi_1^{p^{m+n-2}}(x, y)$,

b) $l_3(x, y) = x^{p^{m-1}} y^{p^{m+n}}$, $g_3(x, y) = \Phi_1^{p^{m-2}}(x, y)$,

c) $l_4(x, y) = x^{p^m} y^{p^{m+n-2}}$, $g_4(x, y) = x^{p^{m+n-2}} y^{p^{m+n-1}}$,

d) $l_5(x, y) = x^{p^{m-2}} y^{p^{m+n}}$, $g_5(x, y) = x^{p^{m-2}} y^{p^{m-1}}$.

Since for any $a, b \in \mathsf{k}$ the following identities hold

$$\sum_{h=1}^{p-1} \frac{(p-1)!}{(h-1)!(p-h)!} \, a^{h-1} b^{p-h} = (a+b)^{p-1} - a^{p-1}$$

$$\sum_{h=1}^{p-1} \frac{(p-1)!}{(h)!(p-h-1)!} \, a^{h} b^{p-h-1} = (a+b)^{p-1} - b^{p-1},$$

for any $i = 2, 3, 4, 5$, the polynomial $f_i(x, y, z)$ is equal to

$$\left(l_i(y, z) + l_i(x, y+z) \right) \left(g_i(y, z) + g_i(x, y+z) \right)^{p-1} +$$

$$-l_i(y, z) g_i^{p-1}(y, z) - l_i(x, y+z) g_i^{p-1}(x, y+z) +$$

$$-\left(l_i(x, y) + l_i(x+y, z) \right) \left(g_i(x, y) + g_i(x+y, z) \right)^{p-1} +$$

$$l_i(x, y) g_i^{p-1}(x, y) + l_i(x+y, z) g_i^{p-1}(x+y, z).$$

Applying the fact that every l_i and every g_i is a factor system, a direct computation shows that

$$\begin{aligned}
f_i(x,y,z) &= -l_i(y,z)g_i^{p-1}(y,z) - l_i(x,y+z)g_i^{p-1}(x,y+z) + \\
&\quad + l_i(x,y)g_i^{p-1}(x,y) + l_i(x+y,z)g_i^{p-1}(x+y,z) \\
&= \delta^2\big(l_i(x,y)g_i^{p-1}(x,y)\big)(x,y,z)
\end{aligned}$$

The arguments about the polynomials $h_i(x,y,z)$ give now the following

4.2.5 Lemma. *Let $d = p^m + p^{m+n}$, with $n > 0$, and let $T_d(x,y,z)$ be the homogeneous component of degree d of the polynomial $T(x,y,z)$. There exists a polynomial $\gamma \in k[x,y]$ such that*

$$T_d(x,y,z) = \varepsilon_m \cdot b_{m-1,n}^p\left(z^{p^{m+n}}\Phi_1^{p^{m-1}}(x,y) - z^{p^m}\Phi_1^{p^{m+n-1}}(x,y)\right) + \delta^2\gamma(x,y,z),$$

where $\varepsilon_0 = 0$ and $\varepsilon_m = 1$ for $m > 0$. $\square$

Recalling that $Q(x,y,z) = \sum_{i=0}^{s} c_i T(x,y,z)^{p^i}$, for $d = p^m + p^{m+n}$ we have

$$Q_d(x,y,z) = \sum_{i \le m,s} c_i \cdot T_{d_i}(x,y,z)^{p^i},$$

where T_{d_i} is the homogeneous component of degree $d_i = p^{m-i} + p^{m+n-i}$ of the polynomial T. By the above lemma we get the following

4.2.6 Proposition. *Let $d = p^m + p^{m+n}$, with $n > 0$, and let $Q_d(x,y,z)$ be the homogeneous component of degree d of the polynomial $Q(x,y,z)$. Then there exists a polynomial $\gamma \in k[x,y]$ such that*

$$Q_d(x,y,z) = a\big(z^{p^{m+n}}\Phi_1^{p^{m-1}}(x,y) - z^{p^m}\Phi_1^{p^{m+n-1}}(x,y)\big) + \delta^2\gamma(x,y,z),$$

where $a = 0$ for $m = 0$ and $a = \sum_{i \le m-1,s} c_i(b_{m-i-1,n})^{p^{i+1}}$ for $m \ge 1$. $\square$

We have $b_{m-i-1,n} = 0$ for any $0 \le i \le s$, because $\deg \sum_j g_j(xy^{p^j}) = p^u + p^{u+v}$ for every pair (m,n) such that $p^m + p^{m+n} > p^{u+s+1} + p^{u+v+s+1}$. In fact, if $m+n > u+v+s+1$, then for any $0 \le i \le s$ it follows $m-i-1+n > u+v$. Hence $p^{m-i-1} + p^{m-i-1+n} > p^u + p^{u+v}$, while we assumed that the set $S = \{(j,k) : b_{jk} \ne 0\}$ is such that for every $(j,k) \subset S$ we have $p^j + p^{j+k} \le p^u + p^{u+v}$. Similarly if $m > u+s+1$ and $m+n = u+v+s+1$, then for every $0 \le i \le s$ we have $m-i-1 > u$, $m-i-1+n \ge u+v$, hence $p^{m-i-1} + p^{m-i-1+n} > p^u + p^{u+v}$. Thus, this discussion together with the preceding proposition yields

4.2.7 Corollary. *Let $d = p^m + p^{m+n}$ with $p^m + p^{m+n} > p^{u+s+1} + p^{u+v+s+1}$. Then there exists a polynomial $\gamma \in k[x,y]$ such that $Q_d(x,y,z) = \delta^2\gamma(x,y,z)$.* $\square$

It remains to evaluate the homogeneous components of the polynomial $\sigma_z\left(\sum_j g_j(xy^{p^j})\right)$ appearing in equation (4.7).

Let S be the subspace of $k[x_1, x_2, x_3]$ consisting of polynomials in which for every monomial one has $\deg x > 0, \deg y > 0, \deg z > 0$. Let $\Omega_3 : S \to S$ be the operator defined by:

$$f(x_1, x_2, x_3) \mapsto \sum_{\sigma \in S_3} (-1)^{sign(\sigma)} f(x_{\sigma(1)}, x_{\sigma(2)}, x_{\sigma(3)}).$$

We want to show that $\Omega_3\left(\sigma_z\left(\sum_j g_j(xy^{p^j})\right)\right) = 0$.

Taking the element $g = (z, 0, 0) \in G$, and considering the inner automorphism induced by g, we have

$$g(0, x_1, x_2)g^{-1} = (0, x_1, x_2 - \sigma_z(x_1)).$$

Furthermore, for suitable mappings $\tau, \psi_z : k \longrightarrow k$ we obtain

$$g^{-1} = (-z, -\alpha(z, -z), \tau(z))$$

and

$$g(x_0, 0, 0)g^{-1} = (x_0, \alpha(z, x_0) + \alpha(x_0 + z, -z) - \alpha(z, -z), \psi_z(x_0)).$$

Put for short $\alpha = \alpha_1 + \alpha_2$, where $\alpha_1(x, y) = \sum_j g_j(xy^{p^j})$, $\alpha_2(x, y) = f\Phi_1(x, y)$, and $V(x_0, x_1; y_0, y_1) = h\Phi_1(x_1, y_1) + h\Phi_1(x_1 + y_1, \alpha(x_0, y_0)) + \sigma_{y_0}(x_1) + \beta(x_0, y_0)$. As $\alpha_2(x, y)$ is symmetric we get $\alpha_2(z, x_0) + \alpha_2(x_0 + z, -z) - \alpha_2(z, -z) = 0$, whereas $\alpha_1(x, y)$ being bi-additive yields $\alpha_1(z, x_0) + \alpha_1 b(x_0 + z, -z) - \alpha_1(z, -z) = \alpha_1(z, x_0) - \alpha_1(x_0, z)$. So $(x_0, 0, 0)^g$ is equal to $(x_0, \varphi_z(x_0), \psi_z(x_0))$, where we put for short $\varphi_z(x_0) = \alpha_1(z, x_0) - \alpha_1(x_0, z)$.

Since $(x_0, x_1, x_2)^g = (x_0, 0, 0)^g(0, x_1, x_2)^g$ we obtain

$$(x_0, x_1, x_2)^g = \left(x_0, x_1 + \varphi_z(x_0), x_2 + \psi_z(x_0) - \sigma_z(x_1) + h\Phi_1(x_1, \varphi_z(x_0))\right).$$

Furthermore, as $(x_0, x_1, x_2)^g(y_0, y_1, y_2)^g = ((x_0, x_1, x_2)(y_0, y_1, y_2))^g$, we see that the function ψ_z fulfils the equation

$$\psi_z(x_0 + y_0) - \psi_z(x_0) - \psi_z(y_0) =$$
$$V(x_0, x_1 + \varphi(x_0); y_0, y_1 + \varphi(y_0)) - V(x_0, x_1; y_0, y_1) +$$
$$+ h\Phi_1(x_1, \varphi_z(x_0)) + h\Phi_1(y_1, \varphi_z(y_0)) + \sigma_z(\alpha(x_0, y_0)) -$$
$$- h\Phi_1(x_1 + y_1 + \alpha(x_0, y_0), \varphi_z(x_0 + y_0)).$$

Since $\varphi_z : \mathbf{G}_a \longrightarrow \mathbf{G}_a$ is a homomorphism, for $x_1 = y_1 = 0$ we obtain the identity

$$\psi_z(x + y) - \psi_z(x) - \psi_z(y) =$$
$$h\Phi_1(\varphi_z(x), \varphi_z(y)) + \sigma_y(\varphi_z(x)) + \sigma_z(\alpha(x, y)).$$

As on the left-hand side we have a symmetric polynomial in x, y, the equation

$$\sigma_y\big(\varphi_z(x)\big) + \sigma_z\big(\alpha(x, y)\big) = \sigma_x\big(\varphi_z(y)\big) + \sigma_z\big(\alpha(y, x)\big)$$

holds, that is

$$0 = \sigma_y\big(\alpha_1(z, x)\big) - \sigma_y\big(\alpha_1(x, z)\big) + \sigma_z\big(\alpha_1(x, y)\big) +$$

$$-\sigma_x\big(\alpha_1(z, y)\big) + \sigma_x\big(\alpha_1(y, z)\big) - \sigma_z\big(\alpha_1(y, x)\big) = \Omega_3\Big(\sigma_z\big(\alpha_1(x, y)\big)\Big).$$

As the operator Ω_3 is homogeneous and the homogeneous components of $\sigma_z\big(\sum_j g_j(xy^{p^j})\big)$ coincide with the homogeneous components of the polynomial $R(x, y, z) = \sigma_z\big(\alpha(x, y)\big)$ having degree equal to the sum of three p-powers, we can state the following

4.2.8 Proposition. $\Omega_3\big(R_d(x, y, z)\big) = 0$ for any d equal to the sum of three p-powers. $\qquad\square$

For the reader's convenience we collect here some of the polynomials we will be using to prove that no polynomial β satisfies equation (4.7).

$$Q(x, y, z) = \sum_{i=0}^{s} c_i \Phi_1^{p^i}\big(\alpha(y, z), \alpha(x, y + z)\big) - \sum_{i=0}^{s} c_i \Phi_1^{p^i}\big(\alpha(x, y), \alpha(x, y + z)\big),$$

$$R(x, y, z) = \sigma_z\big(\alpha(x, y)\big),$$

$$P(x, y, z) = Q(x, y, z) - R(x, y, z),$$

where

$$\alpha(x, y) = \sum_{i=0}^{t} a_i \Phi_1^{p^i}(x, y) + \sum_{j,k} b_{jk} x^{p^j} y^{p^{j+k}},$$

$$\deg \sum_{j,k} b_{jk} x^{p^j} y^{p^{j+k}} = p^u + p^{u+v},$$

$\sigma_z(x)$ is a p-polynomial in the variables x and z.

4.2.9 Theorem. *Let* k *be a field of characteristic* $p > 2$. *Then there exists no polynomial* $\beta(x, y) \in \mathsf{k}[x, y]$ *such that* $\delta^2\beta(x, y, z) = P(x, y, z)$.

Proof. By contradiction, let $\beta(x, y) \in \mathsf{k}[x, y]$ be such that $\delta^2\beta(x, y, z) = P(x, y, z)$. Then for every homogeneous component $P_d(x, y, z)$ of $P(x, y, z)$ we have

$$\delta^2\beta_d(x, y, z) = P_d(x, y, z) \qquad\qquad (4.11)$$

where $\beta_d(x, y)$ is the homogeneous component of degree d of $\beta(x, y)$.

We remark that the homogeneous components of even degree of the polynomial $R(x, y, z)$ coincide with those of the polynomial $\sigma_z\big(\sum_{i=0}^{t} a_i \Phi_1^{p^i}(x, y)\big)$, whereas the homogeneous components of odd degree coincide with those of the polynomial $\sigma_z\left(\sum_{j,k} b_{jk} x^{p^j} y^{p^{j+k}}\right)$.

By Proposition 4.2.6 for $d = p^{u+s+1} + p^{u+v+s+1}$ there exists $\gamma(x,y) \in k[x,y]$ such that

$$P_d(x,y,z) = Q_d(x,y,z) - R_d(x,y,z) =$$

$$= c_s(b_{uv})^{p^{s+1}}\left(z^{p^{u+v+s+1}}\Phi_1^{p^{u+s}}(x,y) - z^{p^{u+s+1}}\Phi_1^{p^{u+v+s}}(x,y)\right) + \delta^2\gamma(x,y,z) +$$

$$-R_d(x,y,z).$$

According to (4.11) it follows that

$$R_d(x,y,z) - c_s(b_{uv})^{p^{s+1}}\left(z^{p^{u+v+s+1}}\Phi_1^{p^{u+s}}(x,y) - z^{p^{u+s+1}}\Phi_1^{p^{u+v+s}}(x,y)\right) =$$

$$\delta^2(\gamma - \beta_d)(x,y,z).$$

Hence by Proposition 4.2.3 we have

$$R_d(x,y,z) = c_s(b_{uv})^{p^{s+1}}\left(z^{p^{u+v+s+1}}\Phi_1^{p^{u+s}}(x,y) - z^{p^{u+s+1}}\Phi_1^{p^{u+v+s}}(x,y)\right).$$

As we are dealing with the homogeneous component R_d of the polynomial $\sigma_z\left(\sum_{i=0}^t a_i \Phi_1^{p^i}(x,y)\right)$, there exist two non-trivial p-polynomials $A_1(x), A_2(x)$, such that

$$\sigma_z(x) = z^{p^{u+v+s+1}} A_1(x) + z^{p^{u+s+1}} A_2(x) + S(z,x), \qquad (4.12)$$

where $S(z,x) = \sum_h z^{p^h} A_h(x)$, with $h \neq u+v+s+1, u+s+1$, and $A_h(x)$ a (possibly trivial) p-polynomial.

We claim that $\deg A_1 = p^{u+s-t}$. In fact, if $\deg A_1 = p^n$ with $n < u+s-t$, then the polynomial $z^{p^{u+v+s+1}} A_1\left(\sum_{i=0}^t a_i \Phi_1^{p^i}(x,y)\right)$ has degree $p^{u+v+s+1} + p^{t+n+1} < p^{u+v+s+1} + p^{u+s+1}$.

On the other hand the polynomial $z^{p^{u+v+s+1}} A_1\left(\sum_{i=0}^t a_i \Phi_1^{p^i}(x,y)\right)$ has among its terms the following one:

$$c_s(b_{uv})^{p^{s+1}} z^{p^{u+v+s+1}} \Phi_1^{p^{u+s}}(x,y).$$

Furthermore, if $\deg A_1 = p^n$ with $n > u+s-t$, the polynomial $z^{p^{u+v+s+1}} A_1\left(\sum_{i=0}^t a_i \Phi_1^{p^i}(x,y)\right)$ has degree $d = p^{u+v+s+1} + p^{t+n+1} > p^{u+v+s+1} + p^{u+s+1}$. Hence by (4.11) and Corollary 4.2.7 it follows that $R_d(x,y,z) \in \delta^2(k[x,y])$, in contradiction to Proposition 4.2.3. Analogously one sees that $\deg A_2 = p^{u+v+s-t}$.

Denote by r_i the leading coefficient of $A_i(x)$, $i = 1,2$. Then the homogeneous component of maximal degree of the polynomial $z^{p^{u+v+s+1}} A_1\left(\sum_{j,k} b_{jk} x^{p^j} y^{p^{j+k}}\right)$ is equal to $r_1 z^{p^{u+v+s+1}} x^{p^{2u+s-t}} y^{p^{2u+v+s-t}}$ and for its degree d_1 one has $d_1 = p^{2u+s-t} + p^{2u+v+s-t} + p^{u+v+s+1}$, whereas the homogeneous component of maximal degree of the polynomial $z^{p^{u+s+1}} A_2\left(\sum_{j,k} b_{jk} x^{p^j} y^{p^{j+k}}\right)$ is equal to $r_2 z^{p^{u+s+1}} x^{p^{2u+v+s-t}} y^{p^{2u+2v+s-t}}$ and its degree d_2 is given by $d_2 = p^{u+s+1} + p^{2u+v+s-t} + p^{2u+2v+s-t}$.

We will show that the existence of such homogeneous components leads to a contradiction. We will do it treating separately each of the three cases $t \geq u + v$, $t = u + v - 1$ and $t < u + v - 1$.

i) Let $t \geq u + v$.

In this case we have $2u + s - t < 2u + v + s - t < u + v + s + 1$. Hence by Proposition 4.2.8 we have $\Omega_3\big(R_{d_1}(x, y, z)\big) = 0$, i.e. the homogeneous component $R_{d_1}(x, y, z)$ must be equal to

$$r_1 z^{p^{u+v+s+1}} x^{p^{2u+s-t}} y^{p^{2u+v+s-t}} + h_1 z^{p^{2u+s-t}} x^{p^{2u+v+s-t}} y^{p^{u+v+s+1}} +$$

$$k_1 z^{p^{2u+v+s-t}} x^{p^{2u+s-t}} y^{p^{u+v+s+1}} \tag{4.13}$$

with $r_1 + h_1 - k_1 = 0$. In particular, as $r_1 \neq 0$, at least one h_1 and k_1 must be non-zero. If $h_1 \neq 0$, then there exists a non-trivial p-polynomial $B(x)$ such that

$$\sigma_z(x) = z^{p^{u+v+s+1}} A_1(x) + z^{p^{u+s+1}} A_2(x) + z^{p^{2u+s-t}} B(x) + S'(z, x).$$

Moreover, a necessary condition for $h_1 z^{p^{2u+s-t}} x^{p^{2u+v+s-t}} y^{p^{u+v+s+1}}$ to appear as a term in $z^{p^{2u+s-t}} B\big(\sum_{j,k} b_{jk} x^{p^j} y^{p^{j+k}}\big)$ is that $\deg B \geq p^{s+1}$ (recalling that $\deg \sum_{j,k} b_{jk} x^{p^j} y^{p^{j+k}} = p^u + p^{u+v}$). Hence, putting $p^n = \deg B$, the polynomial $z^{p^{2u+s-t}} B\big(\sum_{i=0}^{t} a_i \Phi_1^{p^i}(x, y)\big)$ has degree

$$d = p^{2u+s-t} + p^{t+n+1} \geq p^{2u+s-t} + p^{u+v+s+2} > p^{u+s+1} + p^{u+v+s+1}.$$

It follows from (4.11) and Corollary 4.2.7 that $R_d(x, y, z) \in \delta^2(k[x, y])$, in contradiction to Proposition 4.2.3. Analogously one proves that $k_1 \neq 0$ leads to a contradiction, as well.

ii) Let $t = u + v - 1$.

In this case we have again $2u + s - t < 2u + v + s - t < u + v + s + 1$. Hence the component $R_{d_1}(x, y, z)$ must be equal to the polynomial (4.13) with h_1, k_1 not both zero. If $h_1 \neq 0$ we obtain again

$$\sigma_z(x) = z^{p^{u+v+s+1}} A_1(x) + z^{p^{u+s+1}} A_2(x) + z^{p^{2u+s-t}} B(x) + S'(z, x),$$

where $\deg B \geq p^{s+1}$. In particular, if $\deg B = p^n > p^{s+1}$, then the polynomial $z^{p^{2u+s-t}} B\big(\sum_{i=0}^{t} a_i \Phi_1^{p^i}(x, y)\big)$ has degree

$$d = p^{2u+s-t} + p^{t+n+1} = p^{u-v+s+1} + p^{u+v+n} > p^{u+s+1} + p^{u+v+s+1}.$$

Now by (4.11) and Corollary 4.2.7 it follows that $R_d(x, y, z) \in \delta^2(k[x, y])$, in contradiction to Proposition 4.2.3.

Hence $\deg B = p^{s+1}$ and $z^{p^{2u+s-t}} B(\sum_{i=0}^{t} a_i \Phi_1^{p^i}(x, y))$ has degree $p^{u-v+s+1} + p^{u+v+s+1}$. By (4.11) and Propositions 4.2.3, 4.2.6 it follows that $b_{u-v,2v} \neq 0$. Analogously, if $k_1 \neq 0$ we have

$$\sigma_z(x) = z^{p^{u+v+s+1}} A_1(x) + z^{p^{u+s+1}} A_2(x) + z^{p^{2u+v+s-t}} B(x) + S'(z, x),$$

where $\deg B \geq p^{s+1}$. Using the same arguments as before one proves that $B(x)$ has precisely degree p^{s+1}. Since $k_1 z^{p^{2u+v+s-t}} x^{p^{2u+s-t}} y^{p^{u+v+s+1}}$ is a term in $z^{p^{2u+v+s-t}} B\left(\sum_{j,k} b_{jk} x^{p^j} y^{p^{j+k}}\right)$ and $2u + s - t = u - v + s + 1$, it follows that $b_{u-v,2v} \neq 0$ in this case as well.

Now by a recursive procedure we show that $b_{u-v,2v} \neq 0$ implies $b_{u-2v,3v} \neq 0$. In fact, by (4.12), since $\deg A_1 = p^{u+s-t}$, we have that $b^{p^{u+s-t}} z^{p^{u+v+s+1}} x^{p^{2u+s-v-t}} y^{p^{2u+v+s-t}}$ is a term in $z^{p^{u+v+s+1}} A_1\left(\sum_{j,k} b_{jk} x^{p^j} y^{p^{j+k}}\right)$, where we put for short $b = b_{u-v,2v}$.

Hence, for $d' = p^{2u+s-v-t} + p^{2u+v+s-t} + p^{u+v+s+1}$, the homogeneous component $R_{d'}(x, y, z)$ is non-trivial. Since we have

$$2u + s - v - t < 2u + v + s - t < u + v + s + 1,$$

by Proposition 4.2.8 it follows that the component $R_{d'}(x, y, z)$ must be equal to

$$b^{p^{u+s-v-t}} z^{p^{u+v+s+1}} x^{p^{2u+s-v-t}} y^{p^{2u+v+s-t}} + h z^{p^{2u+s-v-t}} x^{p^{2u+v+s-t}} y^{p^{u+v+s+1}} +$$
$$k z^{p^{2u+v+s-t}} x^{p^{2u+s-v-t}} y^{p^{u+v+s+1}}$$

with $b^{p^{u+s-v-t}} + h - k = 0$.

If $h \neq 0$, then we have

$$\sigma_z(x) = z^{p^{u+v+s+1}} A_1(x) + z^{p^{u+s+1}} A_2(x) + z^{p^{2u+s-v-t}} B(x) + S'(z, x)$$

where $\deg B \geq p^{s+1}$. In particular, if $\deg B = p^n > p^{s+1}$, then the polynomial $z^{p^{2u+s-v-t}} B\left(\sum_{i=0}^t a_i \Phi_1^{p^i}(x, y)\right)$ has degree

$$d = p^{2u+s-v-t} + p^{t+n+1} = p^{u-2v+s+1} + p^{u+v+n} > p^{u+s+1} + p^{u+v+s+1},$$

forcing, by (4.11) and Corollary 4.2.7, $R_d(x, y, z) \in \delta^2(k[x, y])$, a contradiction to Proposition 4.2.3.

Thus $\deg B = p^{s+1}$ and $z^{p^{2u+s-v-t}} B\left(\sum_{i=0}^t a_i \Phi_1^{p^i}(x, y)\right)$ has degree $p^{u-2v+s+1} + p^{u+v+s+1}$. It follows, by (4.11) and Propositions 4.2.3, 4.2.6, that $b_{u-2v,3v} \neq 0$.

Analogously, if $k \neq 0$ we have

$$\sigma_z(x) = z^{p^{u+v+s+1}} A_1(x) + z^{p^{u+s+1}} A_2(x) + z^{p^{2u+v+s-t}} B(x) + S'(z, x),$$

where $\deg B \geq p^{s+1}$. As before one proves that $B(x)$ has degree p^{s+1}. Since $k z^{p^{2u+v+s-t}} x^{p^{2u+s-v-t}} y^{p^{u+v+s+1}}$ is a term in $z^{p^{2u+v+s-t}} B(\sum_{j,k} b_{jk} x^{p^j} y^{p^{j+k}})$ and $2u + s - v - t = u - 2v + s + 1$ holds, it follows that $b_{u-2v,3v} \neq 0$. After finitely many steps recursively we reach a contradiction.

iii) Let $t < u + v - 1$.

In this case we have $u + s + 1 < 2u + v + s - t < 2u + 2v + s - t$. As $d_2 = p^{u+s+1} + p^{2u+v+s-t} + p^{2u+2v+s-t}$ holds, by Proposition 4.2.8 the homogeneous component $R_{d_2}(x, y, z)$ must be equal to

$$r_2 z^{p^{u+s+1}} x^{p^{2u+v+s-t}} y^{p^{2u+2v+s-t}} + h_2 z^{p^{2u+v+s-t}} x^{p^{u+s+1}} y^{p^{2u+2v+s-t}} +$$
$$k_2 z^{p^{2u+2v+s-t}} x^{p^{u+s+1}} y^{p^{2u+v+s-t}}$$

with $r_2 - h_2 + k_2 = 0$. Hence one of h_2, k_2 is not zero.

If $k_2 \neq 0$, then there should exist a non-trivial p-polynomial $B(x)$ such that

$$\sigma_z(x) = z^{p^{u+v+s+1}} A_1(x) + z^{p^{u+s+1}} A_2(x) + z^{p^{2u+2v+s-t}} B(x) + S'(z, x).$$

Putting $p^n = \deg B$ and $d = p^{2u+2v+s-t} + p^{t+n+1}$, the homogeneous component $R_d(x, y, z)$ is non-trivial. Since $2u + 2v + s - t > u + v + s + 1$, by Corollary 4.2.7 and by (4.11) we have that $R_d(x, y, z) \in \delta^2(\mathsf{k}[x, y])$, a contradiction to Proposition 4.2.3.

On the other hand, if $h_2 \neq 0$, then there should exist a non-trivial p-polynomial $B(x)$, such that

$$\sigma_z(x) = z^{p^{u+v+s+1}} A_1(x) + z^{p^{u+s+1}} A_2(x) + z^{p^{2u+v+s-t}} B(x) + S'(z, x).$$

Moreover, a necessary condition for $h_2 z^{p^{2u+v+s-t}} x^{p^{u+s+1}} y^{p^{2u+2v+s-t}}$ to be a term in $z^{p^{2u+v+s-t}} B(\sum_{j,k} b_{jk} x^{p^j} y^{p^{j+k}})$ is that $\deg B \geq p^{u+v+s-t}$. Putting $p^n = \deg B$, the polynomial $z^{p^{2u+v+s-t}} B(\sum_{i=0}^{t} a_i \Phi_1^{p^i}(x, y))$ has degree

$$d = p^{2u+v+s-t} + p^{t+n+1} \geq p^{2u+v+s-t} + p^{u+v+s+1} > p^{u+s+1} + p^{u+v+s+1}.$$

From (4.11) and Corollary 4.2.7 it follows that $R_d(x, y, z) \in \delta^2(\mathsf{k}[x, y])$ in contrast to Proposition 4.2.3, which is the final contradiction of this proof. $\square$

4.2.10 Corollary. *Let G be a non-commutative chain defined over a perfect field k of characteristic $p > 2$. If G has dimension $n > 2$, then the commutator subgroup G' has co-dimension at least two and the centre $_3G$ has dimension at least two.* $\square$

Proof. According to the results of this section the assertion on the dimension of G' is true for $n = 3$, and for $n > 3$ it follows by induction. Theorem 3.2.10 yields the assertion on the dimension of $_3G$ directly. $\square$

As immediate consequences we have:

4.2.11 Corollary. *Let G be a non-commutative n-dimensional chain defined over a perfect field k of characteristic $p > 2$. If $n \leq 4$, then G has nilpotency class two.* $\square$

4.2.12 Corollary. *Let G be a non-commutative chain defined over a perfect field k of characteristic $p > 2$. If G has dimension $n > 2$, then G cannot contain a two-dimensional non-commutative chain.* $\square$

4.3 Chains with One-Dimensional Commutator Subgroup

In this section we characterise, up to k-isogenies, the chains $\mathfrak{C}_n(f, g_1, \ldots, g_s)$ as the only chains, defined over a perfect field k, having a one-dimensional commutator subgroup and consequently a central maximal connected subgroup.

Also the existence of these p-polynomials $f, g_1, \ldots, g_s \in k[\mathbf{F}]$ gives us a classification of the k-rational points of chains in the case that k is an infinite perfect field (see the beginning of Section 4.2).

4.3.1 Theorem. *Let G be a unipotent n-dimensional chain defined over a perfect field* k. *If* $\dim G' = 1$, *then there exists a surjective algebraic k-homomorphism* $\psi : \mathfrak{C}_n(f, g_1, \ldots, g_s) \longrightarrow G$ *with finite kernel, for suitable* $f, g_1, \ldots, g_s \in k[\mathbf{F}]$.

Proof. By [18], II, § 3, n. 4, Théorème 4.6, p. 197, we may assume $n > 2$. As $\dim G' = 1$, by Theorem 3.2.8 we have $\dim {}_3G = n - 1$. Furthermore, as G/G' is a commutative $(n-1)$-dimensional chain, there exists a surjective algebraic k-homomorphism $\varphi : \mathfrak{W}_{n-1} \longrightarrow G/G'$ with finite kernel. Let G^* be the extension of $\mathfrak{W}_1$ by $\mathfrak{W}_{n-1}$ which makes the following diagram commutative (see Remark 2.1.11):

$$
\begin{array}{ccccccccc}
1 & \longrightarrow & \mathfrak{W}_1 & \longrightarrow & G & \longrightarrow & G/G' & \longrightarrow & 1 \\
& & i\uparrow & & \varphi^*\uparrow & & \varphi\uparrow & & \\
1 & \longrightarrow & \mathfrak{W}_1 & \longrightarrow & G^* & \longrightarrow & \mathfrak{W}_{n-1} & \longrightarrow & 1
\end{array}
$$

As G^* is a k-group and the k-homomorphism φ^* is surjective with finite kernel, up to this homomorphism we can take a coordinate system for the n-dimensional unipotent group G such that:

$$G' = \{(x_0, x_1, \ldots, x_{n-1}) : x_0 = \cdots = x_{n-2} = 0\}, \quad \text{and}$$

$$_3{}^\circ G = \{(x_0, x_1, \ldots, x_{n-1}) : x_0 = 0\},$$

with G/G' isomorphic to $\mathfrak{W}_{n-1}$. As $_3{}^\circ G/G'$ is isomorphic to $\mathfrak{W}_{n-2}$, there exists a non-zero p-polynomial $f \in k[\mathbf{F}]$ and a regular k-function $c : \mathfrak{W}_{n-2} \longrightarrow \mathfrak{W}_1$ such that the factor system corresponding to the extension

$$1 \longrightarrow \mathfrak{W}_1 \longrightarrow {}_3{}^\circ G \longrightarrow \mathfrak{W}_{n-2} \longrightarrow 1$$

is $f\Phi_{n-2} + \delta^1(c)$. As in Remark 4.2.1, up to the k-isomorphism

$$(x_0, x_1, \cdots, x_{n-1}) \mapsto (x_0, x_1, \cdots, x_{n-1} - c(x_1, \cdots, x_{n-2})),$$

we can assume that to the extension $1 \longrightarrow \mathfrak{W}_1 \longrightarrow {}_3{}^\circ G \longrightarrow \mathfrak{W}_{n-2} \longrightarrow 1$ there corresponds the factor system $f\Phi_{n-2}$.

Consider the following central extensions corresponding to G

i) $1 \longrightarrow {}_3{}^\circ G \longrightarrow G \longrightarrow \mathfrak{W}_1 \longrightarrow 1.$

Let $(\theta_1, \ldots, \theta_{n-1}) : \mathfrak{W}_1 \times \mathfrak{W}_1 \longrightarrow {}_3{}^\circ G$ the factor system corresponding to the section $x_0 \mapsto (x_0, 0, \cdots, 0)$ of the extension i).

In order to determine the polynomials θ_i, it is useful to take the canonical extension

ii) $1 \longrightarrow \mathfrak{W}_{n-1} \xrightarrow{\ v\ } \mathfrak{W}_n \xrightarrow{\ \mathcal{R}^{n-1}\ } \mathfrak{W}_1 \longrightarrow 1$

and to write $(\Psi_1, \ldots, \Psi_{n-1}) : \mathfrak{W}_1 \times \mathfrak{W}_1 \longrightarrow \mathfrak{W}_{n-1}$ for the factor system corresponding to the section $x_0 \mapsto (x_0, 0, \cdots, 0)$ of the extension $ii)$. With the same argument as Remark 2.1.15 we can compare the addition formulas in the ith-entry for any $i = 1, 2, \ldots, n-2$, finding that the mapping θ_i is equal to the mapping Ψ_i for any such $i = 1, 2, \ldots, n-2$. Moreover, applying the associative property to the extension $i)$ we find that the map

$$f\Phi_{n-2}(\overline{x}_{n-2}; \overline{y}_{n-2}) + \theta_{n-1}(x_0, y_0) +$$

$$f\Phi_{n-2}(x_1+y_1, \ldots, x_{n-2}+y_{n-2} + \Phi_{n-3}(\overline{x}_{n-3}; \overline{y}_{n-3}); \Psi_1(x_0, y_0), \ldots, \Psi_{n-2}(x_0, y_0)),$$

is a factor system $\Theta : \mathfrak{W}_{n-1} \times \mathfrak{W}_{n-1} \longrightarrow \mathfrak{W}_1$, where we put for short $(x_1, \ldots, x_r) = \overline{x}_r$ for any $r \geq 1$.

Since for the n-dimensional Witt group $\mathfrak{W}_n$ we have

$$\Phi_{n-1}(x_0, \overline{x}_{n-2}; y_0, \overline{y}_{n-2}) = \Psi_{n-1}(x_0, y_0) + \Phi_{n-2}(\overline{x}_{n-2}; \overline{y}_{n-2}) +$$

$$\Phi_{n-2}(x_1+y_1, \ldots, x_{n-2}+y_{n-2} + \Phi_{n-3}(\overline{x}_{n-3}; \overline{y}_{n-3}); \Psi_1(x_0, y_0), \ldots, \Psi_{n-2}(x_0, y_0)),$$

substituting this in Θ we find

$$\Theta(x_0, \overline{x}_{n-2}; y_0, \overline{y}_{n-2}) =$$

$$f\Phi_{n-1}(x_0, \overline{x}_{n-2}; y_0, \overline{y}_{n-2}) + \theta_{n-1}(x_0, y_0) - f\Psi_{n-1}(x_0, y_0).$$

It follows that the mapping $\beta : \mathfrak{W}_1 \times \mathfrak{W}_1 \longrightarrow \mathfrak{W}_1$

$$\beta(x_0, y_0) = \theta_{n-1}(x_0, y_0) - f\Psi_{n-1}(x_0, y_0)$$

is a factor system. Thus we find p-polynomials $h, g_1, \ldots, g_s \in k[\mathbf{F}]$, and a regular k-function $l : \mathfrak{W}_1 \longrightarrow \mathfrak{W}_1$, such that

$$\Theta(x_0, \overline{x}_{n-2}; y_0, \overline{y}_{n-2}) =$$

$$f\Phi_{n-1}(x_0, \overline{x}_{n-2}; y_0, \overline{y}_{n-2}) + h\Phi_1(x_0, y_0) + \sum_{i=1}^{s} g_i(x_0 y_0^{p^i}) + \delta^1 l(x_0, y_0).$$

As the mapping $\mathfrak{C}_n(f, g_1, \ldots, g_s) \longrightarrow G$,

$$(x_0, \overline{x}_{n-2}, x_{n-1}) \mapsto (x_0, \overline{x}_{n-2}, x_{n-1} + h(x_1) + l(x_0)),$$

is a k-isomorphism, we are done.   □

4.3.2 Corollary. *Let G be a three-dimensional non-commutative unipotent chain defined over a perfect field k of characteristic $p > 2$. Then there exists a surjective algebraic k-homomorphism $\psi : \mathfrak{C}_3(f, g_1, \ldots, g_s) \longrightarrow G$ with finite kernel, for suitable $f, g_1, \ldots, g_s \in k[\mathbf{F}]$.*

Proof. The assertion follows from Corollary 4.2.10 and Theorem 4.3.1.   □

4.3.3 Corollary. *Let G be a non-commutative chain, defined over a perfect field* k *of characteristic $p > 2$. If G has dimension $n > 2$, then G^p is a $(n-1)$-dimensional chain.*

Proof. For $n = 3$ it is true by Corollary 4.3.2. For $n > 3$ let K be the unique one-dimensional connected algebraic subgroup of G. Since by Corollary 4.2.10 the center $_3G$ has dimension $d \geq 2$, the center $_3G$ contains a commutative two-dimensional chain of exponent p^2 which forces $K \leq G^p$. The assertion follows now by induction on n considering the factor group G/K. $\square$

We have already seen in Remark 4.1.6 that there exist chains of arbitrary dimension with a two-dimensional commutator subgroup. We can even give examples of chains with a three-dimensional commutator subgroup. These examples show also that the maximal connected algebraic subgroup of a chain is not necessarily a commutative group.

4.3.4 Examples. Let k be a perfect field of characteristic three or five.

Let $n \geq 6$. We denote by $\mathbf{x}_i$ the i-tuple $(x_0, \cdots x_i)$ and by $\Phi_{i+1}(\mathbf{x}_i; \mathbf{y}_i)$ the polynomials defining the Witt group $\mathfrak{W}_n$ by the operation

$$(x_0, \cdots, x_{n-1}) + (y_0, \cdots, y_{n-1}) = (z_0, \cdots, z_{n-1})$$

where $z_0 = x_0 + y_0$ and $z_i = x_i + y_i + \Phi_i(\mathbf{x}_{i-1}; \mathbf{y}_{i-1})$ for $i = 1, \cdots, n-1$.

Let $\Psi : \mathfrak{W}_{n-3} \times \mathfrak{W}_{n-3} \longrightarrow \mathfrak{W}_3$ be the maps defined by

$$\Psi(\mathbf{x}_{n-4}, \mathbf{y}_{n-4}) = (\beta_1(\mathbf{x}_{n-4}, \mathbf{y}_{n-4}), \beta_2(\mathbf{x}_{n-4}, \mathbf{y}_{n-4}), \beta_3(\mathbf{x}_{n-4}, \mathbf{y}_{n-4})),$$

where for $p = 3$ one has

$$\beta_1(\mathbf{x}_{n-4}, \mathbf{y}_{n-4}) = x_0 y_0^3,$$

$$\beta_2(\mathbf{x}_{n-4}, \mathbf{y}_{n-4}) = x_0^9 y_0^3 + 2x_1^3 y_0^3 + 2x_0^6 y_0^6 + x_0^3 y_0^9 + x_1 y_0^9,$$

$$\begin{aligned}
\beta_3(\mathbf{x}_{n-4}, \mathbf{y}_{n-4}) = {} & x_0^{33} y_0^3 + 2x_0^6 x_1^9 y_0^3 + x_0^{30} y_0^6 + 2x_0^3 x_1^9 y_0^6 + x_0^{18} x_1^3 y_0^9 + 2x_0^9 x_1^6 y_0^9 + \\
& 2x_2^3 y_0^9 + 2x_0^{24} y_0^{12} + x_0^{15} x_1^3 y_0^{12} + x_0^6 x_1^6 y_0^{12} + 2x_0^{21} y_0^{15} + 2x_0^{18} x_1 y_0^{15} + \\
& 2x_0^9 x_1^4 y_0^{15} + 2x_0^3 x_1^6 y_0^{15} + 2x_1^7 y_0^{15} + 2x_0^{15} x_1 y_0^{18} + x_0^9 x_1^3 y_0^{18} + x_0^6 x_1^4 y_0^{18} + \\
& 2x_0^{15} y_0^{21} + 2x_0^9 x_1^2 y_0^{21} + x_0^6 x_1^3 y_0^{21} + 2x_0^3 x_1^4 y_0^{21} + x_1^5 y_0^{21} + 2x_0^{12} y_0^{24} + \\
& 2x_0^9 x_1 y_0^{24} + x_0^6 x_1^2 y_0^{24} + 2x_0^6 x_1 y_0^{27} + 2x_0^3 x_1^2 y_0^{27} + x_2 y_0^{27} + x_0^6 y_0^{30} + \\
& x_0^3 y_0^{33} + x_1^3 y_1^9 + 2x_0^6 y_0^3 y_1^9 + 2x_0^3 y_0^6 y_1^9,
\end{aligned}$$

whereas for $p = 5$ one has

$$\beta_1(\mathbf{x}_{n-4}, \mathbf{y}_{n-4}) = x_0 y_0^5,$$

$$\beta_2(\mathbf{x}_{n-4}, \mathbf{y}_{n-4}) = x_0^{25} y_0^5 + 4x_1^5 y_0^5 + 3x_0^{20} y_0^{10} + 4x_0^{15} y_0^{15} + 3x_0^{10} y_0^{20} + x_0^5 y_0^{25} + x_1 y_0^{25}$$

$$\beta_3(\mathbf{x}_{n-4}, \mathbf{y}_{n-4}) = x_0^{145}y_0^5 + 4x_0^{20}x_1^{25}y_0^5 + 2x_0^{140}y_0^{10} + 3x_0^{15}x_1^{25}y_0^{10} + 2x_0^{135}y_0^{15} +$$
$$3x_0^{10}x_1^{25}y_0^{15} + x_0^{130}y_0^{20} + 4x_0^5x_1^{25}y_0^{20} + x_0^{100}x_1^5y_0^{25} + 3x_0^{75}x_1^{10}y_0^{25} +$$
$$2x_0^{50}x_1^{15}y_0^{25} + 4x_0^{25}x_1^{20}y_0^{25} + 4x_2^5y_0^{25} + 3x_0^{120}y_0^{30} +$$
$$2x_0^{95}x_1^5y_0^{30} + 2x_0^{70}x_1^{10}y_0^{30} + 2x_0^{45}x_1^{15}y_0^{30} +$$
$$2x_0^{20}x_1^{20}y_0^{30} + 2x_0^{65}x_1^{10}y_0^{35} + 4x_0^{40}x_1^{15}y_0^{35} + x_0^{15}x_1^{20}y_0^{35} +$$
$$2x_0^{110}y_0^{40} + 4x_0^{85}x_1^5y_0^{40} + 4x_0^{60}x_1^{10}y_0^{40} + 2x_0^{10}x_1^{20}y_0^{40} + 4x_0^{100}x_1y_0^{45} +$$
$$x_0^{80}x_1^5y_0^{45} + 4x_0^{75}x_1^6y_0^{45} + 4x_0^{55}x_1^{10}y_0^{45} + 4x_0^{50}x_1^{11}y_0^{45} + 2x_0^{30}x_1^{15}y_0^{45} +$$
$$4x_0^{25}x_1^{16}y_0^{45} + 4x_0^5x_1^{20}y_0^{45} + 4x_1^{21}y_0^{45} + 3x_0^{95}x_1y_0^{50} + 2x_0^{75}x_1^5y_0^{50} +$$
$$x_0^{70}x_1^6y_0^{50} + 4x_0^{45}x_1^{11}y_0^{50} + 2x_0^{20}x_1^{16}y_0^{50} + 2x_0^{95}y_0^{55} + 3x_0^{70}x_1^5y_0^{55} +$$
$$x_0^{65}x_1^6y_0^{55} + 4x_0^{45}x_1^{10}y_0^{55} + 3x_0^{40}x_1^{11}y_0^{55} + 4x_0^{20}x_1^{15}y_0^{55} + x_0^{15}x_1^{16}y_0^{55} +$$
$$4x_0^{90}y_0^{60} + x_0^{85}x_1y_0^{60} + x_0^{65}x_1^5y_0^{60} + 2x_0^{60}x_1^6y_0^{60} + 2x_0^{15}x_1^{15}y_0^{60} +$$
$$2x_0^{10}x_1^{16}y_0^{60} + x_0^{85}y_0^{65} + 4x_0^{80}x_1y_0^{65} + 3x_0^{75}x_1^2y_0^{65} + 3x_0^{60}x_1^5y_0^{65} +$$
$$2x_0^{55}x_1^6y_0^{65} + x_0^{50}x_1^7y_0^{65} + 4x_0^{35}x_1^{10}y_0^{65} + 4x_0^{30}x_1^{11}y_0^{65} + 4x_0^{25}x_1^{12}y_0^{65} +$$
$$2x_0^{10}x_1^{15}y_0^{65} + 4x_0^5x_1^{16}y_0^{65} + 2x_1^{17}y_0^{65} + 4x_0^{80}y_0^{70} + 3x_0^{75}x_1y_0^{70} +$$
$$2x_0^{70}x_1^2y_0^{70} + x_0^{55}x_1^5y_0^{70} + x_0^{45}x_1^7y_0^{70} + 4x_0^{30}x_1^{10}y_0^{70} + 2x_0^{20}x_1^{12}y_0^{70} +$$
$$2x_0^{70}x_1y_0^{75} + 2x_0^{65}x_1^2y_0^{75} + 3x_0^{50}x_1^5y_0^{75} + 2x_0^{45}x_1^6y_0^{75} + 2x_0^{40}x_1^7y_0^{75} +$$
$$2x_0^{25}x_1^{10}y_0^{75} + 3x_0^{20}x_1^{11}y_0^{75} + x_0^{15}x_1^{12}y_0^{75} + 4x_0^{70}y_0^{80} + 4x_0^{65}x_1y_0^{80} +$$
$$4x_0^{60}x_1^2y_0^{80} + 2x_0^{45}x_1^5y_0^{80} + 2x_0^{20}x_1^{10}y_0^{80} + 4x_0^{15}x_1^{11}y_0^{80} + 2x_0^{10}x_1^{12}y_0^{80} +$$
$$x_0^{65}y_0^{85} + 2x_0^{60}x_1y_0^{85} + 4x_0^{55}x_1^2y_0^{85} + 3x_0^{50}x_1^3y_0^{85} + x_0^{40}x_1^5y_0^{85} +$$
$$2x_0^{35}x_1^6y_0^{85} + x_0^{30}x_1^7y_0^{85} + 4x_0^{25}x_1^8y_0^{85} + 3x_0^{15}x_1^{10}y_0^{85} + 4x_0^{10}x_1^{11}y_0^{85} +$$
$$4x_0^5x_1^{12}y_0^{85} + 3x_1^{13}y_0^{85} + 4x_0^{60}y_0^{90} + 4x_0^{55}x_1y_0^{90} + 3x_0^{45}x_1^3y_0^{90} +$$
$$4x_0^{35}x_1^5y_0^{90} + 2x_0^{30}x_1^6y_0^{90} + 2x_0^{20}x_1^8y_0^{90} + 2x_0^{55}y_0^{95} + 2x_0^{50}x_1y_0^{95} +$$
$$4x_0^{45}x_1^2y_0^{95} + x_0^{40}x_1^3y_0^{95} + x_0^{25}x_1^6y_0^{95} + 2x_0^{20}x_1^7y_0^{95} + x_0^{15}x_1^8y_0^{95} +$$
$$3x_0^{45}x_1y_0^{100} + 2x_0^{25}x_1^5y_0^{100} + x_0^{20}x_1^6y_0^{100} + x_0^{15}x_1^7y_0^{100} + 2x_0^{10}x_1^8y_0^{100} +$$
$$4x_0^{40}x_1y_0^{105} + 4x_0^{35}x_1^2y_0^{105} + 3x_0^{30}x_1^3y_0^{105} + 4x_0^{25}x_1^4y_0^{105} + x_0^{20}x_1^5y_0^{105} +$$
$$4x_0^{15}x_1^6y_0^{105} + x_0^{10}x_1^7y_0^{105} + 4x_0^5x_1^8y_0^{105} + x_1^9y_0^{105} + 2x_0^{40}y_0^{110} +$$
$$x_0^{35}x_1y_0^{110} + 4x_0^{30}x_1^2y_0^{110} + 2x_0^{20}x_1^4y_0^{110} + 2x_0^{25}x_1^2y_0^{115} + x_0^{20}x_1^3y_0^{115} +$$
$$x_0^{15}x_1^4y_0^{115} + 3x_0^{30}y_0^{120} + 3x_0^{25}x_1y_0^{120} + 2x_0^{20}x_1^2y_0^{120} + 3x_0^{15}x_1^3y_0^{120} +$$
$$2x_0^{10}x_1^4y_0^{120} + 4x_0^{20}x_1y_0^{125} + 3x_0^{15}x_1^2y_0^{125} + 3x_0^{10}x_1^3y_0^{125} + 4x_0^5x_1^4y_0^{125}$$
$$x_2y_0^{125} + x_0^{20}y_0^{130} + 2x_0^{15}y_0^{135} + 2x_0^{10}y_0^{140} + x_0^5y_0^{145} +$$
$$x_1^5y_1^{25} + 4x_0^{20}y_0^5y_1^{25} + 3x_0^{15}y_0^{10}y_1^{25} + 3x_0^{10}y_0^{15}y_1^{25} + 4x_0^5y_0^{20}y_1^{25}.$$

By an elementary but tedious calculation, we obtain in both cases $\delta^2\Psi = (0,0,0)$ which shows that Ψ is a factor system.

The multiplication on the n-dimensional affine space given by

$$(x_0, \cdots, x_{n-1}) \cdot (y_0, \cdots, y_{n-1}) = (z_0, \cdots, z_{n-1}),$$

where $z_0 = x_0 + y_0$ and

$$
\begin{cases}
z_i &= x_i + y_i + \Phi_i(\mathbf{x}_{i-1}; \mathbf{y}_{i-1}) \quad \text{for } i = 1, \cdots, n-4; \\[2mm]
z_{n-3} &= x_{n-3} + y_{n-3} + \Phi_{n-3}(\mathbf{x}_{n-4}; \mathbf{y}_{n-4}) + \beta_1(\mathbf{x}_{n-4}, \mathbf{y}_{n-4}) \\[2mm]
z_{n-2} &= x_{n-2} + y_{n-2} + \Phi_{n-2}(\mathbf{x}_{n-3}; \mathbf{y}_{n-3}) + \beta_2(\mathbf{x}_{n-4}, \mathbf{y}_{n-4}) + \\
& \quad \Phi_1(x_{n-3} + y_{n-3} + \Phi_{n-3}(\mathbf{x}_{n-4}; \mathbf{y}_{n-4}), \beta_1(\mathbf{x}_{n-4}, \mathbf{y}_{n-4})) \\[2mm]
z_{n-1} &= x_{n-1} + y_{n-1} + \Phi_{n-1}(\mathbf{x}_{n-2}; \mathbf{y}_{n-2}) + \beta_3(\mathbf{x}_{n-4}, \mathbf{y}_{n-4}) + \\
& \quad \Phi_2(x_{n-3} + y_{n-3} + \Phi_{n-3}(\mathbf{x}_{n-4}; \mathbf{y}_{n-4}), x_{n-2} + y_{n-2} + \Phi_{n-2}(\mathbf{x}_{n-3}; \mathbf{y}_{n-3}), \\
& \quad \beta_1(\mathbf{x}_{n-4}, \mathbf{y}_{n-4}), \beta_2(\mathbf{x}_{n-4}, \mathbf{y}_{n-4}))
\end{cases}
$$

defines an n-dimensional chain G. Namely, the commutator subgroup G' of G is the three-dimensional subgroup

$$
G' = \{(x_0, \cdots, x_{n-1}) \in G : x_0 = x_1 = \cdots = x_{n-4} = 0\}.
$$

Since G/G' is isomorphic to a Witt group, G has a unique maximal connected subgroup M, which is defined by $x_0 = 0$. As M is a chain $\mathfrak{C}_{n-1}$ with one-dimensional commutator subgroup, also G is a chain. $\qquad\square$

Now we want to examine to which extent the chains $\mathfrak{C}_n$ are rigid. For this reason we investigate the automorphism group of $\mathfrak{C}_n$.

The following result is a direct consequence of [89], Lemma 6, p. 175.

4.3.5 Proposition. *Let Υ be an endomorphism of $\mathfrak{W}_n$ having the form $\Upsilon(x_0, \ldots, x_{n-1}) = (\alpha x_0, \psi_1(x_0, x_1), \ldots, \psi_{n-1}(x_0, \ldots, x_{n-1}))$, where $0 \neq \alpha$ in the algebraic closure $\bar{\mathsf{k}}$ of the field k of definition of $\mathfrak{W}_n$ and the ψ_i are maps from $\mathfrak{W}_{i+1}$ into $\mathfrak{W}_1$. Then the factor system*

$$
\Phi_{n-1}(\Upsilon(x_0, \ldots, x_{n-2}); \Upsilon(y_0, \ldots, y_{n-2})) - \alpha^{p^{n-1}} \Phi_{n-1}(x_0, \ldots, x_{n-2}; y_0, \ldots, y_{n-2})
$$

is a coboundary. $\qquad\square$

4.3.6 Proposition. *Let $n \geq 1$ and let Θ be an algebraic automorphism of $\mathfrak{W}_n$. Then*

$$
\Theta(x_0, \cdots, x_{n-1}) = (\alpha x_0, \alpha^p x_1 + \psi_1(x_0), \cdots, \alpha^{p^{n-1}} x_{n-1} + \psi_{n-1}(x_0, \cdots, x_{n-2}))
$$

for a suitable $0 \neq \alpha$ in the algebraic closure $\bar{\mathsf{k}}$ of the field k of definition of $\mathfrak{W}_n$ and for $n-1$ maps $\psi_i : \mathfrak{W}_i \longrightarrow \mathfrak{W}_1$ satisfying, for any $i = 1, \cdots, n-1$ and for any $\mathbf{x}_i = (x_0, \cdots, x_i), \mathbf{y}_i = (y_0, \cdots, y_i) \in \mathfrak{W}_{i+1}$, the condition

$$
\Phi_i(\alpha x_0, \cdots, \alpha^{p^{i-1}} x_{i-1} + \psi_{i-1}(\mathbf{x}_{i-2}), \alpha y_0, \cdots, \alpha^{p^{i-1}} y_{i-1} + \psi_{i-1}(\mathbf{y}_{i-2})) -
$$

$$
\alpha^{p^i} \Phi_i(\mathbf{x}_{i-1}, \mathbf{y}_{i-1}) = \psi_i(\mathbf{x}_{i-1} + \mathbf{y}_{i-1}) - \psi_i(\mathbf{x}_{i-1}) - \psi_i(\mathbf{y}_{i-1}). \qquad (4.14)
$$

Proof. We prove the claim by induction on n. For $n = 1$ the assertion follows from the fact that the group of the algebraic automorphism of $\mathfrak{W}_1$ is isomorphic to $\mathbf{G}_m$. Let $n > 1$. Since any connected algebraic subgroup of $\mathfrak{W}_n$ is characteristic, the automorphism Θ induces in any factor group of $\mathfrak{W}_n$ an algebraic automorphism. Therefore by induction there exist $\gamma \neq 0$ and a map $\varphi : \mathfrak{W}_{n-1} \longrightarrow \mathfrak{W}_1$ such that

$$\Theta(\mathbf{x}_{n-1}) = (\alpha x_0, \alpha^p x_1 + \psi_1(x_0), \cdots, \alpha^{p^{n-2}} x_{n-2} + \psi_{n-2}(\mathbf{x}_{n-3}), \gamma x_{n-1} + \varphi(\mathbf{x}_{n-2}))$$

for a suitable $0 \neq \alpha \in \bar{\mathsf{k}}$ and $n-2$ maps $\psi_1, \cdots, \psi_{n-2}$ any of which satisfies the condition (4.14). Using the fact that Θ is a homomorphism we get the following relation

$$\Phi_{n-1}(\alpha x_0, \cdots, \alpha^{p^{n-2}} x_{n-2} + \psi_{n-2}(\mathbf{x}_{n-3}), \alpha y_0, \cdots, \alpha^{p^{n-2}} y_{n-2} + \psi_{n-2}(\mathbf{y}_{n-3}))$$

$$-\gamma \Phi_{n-1}(\mathbf{x}_{n-2}, \mathbf{y}_{n-2}) = \varphi(\mathbf{x}_{n-2} + \mathbf{y}_{n-2}) - \varphi(\mathbf{x}_{n-2}) - \varphi(\mathbf{y}_{n-2}),$$

from which it follows, by Proposition 4.3.5, that $\gamma = \alpha^{p^{n-1}}$. $\qquad\square$

4.3.7 Proposition. *Let* $G = \mathfrak{C}_n(f, g_1, \cdots, g_t)$ *be a non-commutative chain, with* $f(u) = \sum_{j=0}^t b_j u^{p^j}$ *and* $\sum_i g_i(xy^{p^i}) = \sum_{(h,k) \in I} \lambda_{hk} x^{p^h} y^{p^{h+k}}$ *for a given subset* I *of* $\mathbb{N} \times \mathbb{N}$. *If* Θ *is an automorphism of* G, *then for any* $\mathbf{x}_{n-1} = (x_0, \cdots, x_{n-1}) \in G$ *we have*

$$\Theta(\mathbf{x}_{n-1}) = (\alpha x_0, \alpha^p x_1 + \psi_1(x_0), \cdots, \alpha^{p^{n-2}} x_{n-2} + \psi_{n-2}(\mathbf{x}_{n-3}), \beta x_{n-1} + \varphi(\mathbf{x}_{n-2}))$$

for suitable $\alpha, \beta \in \bar{\mathsf{k}}$ *and* $n-1$ *maps* $\psi_1, \cdots, \psi_{n-2}, \varphi$ *which satisfy the following conditions:*

1) *any of the map* $\psi_i : \mathfrak{W}_i \longrightarrow \mathfrak{W}_1$, $i = 1, \cdots, n-2$, *fulfils the condition (4.14) of Proposition 4.3.6,*
2) *the map* $\varphi : \mathfrak{W}_{n-1} \longrightarrow \mathfrak{W}_1$ *is such that*

$$f(\Phi_{n-1}(\alpha x_0, \cdots, \alpha^{p^{n-2}} x_{n-2} + \psi_{n-2}(\mathbf{x}_{n-3}),$$

$$\alpha y_0, \cdots, \alpha^{p^{n-2}} y_{n-2} + \psi_{n-2}(\mathbf{y}_{n-3})) +$$

$$\sum_i g_i((\alpha x_0)(\alpha y_0)^{p^i}) - \beta f(\Phi_{n-1}(\mathbf{x}_{n-2}, \mathbf{y}_{n-2})) - \beta \sum_i g_i(x_0 y_0^{p^i}) =$$

$$\varphi(\mathbf{x}_{n-2} + \mathbf{y}_{n-2}) - \varphi(\mathbf{x}_{n-2}) - \varphi(\mathbf{y}_{n-2})$$

for any $\mathbf{x}_{n-2}, \mathbf{y}_{n-2} \in \mathfrak{W}_{n-1}$,
3) $\alpha \neq 0 \neq \beta$ *and* $\beta = \alpha^{p^{n-1+j}} = \alpha^{p^h + p^{h+k}}$ *for any* $j = 0, \ldots, t$ *and for any* $(h, k) \in I$.

Proof. Since any connected algebraic subgroup of G is characteristic in G, the automorphism Θ induces in any factor group of G an algebraic automorphism. Thus by Proposition 4.3.6 there exist $\beta \neq 0$ and a map $\varphi : \mathfrak{W}_{n-1} \longrightarrow \mathfrak{W}_1$ such that

$$\Theta(\mathbf{x}_{n-1}) = (\alpha x_0, \alpha^p x_1 + \psi_1(x_0), \cdots, \alpha^{p^{n-2}} x_{n-2} + \psi_{n-2}(\mathbf{x}_{n-3}), \beta x_{n-1} + \varphi(\mathbf{x}_{n-2}))$$

for a suitable $\alpha \neq 0$ and $n - 2$ maps $\psi_1, \cdots, \psi_{n-2}$ any of which satisfies the condition (4.14). In particular, since Θ is an homomorphism the map φ is forced to satisfy condition 2) and consequently the polynomial $\sum_i g_i((\alpha x_0)(\alpha y_0)^{p^i}) - \beta \sum_i g_i(x_0 y_0^{p^i})$ is forced to be symmetric. Therefore $\beta = \alpha^{p^h + p^{h+k}}$ for any $(h, k) \in I$ and

$$f(\Phi_{n-1}(\alpha x_0, \cdots, \alpha^{p^{n-2}} x_{n-2} + \psi_{n-2}(\mathbf{x}_{n-3}),$$

$$\alpha y_0, \cdots, \alpha^{p^{n-2}} y_{n-2} + \psi_{n-2}(\mathbf{y}_{n-3})) -$$

$$\beta f(\Phi_{n-1}(\mathbf{x}_{n-2}, \mathbf{y}_{n-2})) = \varphi(\mathbf{x}_{n-2} + \mathbf{y}_{n-2}) - \varphi(\mathbf{x}_{n-2}) - \varphi(\mathbf{y}_{n-2}),$$

for any $\mathbf{x}_{n-2}, \mathbf{y}_{n-2} \in \mathfrak{W}_{n-1}$. By Proposition 4.3.5 it follows now that $\beta = \alpha^{p^{n-1+j}}$ for any $j = 0, \ldots, t$. $\qquad \square$

By the above proposition and by the fact that any algebraic homomorphism from $\mathfrak{W}_n$ to $\mathfrak{W}_1$ has an $(n-1)$-dimensional algebraic subgroup as the kernel, we get the following

4.3.8 Corollary. *Let Θ be an algebraic automorphism of the chain $G = \mathfrak{C}_n(f, g_1, \cdots, g_t)$. If Θ fixes element-wise the factor group G/G', then Θ has the form*

$$\Theta(x_0, \ldots, x_{n-1}) = (x_0, \ldots, x_{n-1} + \varphi(x_0))$$

for a suitable p-polynomial φ. $\qquad \square$

4.3.9 Remark. We want to examine condition 3) of Proposition 4.3.7. If $n > 2$ we have for any j, h, k

$$\alpha^{p^{n-1+j}} = \alpha^{p^h + p^{h+k}},$$

hence the set of algebraic automorphisms of $G/\mathfrak{z}°(G)$ which lift to an algebraic automorphism of the non-commutative chain $G = \mathfrak{C}_n(f, g_1, \cdots, g_t)$ is finite. For instance, for $f(t) = t$ and $\sum_i g_i(xy^{p^i}) = xy^{p^{n-1}}$, we have that this set consists just of the identity, because the equation $\alpha^{p^{n-1}} = \alpha^{1+p^{n-1}}$ has $\alpha = 1$ as unique non-zero solution.

Let G be non-commutative and of dimension two. If $f \neq 0$ the same condition holds, that is for any j, h, k we have $\alpha^{p^{1+j}} = \alpha^{p^h + p^{h+k}}$. Hence the set of algebraic automorphisms of $G/\mathfrak{z}°(G)$ which lift to one of the non-commutative two-dimensional chain $G = \mathfrak{C}_2(f, g_1, \cdots, g_t)$ is finite. In contrast, if $f = 0$ and I consists of a unique pair (h, k), condition 3) reduces only to $\beta = \alpha^{p^h + p^{h+k}}$. Therefore the set of algebraic automorphisms of $G/\mathfrak{z}°(G)$ which lift to those of G is infinite.

If G is a two-dimensional commutative chain, then the automorphism group of G is infinite if and only if f consists only of one monomial. $\qquad \square$

4.3.10 Theorem. *Let G be a solvable, but not nilpotent connected affine algebraic group defined over a perfect field of characteristic $p > 0$. Let U be the unipotent radical of G and T a maximal torus of G. If U is a chain, then the following holds:*

1) *$G' = U$ and for any connected algebraic subgroup $L \leq U$ we have $[T, L] = L$;*
2) *the connected component of the centre of G is trivial if and only if a maximal torus T of G has dimension one;*
3) *if $p > 2$, then either U is commutative and any factor group U_i/U_{i-2}, where U_i is the i-dimensional subgroup of U and $U_n = U$, is isogenous to a group defined on the affine plane A by the product*

$$(x_0, x_1)(y_0, y_1) = (x_0 + y_0, x_1 + y_1 + \Phi_1^{p^d}(x_0, y_0)),$$

for a suitable $d \in \mathbb{N}$,
or U has dimension two and is isomorphic to a group defined on A by the product

$$(x_0, x_1)(y_0, y_1) = (x_0 + y_0, x_1 + y_1 + x_0^{p^r} y_0^{p^{r+s}})$$

for suitable $r, s \in \mathbb{N}$.

Proof. 1) It suffices to prove that $[T, L] = L$ for any connected algebraic subgroup $L \leq U$. If U is one-dimensional the assertion follows since G is not nilpotent. Let U be a counter-example of minimal dimension and let $L \neq 1$ be a connected algebraic subgroup of U.

Assume that $[T, L] = 1$ and let K be the connected one-dimensional subgroup of U. As U is a chain, we have $K \leq L$. Since U/K is a chain, by induction we have either $[T, U] = K$ or $[T, U] = U$. If $[T, U] = K$, then for any given $u \in U$ the map $\sigma_u : T \longrightarrow K$ defined by $\sigma_u(t) = [t, u]$ is such that $\sigma_u(t_1 t_2) = \sigma_u(t_1)\sigma_u(t_2)$, because $\sigma_u(t) \in K \leq L$ and $[T, L] = 1$. Since no proper homomorphism exists from T to K, we have $\sigma_u = 1$, that is $[T, U] = 1$ and G is nilpotent, a contradiction.

Hence $[T, L] \neq 1$, that is the group LT is not nilpotent. If $L \neq U$, by minimality we have $[T, L] = L$. In particular, for the maximal connected algebraic subgroup M of U we have $[T, M] = M$. Let H be the maximal connected algebraic subgroup of M. Thus U/H is isomorphic to a chain $\mathfrak{C}_2(\mathcal{S})$ and, by Proposition 4.3.7, it follows that $[T, U] = U$.

2) If T has dimension one, then $\mathfrak{z}°G = U_1 \leq U$, and by the previous point 1) we have $U_1 = 1$.

Conversely, let M be the maximal connected algebraic subgroup of U. If T has dimension greater than one, then the character $\chi : T \longrightarrow \mathsf{Aut}(U/M)$ has a non-trivial kernel. By the previous point 1), the connected component T_1 of the kernel of χ is such that $[T_1, U] = 1$, hence $T_1 \leq \mathfrak{z}°G$.

3) Let U have dimension n. Up to isomorphism, we may assume that the elements of U are points of the n-dimensional affine space and the subgroup U_i

of dimension i has the shape $\{(x_0, \cdots, x_{n-1}) \in U : x_0 = \cdots = x_{n-i-1} = 0\}$. The factor groups U_i/U_{i-2} are of type $\mathfrak{C}_2$ ([18], II, § 3, n. 4, Théorème 4.6, p. 197). Considering the action of T on these factor groups for $i = 2, \cdots, n$ we see using Proposition 4.3.7 that the action of T on U_i/U_{i-i} is not trivial for $i = n, \cdots, 1$. From this it follows that any factor group G/U_i is not nilpotent.

First we assume that U is not commutative. If $n > 2$, then U' has dimension one, since otherwise for the one-dimensional connected algebraic subgroup H of U the factor group U/H would have dimension two by minimality, a contradiction to Corollary 4.2.10. Hence by Theorem 4.3.1 the chain U is isogenous to $\mathfrak{C}_n(f, g_1, \cdots, g_s)$ for suitable p-polynomials $f, g_1 \cdots, g_t$, where $f \neq 0$ and not all g_i are zero. Now it follows from Remark 4.3.9 that the group of algebraic automorphism of U is finite, a contradiction to the fact that the dimension of U is at least three.

If U is not commutative but has dimension two, then U is isogenous to $\mathfrak{C}_2(f, g_1, \cdots, g_t)$, where $g_1, \cdots, g_t$ are not all zero. But by Remark 4.3.9 a torus can act non-trivially on U if and only if the product in U is given by

$$(x_0, x_1)(y_0, y_1) = (x_0 + y_0, x_1 + y_1 + x_0^{p^r} y_0^{p^{r+s}})$$

for suitable $r, s \in \mathbb{N}$.

If U is commutative, then any factor group U_i/U_{i-2}, where U_i is the i-dimensional connected algebraic subgroup of U, is isomorphic to the group defined on the two-dimensional affine space by the multiplication

$$(x_0, x_1)(y_0, y_1) = (x_0 + y_0, x_1 + y_1 + f(\Phi_1(x_0, y_0)))$$

for a suitable p-polynomial f. Since the action of T on any U_i/U_{i-1} is not trivial for $i = n, \cdots, 1$, it follows from Remark 4.3.9 that f consists only of a monomial and the claim follows.   □

4.3.11 Corollary. *Any connected algebraic group of algebraic automorphisms of a non-commutative unipotent chain of dimension greater than two defined over a perfect field of characteristic greater than two is unipotent.*   □

With our classification of unipotent chains over perfect fields k of characteristic greater than two with a one-dimensional commutator subgroup we can determine all non-commutative k-groups having the centre as maximal connected subgroup.

Let $G_1, \cdots, G_n$ be connected unipotent groups containing connected normal algebraic subgroups M_i $(i = 1, \cdots, n)$ such that all factor groups G_i/M_i are isomorphic to the connected unipotent group H. Then the group G_i is biregularly k-isomorphic to a suitable explicit extension of M_i by H, hence it may be described as the set of pairs (x, y_i) where $x \in H, y_i \in M_i$. The *direct product* $G = (G_1 \curlywedge \cdots \curlywedge G_n)_{[M_1, \cdots, M_n]}$ *of the groups* $G_1, \ldots, G_n$ *with amalgamated factor group* H is the subgroup

$$G = \left\{ ((x, y_1), \cdots, (x, y_n)) : x \in H, y_i \in M_i \right\}$$

of the direct product $G_1 \times \cdots \times G_n$. We note that if we have $M_i = \mathfrak{z}^\circ G_i$ for any $i = 1, \cdots, n$, then

$$\mathfrak{z}^\circ(G_1 \curlywedge \cdots \curlywedge G_n)_{[M_1, \cdots, M_n]} = M_1 \times \cdots \times M_n.$$

4.3.12 Theorem *Let G be a connected algebraic group over a perfect field of positive characteristic. The connected component of the centre $\mathfrak{z}^\circ G$ is maximal connected in G if and only if G is isogenous to the direct product of a commutative group A with a group*

$$B = (V_1 \curlywedge \cdots \curlywedge V_t)_{[\mathfrak{W}_{m_1}, \cdots, \mathfrak{W}_{m_t}]}$$

where V_i is isogenous either to $\mathfrak{C}_{m_i+1}(\mathbb{S})$ or to $(\mathfrak{W}_{m_i} \curlyvee \mathfrak{C}_2)_{\mathfrak{W}_1}$.

Proof. Let $G = L(G)D(G)$ be the decomposition as in Theorem 1.3.2. Since $D(G)$ is central in G, the affine group $L(G)$ has centre of co-dimension one. The nilpotent group $L(G)$ is the direct product of its unipotent radical U with a torus. The connected component $\mathfrak{z}^\circ U$ of the centre of U has co-dimension one in U. The group $\mathfrak{z}^\circ U$ is isogenous to the direct product of Witt groups $\mathfrak{W}_{m_i}$ ($i = 1, \ldots, t$) and U is isogenous to an extension V of $W = \mathfrak{W}_{m_1} \times \cdots \times \mathfrak{W}_{m_t}$ by $\mathfrak{W}_1$. The factor system $\Theta : \mathfrak{W}_1 \times \mathfrak{W}_1 \longrightarrow W$ of this extension is given by $\Theta(x, y) = (\theta_1(x, y), \ldots, \theta_t(x, y))$, where $\theta_i : \mathfrak{W}_1 \times \mathfrak{W}_1 \longrightarrow \mathfrak{W}_{m_i}$ is a factor system. Let V_i be the central extension of $\mathfrak{W}_{m_i}$ by $\mathfrak{W}_1$ with respect to the factor system θ_i. Therefore V is the direct product $V = (V_1 \curlywedge \cdots \curlywedge V_n)_{[\mathfrak{W}_{m_1}, \cdots, \mathfrak{W}_{m_n}]}$ of the groups $V_1, \ldots, V_n$ with amalgamated factor group $\mathfrak{W}_1$.

If θ_i is a coboundary, then $\mathfrak{W}_{m_i}$ is a summand in V. If θ_i is non-trivial and symmetric, then V_i is a commutative chain. If θ_i is not symmetric, then the centre of V_i has co-dimension one in V_i and the commutator subgroup V_i' of V_i has dimension one by Theorem 3.2.8. The factor group V_i/V_i' is either a chain or isogenous to the direct product of $\mathfrak{W}_1$ with $\mathfrak{z}^\circ V_i/V_i'$.

In the first case, V_i is a non-commutative chain since it has a unique maximal connected algebraic subgroup which is a chain. In the second case, the image of the factor system θ_i is contained in V_i', by Remark 2.1.15. In this case, V_i is the direct product $V_i = (\mathfrak{W}_{m_i} \curlyvee \mathfrak{C}_2)_{\mathfrak{W}_1}$ of the Witt group $\mathfrak{W}_{m_i}$ and the non-commutative chain $\mathfrak{C}_2$ with amalgamated central subgroup $\mathfrak{W}_1$ (see Proposition 2.1.3). $\qquad \square$

5

Groups with Few Types of Isogenous Factors

In this section we study some properties fulfilled by any chain, and we investigate whether they characterise the chains.

5.1 A Useful Theorem

If a connected algebraic group G has an algebraic subgroup H which is a chain and if G/H is also a chain, then G is not necessarily a chain. But if we can tie together the chains H and G/H properly, then G will be a chain, as well:

5.1.1 Theorem. *Let G be a connected algebraic group and let H be a normal connected algebraic subgroup of G such that both H and G/H are chains. If two connected algebraic subgroups S, T exist such that $S < H < T$, $S \neq H \neq T$ and such that T/S is a chain, then G is a chain.*

Proof. Let G be a minimal counter-example to the theorem. Assume first that G is affine. As H and G/H are chains, they are nilpotent as mentioned at the beginning of Section 4, so G is solvable.

If H is maximal connected in G, then $T = G$. Let M be another maximal connected subgroup of G. Then $H \cap M$ has co-dimension 1, both in H and M. As H is a chain, this forces $S < M$. Because $G/S = T/S$ is a chain, we have $M = H$. Therefore H is the unique maximal connected subgroup of G and it is a chain, so we are done.

Similarly, if H is not maximal connected in G, we may assume $T \neq G$ and find a maximal connected subgroup U, containing T. By minimality U is a chain. Let V be another maximal connected subgroup of G. Then $U \cap V$ has co-dimension 1, both in U and V. Because U is a chain, this forces $H < V$. As G/H is a chain, we have $U = V$, so U is the unique maximal connected subgroup of G; thus G is a chain.

Let now G be non-affine and assume first that H is affine. Then G/H is a non-affine chain. By Theorem 4.1.3 we find a unique maximal connected subgroup G_1 of G, containing H, such that G_1/H is affine, has at most dimension one and G/G_1 is a simple abelian variety. Since H is affine, the group G_1 is affine. By the above argument G_1 is a chain. If the chain H is a one-dimensional torus, then the chain G_1 is equal to H. It follows that $G = T$ and $S = 1$ and G is a chain. If the chain H is unipotent, then the chain G_1 is unipotent. If $G_1 \neq H$, then the chain G/H contains the unipotent chain G_1/H, hence, by Theorem 4.1.3, the characteristic of the ground field is zero. It follows that $G_1 = H$, a contradiction. Therefore $G_1 = H$ and again $G = T$. Since the non-affine chain G/S contains the unipotent chain H/S, the characteristic of the ground field is zero and the dimension of H/S is one. This means that $S = 1$ and, together with T/S, the group G is a chain, contradicting the fact that G should be a counter-example.

Assume finally that H is not affine. Again by Theorem 4.1.3 we find a unique maximal connected subgroup H_1 of H, which is an affine subgroup of dimension at most one, and H/H_1 is a simple abelian variety. We have $S \leq H_1 < H < T$. Since T/S is a chain which is not affine, this contradicts Theorem 4.1.3. $\square$

5.2 Uni-Minimal and Uni-Maximal Groups

Affine chains and affine algebraic groups in which every two-dimensional connected algebraic subgroup is a chain are examples of algebraic groups with a unique minimal connected subgroup. Dually, unipotent algebraic groups for which the factor group with respect to the commutator subgroup is a chain are examples of algebraic groups having a unique maximal connected algebraic subgroup. In this section we clarify the structure of algebraic groups having a unique minimal or a unique maximal connected algebraic subgroup. In particular, we show that for the classes of groups under consideration these phenomena arise essentially only in algebraic groups over fields of positive characteristic and in complex non-linear Lie groups.

Let G be a group in a group class $\mathfrak{D}$. A subgroup U of G belonging to the class $\mathfrak{D}$ is minimal with respect to $\mathfrak{D}$ if it contains no non-trivial proper subgroup of $\mathfrak{D}$. A subgroup M of G belonging to the class $\mathfrak{D}$ is maximal with respect to $\mathfrak{D}$ if no proper subgroup of G being in $\mathfrak{D}$ contains M properly.

Any subgroup of G belonging with G to the class $\mathfrak{D}$ has a unique minimal subgroup with respect to $\mathfrak{D}$ if the group G has a unique minimal subgroup belonging to $\mathfrak{D}$.

Let $\mathfrak{D}$ be a class of groups such that the images and preimages of any epimorphism σ with kernel in $\mathfrak{D}$ belong to $\mathfrak{D}$. If a group G in such a class $\mathfrak{D}$ has a unique maximal subgroup with respect to $\mathfrak{D}$, then also G^σ has a unique maximal subgroup with respect to $\mathfrak{D}$.

In most classes $\mathfrak{D}$ of groups it is immediately clear which groups in $\mathfrak{D}$ have precisely one minimal, respectively one maximal, subgroup with respect to $\mathfrak{D}$.

For instance, let G be in the class of finite groups. If G has only one minimal subgroup, then G is a p-group because it coincides with one Sylow subgroup. If $p > 2$, then G is itself cyclic (cf. [46], Satz 7.6 (a), p. 304). If $p = 2$, then G is either cyclic or a generalized quaternion group (cf. [46], Satz 7.6 (b), p. 304 and Satz 11.2, p. 336). If G has only one maximal subgroup M, then G is a cyclic p-group. Namely, there are elements not contained in M but any cyclic subgroup of G is contained in M.

For real Lie groups and for algebraic groups over a field of characteristic zero it is also easy to see which groups in these classes have only one minimal, respectively only one maximal, subgroup.

5.2.1 Proposition. *A finite-dimensional Lie algebra $\mathfrak{g}$ has dimension one if and only if one of the following conditions holds:*

(1) $\mathfrak{g}$ has exactly one minimal subalgebra,
(2) $\mathfrak{g}$ has exactly one maximal subalgebra.

Proof. Since $\mathfrak{g}$ is a vector space we have *(1)*. If $\mathfrak{g}$ has a unique maximal subalgebra $\mathfrak{m}$ and if x is an element of $\mathfrak{g}\backslash\mathfrak{m}$, then x lies in a maximal subalgebra of $\mathfrak{g}$ different from $\mathfrak{m}$. □

5.2.2 Proposition. *Let G be a group belonging to one of the following classes:*

(i) A formal group over some field of characteristic zero,
(ii) a connected affine algebraic group over a field of characteristic zero,
(iii) a connected real Lie group.

If G has a unique maximal or minimal formal, respectively closed connected subgroup, then G has dimension one.

Proof. If G is a formal group, then the assertion follows from Proposition 5.2.1 together with the description of functorial equivalence between formal groups and their Lie algebras in [39], Theorem 14.2.3 p. 81. In case *(ii)* the same argument holds, now using [45], 13.1 Theorem, p. 87.

If G is a real Lie group which has a unique maximal or minimal closed connected subgroup, then its universal covering group $\widetilde{G}$ has the same property. Since the Lie algebras of G and $\widetilde{G}$ coincide and since every analytic subgroup of a simply connected real Lie group is closed, the assertion holds in this case. □

In the class of complex Lie groups, the groups having only one minimal subgroup or only one maximal subgroup are more interesting. In order to deal with them we need some preparation which is given in the following.

5.2.3 Remark. Let G be a connected complex Lie group and let A be the exponential image of a one-dimensional (complex) subspace $\mathfrak{a}$ of the Lie algebra of G. Since the exponential map is complex analytic, the set A a is a not necessarily closed complex analytic subgroup of G. If A is closed in G, then

A is isomorphic to $\mathbb{C}^+$ or $\mathbb{C}^*$ or is a one-dimensional complex torus. If A is not closed in G, then the restriction of the exponential map to $\mathfrak{a}$ is injective. Considering G as a real Lie group one sees that the (topological) closure B of A in G is isomorphic to $\mathbb{R}^\epsilon \times (SO_2(\mathbb{R}))^l$ with $\epsilon \in \{0, 1\}$ and $l > 1$. Then there is a minimal analytic subgroup B^* of G containing B (cf. [55], Proposition 4.1, p. 99). We will call B^* the *complex analytic closure of A*. Let C be the maximal complex torus contained in B. Then the complex Lie group B^*/C is the complex analytic closure of its subgroup B/C. The group B^*/C is isogenous to the direct product of a linear complex torus, a linear complex torus $(\mathbb{C}^*)^r$ and a torus-free toroidal group. Hence B^* is isomorphic to the direct product of a linear complex torus and a toroidal group.

Every complex torus arises as the complex analytic closure of a complex one-parameter subgroup. If a toroidal torus-free complex Lie group H is the only complex Lie subgroup containing the maximal real torus of H, then H is the complex analytic closure of one of its complex analytic one-parameter subgroups. □

5.2.4 Proposition. *Let G be a connected complex Lie group of (complex) dimension $n > 1$ which is not a simple complex torus.*

(a) The group G has a unique maximal closed connected complex subgroup if and only if G is a non-splitting extension of a toroidal group by a simple complex torus group.

(b) The group G has a unique minimal closed connected complex subgroup if and only if G is either a toroidal group of rank $n - 1$ or a non-splitting extension of a simple complex torus by a complex torus.

Proof. Assume first that G has a unique maximal closed connected complex subgroup N. Consider the Lie algebra $\mathfrak{g}$ of G and the Lie algebra $\mathfrak{n}$ of N. Let $\mathfrak{a}$ be a one-dimensional subspace of $\mathfrak{g}$ not contained in $\mathfrak{n}$. The complex analytic closure A of the analytic subgroup corresponding to $\mathfrak{a}$ coincides with G. Since G is a commutative complex Lie group of dimension greater than one, the group G is toroidal (cf. [1], 1.1.5, p. 6) and then G is a non-split extension of a toroidal group N by a simple complex torus group (cf. [1], Section 1.14, p. 12).

Assume now that G has a unique minimal connected subgroup different from G. Let X be the maximal toroidal subgroup of G. Then X is contained in the centre of G ([59], Lemma 6, p. 259). Assume that $G \neq X$ and consider the Lie algebra $\mathfrak{x}$ of X. If $\mathfrak{b}$ is a complex one-dimensional subspace of $\mathfrak{g}$ not contained in $\mathfrak{x}$, then the algebra $\mathfrak{x} + \mathfrak{b}$ is commutative. Thus the closure of the analytic subgroup H given by $\mathfrak{x} + \mathfrak{b}$ is commutative. Since the group H has a decomposition $\mathbb{C}_+{}^m \times \mathbb{C}^{*n} \times X$, it has at least two closed connected complex subgroups. It follows that any commutative connected subgroup of G coincides with X. Hence G is a toroidal group and the rank of G can be either $n - 1$ or n. □

5.2.5 Remark. If in part (a) of Proposition 5.2.4 the subgroup N is a maximal closed connected subgroup of G, then G/N is complex torus, too (cf. [1], Proposition 1.1.17, p. 15). If G/N is polarizable, then it follows from Poincaré's irreducibility theorem that G/N is even a simple torus (cf. [7], Theorem 5.5, p. 52). Thus N necessarily is a maximal closed connected subgroup of G. Examples of three-dimensional toroidal groups having a unique maximal closed connected complex subgroup M can be deduced from [60], Theorem 11.5, p. 227. In these examples, M is the direct product of $\mathbb{C}^*$ with a one-dimensional complex torus.

There are many toroidal groups having a unique minimal closed connected complex subgroup. Examples of (complex) dimension two can be found in [59], Theorem 3 and Remark, § 4, p. 262, p. 263. These examples are a non-splitting torus-free extensions of $\mathbb{C}^*$ by a one-dimensional complex torus.   □

Next we clarify the structure of connected algebraic groups, having a unique minimal, respectively maximal, connected algebraic subgroup. If G is affine, these two properties mean that G has a unique connected algebraic subgroup of dimension one, respectively co-dimension one.

5.2.6 Proposition. *Let G be a connected affine algebraic group of dimension greater than one such that G has a unique minimal connected algebraic subgroup or a unique maximal connected algebraic subgroup. Then G is unipotent.*

Proof. We start assuming that G has unique minimal connected algebraic subgroup. First we suppose that G is solvable. Then G is a semi-direct product of a unipotent group with a torus. Since the dimension is greater than one, the group G has to be unipotent. If G is not solvable, then we consider a Borel subgroup B of G. With G also B has a unique minimal connected algebraic subgroup. As B is a solvable subgroup of G, it has to be unipotent. Hence $G = B$, a contradiction.

Now we treat the case that G has a unique maximal connected algebraic subgroup. Since a semi-simple algebraic group has never a unique maximal connected algebraic subgroup, the group G must be solvable. Then G is the semi-direct product of the unipotent radical G_u with a torus T such that G_u has a series of closed connected subgroups, normal in G, such that each of them has co-dimension one in the next (cf. [8], Theorem 10.6, p. 137). Since the dimension is greater than one, the group G has to be unipotent.   □

5.2.7 Theorem. *Let G be a connected non-affine algebraic group.*

i) If G is an abelian variety and if it has a unique minimal or unique maximal connected algebraic subgroup, then G is a simple abelian variety;

ii) if G has a unique minimal connected algebraic subgroup and if it is not an abelian variety, then $G = D(G)$ and $L(G)$ is one-dimensional;

iii) if G has a unique maximal connected algebraic subgroup and if it is not an abelian variety, then $G = D(G)$ and $G/L(G)$ is a simple abelian variety.

Proof. If G is an abelian variety, then it is isogenous to a direct product of simple abelian varieties (see [64], Corollary 1, p. 174) and thus an abelian variety which has a unique minimal or a unique maximal connected algebraic subgroup is simple.

Now we suppose that the non-affine group G is not an abelian variety. First we treat the case where G has a unique minimal connected algebraic subgroup. In this case $L(G)$ is an affine group having a unique minimal connected algebraic subgroup. Hence it is either a one-dimensional torus or a unipotent group. If the ground field has characteristic zero, then, according to Proposition 5.2.2, $L(G)$ is one-dimensional. It follows that $G = D(G)$. If the characteristic of the ground field is positive, then the connected component of the group $L(G) \cap D(G)$ has to be a torus (see Theorem 1.3.2). It follows from Proposition 5.2.6 that $L(G)$ is a one-dimensional torus and we have again $G = D(G)$.

Now we assume that G has a unique maximal connected algebraic subgroup. Then the factor group $G/L(G)$ is a simple abelian variety and $L(G)$ is the unique maximal connected algebraic subgroup of G. It follows that $D(G)$ is either equal to G or contained in $L(G)$, which is not the case because G is not affine. □

The *Frattini subgroup* $\Phi(G)$ of a connected algebraic group G is defined as the connected component of the intersection of all maximal connected subgroups of G. If G is a unipotent group defined over a perfect field k, then $\Phi(G)$ is a k-subgroup, characterised by the following property (see [26], Proposition 1, p. 14, see also [71], Lemma 7):

5.2.8 Proposition. *Let G be a connected unipotent group and N a connected algebraic subgroup of G. The factor group G/N is a vector group if and only if $N \geq \Phi(G)$.* □

In Proposition 5.2.2 we have seen that a connected affine algebraic group defined over a field of characteristic zero and having a unique minimal or a unique maximal connected algebraic subgroup has dimension one. Hence any connected affine algebraic group with one of these properties and of dimension greater than one is unipotent (cf. Proposition 5.2.6), and it is defined over a field of characteristic $p > 0$. The study of such groups, which we call *uni-minimal*, respectively *uni-maximal*, if they have a unique minimal, respectively a unique maximal, connected algebraic subgroup, is the objective of the main part of this section.

The Frattini subgroup of a connected affine algebraic group is a useful tool for characterisations of connected uni-maximal algebraic groups.

5.2.9 Proposition. *Let G be a connected affine algebraic group of dimension greater than one. Then the following conditions are equivalent:*

1) G is uni-maximal,
2) G/G' is isogenous to a single Witt group,
3) $G/\Phi(G)$ is a vector group of dimension one.

Proof. Let G be uni-maximal. As G is unipotent, the assertions 2) and 3) follow from 1) using Proposition 5.2.8. Conversely, if 2) is satisfied, then G is uni-maximal. Since the Frattini subgroup is the intersection of all connected maximal subgroups of G, the condition 3) yields 1).   □

5.2.10 Proposition. *Let G be a unipotent algebraic k-group defined over a perfect field k. Then there exists a connected algebraic k-subgroup $T \leq G$, containing $\Phi(G)$ as a maximal connected subgroup, such that $T' = G'$.*

Proof. We prove the claim by induction on $n = \dim G$, the assertion being true for $n = 2$. In the general case $n > 2$, we mention the fact that the assertion is true if G is commutative, whereas if G is not commutative, we consider a normal connected one-dimensional algebraic k-subgroup K of $G' \leq \Phi(G)$. By induction, G/K has a connected algebraic k-subgroup T/K, containing $\Phi(G/K)$ as a maximal connected algebraic subgroup, such that $(T/K)' = (G/K)'$. Since $K \leq G' \leq \Phi(G)$, we have $\Phi(G/K) = \Phi(G)/K$. Therefore $\Phi(G)$ is a maximal connected algebraic subgroup of the k-subgroup T of G. Moreover, $K \leq G' \leq \Phi(G) < T$, hence $(T/K)' = T'/K$. On the other hand we have $(G/K)' = G'/K$ and the assertion follows immediately.   □

5.2.11 Proposition. *If the connected affine algebraic group G of dimension $n > 1$ is uni-maximal, then we have:*

1) the maximal connected subgroup M centralises any two-dimensional normal subgroup H which, in particular, is commutative;

2) if the commutator subgroup $[G, M]$ is at most one-dimensional, then M is commutative.

Proof. Proposition 5.2.6 yields that G is unipotent. First we prove the assertion 1). Let H be a two-dimensional normal connected algebraic subgroup of G. By intersecting H with a series of normal connected subgroup of G, each of co-dimension one in the next, we see that H contains always a one-dimensional normal subgroup K of G and the factor group H/K is central in G/K. Hence $[G, H]$ is at most one-dimensional, which yields that $[G, H]$ is central. For any $h \in H$ and for any $x \in G$, the mapping $\sigma_h : G \longrightarrow [G, H]$, defined by $\sigma_h(x) = [h, x]$, is therefore a homomorphism with a kernel of co-dimension at most one in G. Since the unique maximal connected subgroup M is contained in the kernel of σ_h for any $h \in H$, we are done.

To prove the second assertion we consider for any $a \in M$ the mapping $\sigma_a : G \longrightarrow [G, M]$ defined by $\sigma_a(x) = [a, x]$. Since $[G, M]$ is central, the map σ_a is a homomorphism with a kernel of co-dimension at most one in G, and M is again contained in the kernel of σ_a.   □

5.2.12 Remark. The groups $\mathfrak{J}_n(\mathfrak{W}_m)$, introduced in Remark 3.1.12, form a class of uni-maximal groups where property 1) of Proposition 5.2.11 is satisfied, even though one might not see this at first glance. In this class of groups, property 2) is fulfilled only if $n, m \leq 2$.   □

Now we want to characterise affine uni-maximal algebraic groups using homomorphic images. To do this we need the following

5.2.13 Proposition. *Let G be a non-commutative connected affine algebraic group over a field of positive characteristic such that $_3G$ has positive dimension. If every epimorphic image of G with a connected non-trivial kernel is commutative, then $_3°G$ is a chain. In particular, there exists a unique normal one-dimensional connected algebraic subgroup K of G, and $K = G'$.*

Proof. If the centre $_3G$ were not a chain, then two central one-dimensional connected subgroup K_1 and K_2 would exist with $K_1 \cap K_2 = 1$. Since G embeds into the direct product $G/K_1 \times G/K_2$, it would be commutative, contradicting the fact that G' is not equal 1.   □

5.2.14 Proposition. *Let G be a connected non-simple algebraic group of dimension greater than one over a field k of positive characteristic. If every epimorphic image of G with a connected non-trivial kernel is a chain, then G satisfies one of the following conditions:*
i) G is a chain;
ii) $\dim G = 2$;
iii) G is isogenous either to a product of two simple abelian varieties or of a simple abelian variety and a one-dimensional affine group;
iv) G is an algebraic group having no non-trivial affine epimorphic image such that the maximal affine connected subgroup of G is a two-dimensional torus;
v) G is a uni-maximal group containing a normal algebraic subgroup N such that G/N is a non-commutative chain. Moreover, if $G' \leq {}_3G$, then G has dimension three.

Proof. Let $G = L(G)D(G)$ be the decomposition of G as in Theorem 1.3.2 and put $H = L(G) \cap D(G)$. The group G/H is isogenous to $L(G)/H \times D(G)/H$.

If G is not affine and $G \neq D(G)$, then $L(G) \cap D(G)$ is finite. Then G is isogenous to the direct product of a one-dimensional affine group and a simple abelian variety.

Let $G = D(G)$ be not a chain. If $L(G) = 1$, then G is isogenous to the direct product of two simple abelian varieties. If $L(G) \neq 1$, then $L(G)$ is a two-dimensional torus, because the characteristic of k is positive (see Proposition 4.1.3 and Theorem 1.3.2). Hence if G is not affine, then we have *iii*) or *iv*).

If the non-simple group G is affine, then G is an extension of its solvable radical by a semi-simple algebraic group (cf. [8], 11.21, p. 157). From this it follows that G is solvable. Thus G is a semi-direct product of its unipotent radical U and a maximal torus T. If T is non-trivial, then either U is trivial and T is a two-dimensional torus, or U is one-dimensional, as well as T, and we have the case *ii*).

So let now G be unipotent. Of course we assume that G has dimension greater than 2. If G is commutative, then G is isogenous to a single Witt group and we are done. Assume therefore that $G' \neq 1$. Since G/G' is a chain, the group G is uni-maximal. If G/N is commutative for any normal connected algebraic subgroup N of G, then by Proposition 5.2.13 the group G has a unique normal connected one-dimensional algebraic subgroup $K = G'$ and $_3G$ is a chain. If the dimension of $_3G$ is greater than one, then, by Theorem 5.1.1, the group G is a chain. Otherwise $_3^{\circ}G = G'$ and Theorem 3.2.10 applies, forcing G to have dimension smaller than three.

It remains to prove the last assertion of $v)$. Let G be non-commutative, with $G' \leq {}_3G$. Since every epimorphic image of $_3G$ with a non-trivial connected kernel is a chain, $_3G$ is either a chain or a vector group of dimension $d \leq 2$. If $_3G$ were a chain of dimension greater than one, then G would be a chain as well, by Theorem 5.1.1. Therefore $_3G$ is a vector group of dimension $d \leq 2$. By Theorem 3.2.8 we find now that $_3G$ has co-dimension one, hence G has dimension three.   □

5.2.15 Proposition. *Let G be a connected affine algebraic group of dimension greater than one over a field of positive characteristic p. Then the following holds:*

i) the group G is either a two-dimensional vector group or uni-maximal if and only if every non-trivial epimorphic image of G is uni-maximal;

ii) if G is uni-maximal, then either G is a chain or G has a normal subgroup C such that G/C is a non-commutative chain. Moreover, if G has nilpotency class two, then the subgroup C has co-dimension two.

Proof. If G is either a two-dimensional vector group or uni-maximal, then every non-trivial epimorphic image of G is uni-maximal, as well. Conversely, assume that every non-trivial epimorphic image of G is uni-maximal. Let M_1 and M_2 be two maximal connected subgroups of G. If G has dimension greater than two, $M_1 \cap M_2$ contains a non-trivial normal connected algebraic subgroup K. Since G/K is uni-maximal, we find that $M_1/K = M_2/K$, hence $M_1 = M_2$, that is: G is uni-maximal.

Let G be a minimal counter-example to the second assertion. Then G is not a chain. If G/N is a chain for any connected normal subgroup $N \neq 1$ of G, then we are done by Propositions 5.2.14 $v)$. Thus there exists a normal connected subgroup $N \neq 1$ of G such that G/N is not a chain. By minimality of G, the group G/N contains a normal subgroup C/N such that $(G/N)/(C/N)$ is a non-commutative chain. Moreover, if $G' \leq {}_3G$, then the group $(G/N)/(C/N)$ has dimension two. This is a contradiction, because $(G/N)/(C/N)$ is isomorphic to G/C.   □

5.2.16 Remark. If G' is the maximal connected algebraic subgroup of G, then it is clear that for any given maximal connected algebraic subgroup C of G', normal in G, we have that G/C is a non-commutative two-dimensional algebraic group. Non-trivial examples for such subgroups C of G are given when one considers the group $\mathfrak{J}_n(\mathfrak{W}_m)$ introduced in Remark 3.1.12.   □

5.2.17 Theorem. *Let G be a non-commutative unipotent algebraic group over a field of positive characteristic. A sufficient and necessary condition for G to be not uni-maximal is that a maximal connected algebraic subgroup M of G exists such that $M' = G'$.*

Proof. If G is not uni-maximal, then by Proposition 5.2.10 we find a connected algebraic subgroup $T \neq G$, such that $G' = T'$ and the assertion follows for any maximal connected algebraic subgroup M of G containing T.

Conversely, if G is uni-maximal, then by Proposition 5.2.14 there exists a normal algebraic subgroup N of G such that G/N is a non-commutative chain. Therefore, we can restrict attention to the case where G is a non-commutative chain. We prove by induction on $n = \dim G$ that the maximal connected algebraic subgroup M of the non-commutative chain G is such that $M' \neq G'$. If $n = 2$ the assertion is true. Let K be a connected one-dimensional algebraic subgroup of G and consider the chain G/K. If G/K is not commutative, then the assertion follows by induction. If G/K is commutative, then $G' = K$ and by Theorem 3.2.10 the maximal connected algebraic subgroup M is central, hence $M' = 0$. □

In contrast to the situation of vector groups over fields of characteristic zero, the one-dimensional subgroups of a vector group V over a field of positive characteristic are not necessarily given by linear functions. A lemma on p. 101 of [82] claims that any finite subgroup of V is contained in a connected algebraic subgroup of dimension one. This fact can be improved, because any finite set of elements of V lies in a one-dimensional connected algebraic subgroup of G. For the proof we need a technical

5.2.18 Lemma. *The determinant of the matrix*

$$
T = \begin{pmatrix}
t_1 & t_2 & \cdots & t_m \\
t_1^p & t_2^p & \cdots & t_m^p \\
\vdots & \vdots & \vdots & \vdots \\
t_1^{p^{m-1}} & t_2^{p^{m-1}} & \cdots & t_m^{p^{m-1}}
\end{pmatrix}
$$

is a non-zero polynomial in the variables $t_1, \cdots, t_m$ of degree $1 + p + p^2 + \cdots + p^{m-1}$.

Proof. By the Laplace formula one has

$$
|T| = \sum_{\sigma \in S_m} (-1)^{\operatorname{sgn}\sigma} t_1^{p^{\sigma(1)}-1} t_2^{p^{\sigma(2)}-1} \cdots t_m^{p^{\sigma(m)}-1}.
$$

Since any two monomials of $|T|$ are different, the assertion follows. □

5.2.19 Proposition. *Let G be a connected unipotent group over an infinite perfect field k of positive characteristic and $g_1, \cdots, g_m \in G(\mathsf{k})$. There exists a connected algebraic k-subgroup H containing $\Phi(G)$ as a subgroup of co-dimension one such that $g_1, \cdots, g_m \in H$.*

Proof. By Proposition 5.2.8 we may assume that G is an n-dimensional vector group. Fix a coordinate system for G and let $g_i = (a_{i1}, \cdots, a_{in})$ for $i = 1, \cdots, m$. According to Lemma 5.2.18, let $r_1, \cdots, r_m \in \mathsf{k}$ be such that

$$R = \begin{pmatrix} r_1 & r_2 & \cdots & r_m \\ r_1^p & r_2^p & \cdots & r_m^p \\ \vdots & \vdots & \vdots & \vdots \\ r_1^{p^{m-1}} & r_2^{p^{m-1}} & \cdots & r_m^{p^{m-1}} \end{pmatrix}$$

is non-singular. Consider the matrix $C = PR^{-1} = (c_{ij})$, where $P^{\mathsf{t}} = (a_{ij})$. Putting

$$f_i(x) = \sum_{k=0}^{m} c_{ik} x^{p^{k-1}},$$

a direct computation shows that $f_i(r_j) = a_{ij}$. Thus the mapping $r \mapsto (f_1(r), \cdots, f_n(r))$ is an epimorphism from $\mathbf{G}_a$ onto a one-dimensional connected algebraic k-subgroup $H \leq G$ containing the elements g_j.   □

The above Proposition 5.2.19 allows to characterise connected algebraic groups in which every connected algebraic subgroup is commutative. For non-affine groups we have the following fact:

5.2.20 Theorem. *Let G be a connected non-affine k-group defined over a perfect field k such that every proper connected normal algebraic k-subgroup is commutative. Then G is commutative.*

Proof. By Theorem 1.3.2, we have $G = L(G)D(G)$ and $D(G)$ is central in G. Since G is not affine, $L(G)$ is a proper connected normal k-subgroup of G, hence commutative, and the assertion follows.   □

In contrast to Theorem 5.2.20, for affine algebraic groups over a field k of positive characteristic the situation is more complicated, as the following theorem shows.

5.2.21 Theorem. *Let G be a connected non-commutative unipotent k-group over an infinite perfect field k of positive characteristic. If every proper connected normal algebraic k-subgroup is commutative, then G is uni-maximal.*

Proof. Suppose that G is not uni-maximal; then $G/\Phi(G)$ is a vector group of dimension $d \geq 2$. Since $G(\mathsf{k})$ is dense in G (see [8], Corollary 18.3, p. 220), it is not commutative (see [8], 1.7 Proposition, (c), p. 52). So there are two elements $x, y \in G(\mathsf{k})$ such that $xy \neq yx$. By Proposition 5.2.19, the elements x and y lie in a proper connected normal algebraic k-subgroup H of G, a contradiction.   □

This theorem allows the following applications, since the field of definition does not play any rôle there.

5.2.22 Corollary. *Let G be a connected non-commutative unipotent group over a field of positive characteristic. Then any non-commutative connected algebraic subgroup of minimal dimension is uni-maximal.*   □

5.2.23 Corollary. *Let G be a connected unipotent non-commutative algebraic group over a field of positive characteristic. Then any connected algebraic subgroup of G is commutative if and only if the factor group G/G' is a chain and the maximal subgroup of G is commutative.*   □

5.2.24 Corollary. *Let G be a connected unipotent non-commutative algebraic group over a field of positive characteristic. If any connected algebraic subgroup of G is commutative and the maximal connected algebraic subgroup U of G is isogenous to the factor group G/K for a one-dimensional normal algebraic subgroup K of G, then G is a chain and the commutator subgroup has dimension one.*

Proof. By Corollary 5.2.23 the factor group G/G' is a chain and the factor group G/K is uni-maximal. Therefore U, as a uni-maximal commutative group, is isogenous to a Witt group. Now the assertion follows by Theorem 5.1.1.   □

5.2.25 Theorem. *Let G be a connected non-commutative affine group over a field of characteristic $p > 0$. If G is not unipotent and if every proper connected algebraic subgroup of G is commutative, then G is a two-dimensional solvable algebraic affine group.*

Proof. If G were not solvable, no Borel subgroup B of G can be commutative, otherwise $B = G$. Hence the group G is solvable. As G is not unipotent, G contains a maximal torus $T \neq 1$ and G is a semi-direct product of its n-dimensional non-trivial unipotent radical U by T. If T had dimension $d > 1$, then any one-dimensional subtorus of T would centralise U. It follows that T centralises U, which is impossible. Therefore T has dimension one. By [22], Corollary 2.9, p. 531, there exists a bijective algebraic homomorphism $\imath :$ $U \longrightarrow \mathfrak{W}_n$, such that, for any $u \in U, t \in T$, one has $\imath(tut^{-1}) = \lambda_t(\imath(u))$, where for a suitable non-trivial character χ of T we have

$$\lambda_t(x_0, \cdots, x_{n-1}) = (\chi(t)x_0, \chi(t)^p x_1, \cdots, \chi(t)^{p^{n-1}} x_{n-1}).$$

If $n > 1$, then the product of T by the connected one-dimensional algebraic subgroup of U can be commutative only if $\chi(t) = 1$, a contradiction.   □

Now we turn our attention to the property which is dual to uni-maximality and investigate uni-minimal connected affine algebraic k-groups G. If k has characteristic zero, then G has dimension one by Proposition 5.2.2. If the characteristic of k is positive, then the structure of G is more complicated.

5.2.26 Proposition. *Let G be a connected unipotent algebraic group of dimension greater than one. Then the following conditions are equivalent:*

1) G is uni-minimal,
2) any commutative subgroup of G is isogenous to a single Witt group,
3) any vector subgroup of G has dimension one.

Proof. From 1) the conditions 2) and 3) follow immediately. As G is a unipotent group it contains a one-dimensional central subgroup Z. If G satisfies 3) and it is not uni-minimal, then there exists a one-dimensional subgroup different from Z. $\square$

5.2.27 Theorem. *Let G be a connected affine algebraic group of dimension greater than one over a field of characteristic $p > 0$. If every maximal connected subgroup of G is a chain, then G satisfies one of the following conditions:*

i) G is a chain;
ii) $\dim G = 2$;
iii) G is uni-minimal and contains a non-commutative algebraic subgroup.

Proof. Let B a Borel subgroup of G. If $B \neq G$, then B is a chain, but then B is nilpotent and $G = B$, a contradiction. Thus the group G is solvable and splits as a semi-direct product of its unipotent radical U and a torus T. If T is non-trivial, then either U is trivial and T is a two-dimensional torus, or U is one-dimensional as well as T, and we are done.

Let now G be unipotent. Of course we may assume $\dim G \geq 3$. Since two maximal connected algebraic subgroups of G always intersect in a chain, the group G is uni-minimal. If every algebraic subgroup of G were commutative, then G would be uni-maximal by Theorem 5.2.21. But this would force G to be a chain. $\square$

For the next theorem we need the following result, which holds for abstract groups.

5.2.28 Proposition. *Let G be a group such that $G' \leq {}_3G$. For any odd integer k such that $g^k \in {}_3G$ for any $g \in G$, the map $x \mapsto x^k$ is a homomorphism.*

Proof. Since $G' \leq {}_3G$, for any $n \in \mathbb{N}$ we have

$$(xy)^n = [y, x][y^2, x] \cdots [y^{n-1}, x] x^n y^n.$$

This follows by induction on n, since the case $n = 2$ is manifest and we have

$$(xy)^n = (xy)^{n-1}(xy) = [y, x][y^2, x] \cdots [y^{n-2}, x] x^{n-1} y^{n-1}(xy)$$

as well as $x^{n-1} y^{n-1}(xy) = x^n y^n [y^{n-1}, x]$. Moreover, the map $x \mapsto [y, x]$ is a homomorphism, hence $(xy)^k = \left[y^{\frac{k(k-1)}{2}}, x \right] x^k y^k$. Since k is odd, the claim follows. $\square$

5.2.29 Corollary. *Let G be a non-commutative connected algebraic group, defined over a field of characteristic $p > 2$ such that $G' \leq {}_3G$. The factor group $G/{}_3G$ is a vector group if and only if the map $x \mapsto x^p$ is a homomorphism.*

Proof. If $G/{}_3G$ is a vector group, then $g^p \in {}_3G$ for any $g \in G$ and the map $x \mapsto x^p : G \to G$ is a homomorphism by the previous proposition.

Conversely, if the map $x \mapsto x^p$ is a homomorphism, then $(xy)^p = \left[y^{\frac{p(p-1)}{2}}, x\right] x^p y^p = x^p y^p$ for any $x, y \in G$, which means that $g^p \in {}_3G$ for any $g \in G$ and $G/{}_3G$ is a vector group. $\qquad\square$

5.2.30 Theorem. *Let G be a connected affine algebraic group of dimension greater than two over a field of characteristic $p > 2$ such that G is not a chain. If every maximal connected subgroup of G is a chain, then G is a three-dimensional unipotent group containing a unique one-dimensional connected algebraic subgroup $K = G'$ and a two-dimensional non-commutative subgroup. Moreover, either G has exponent p or G contains a unique two-dimensional non-commutative subgroup of exponent p.*

Proof. Let G be a counter-example of minimal dimension. By Theorem 5.2.27 the group G is uni-minimal and contains a non-commutative connected algebraic subgroup. Let K be the unique connected one-dimensional algebraic subgroup of G. The factor group G/K cannot be a chain, because otherwise by Theorem 5.1.1 the group G itself would be a chain. By minimality, G/K is either a two-dimensional group or a three-dimensional group. In the latter case, if any maximal subgroup of G/K were commutative, then by Theorem 5.2.21 G/K would be uni-maximal and by Theorem 5.1.1 G would be a chain. Thus G/K contains a two-dimensional non-commutative subgroup M/K. But by Corollary 4.2.10 this case cannot occur, because M would be a three-dimensional chain with a two-dimensional commutator subgroup M'.

Thus $\dim G = 3$. As any two-dimensional connected subgroup of G is a chain, the group G is uni-minimal. Among these chains there is non-commutative one since otherwise G would be uni-maximal and hence a chain. Moreover the factor group G/K is a two-dimensional vector group.

By Corollary 5.2.29 the map $x \mapsto x^p$ is a homomorphism, the kernel H of which has dimension at least two. If $H = G$, then G has exponent p. If H has dimension two, it is a chain by assumption. Since any commutative two-dimensional chain has exponent p^2, the kernel H is non-commutative. Thus H is the unique two-dimensional connected algebraic subgroup of exponent p. $\qquad\square$

As soon as the dimension of the unipotent group G is at least three, having a unique minimal or a maximal connected subgroup is not a sufficient condition for G to be a chain, as the following examples show:

– Since the centre of the direct product $G = \mathfrak{W}_2 \curlyvee \mathfrak{J}_2$ with amalgamated central subgroup is the subgroup $\mathfrak{W}_2$, the group G has a unique one-dimensional subgroup.

– For any non-commutative two-dimensional unipotent group $\mathfrak{C}_2$, the commutator subgroup G' of the direct product $G = \mathfrak{W}_2 \curlywedge \mathfrak{C}_2$ with amalgamated factor group is a one-dimensional subgroup such that G/G' is isomorphic to $\mathfrak{W}_2$.

5.2.31 Proposition. *Let G be a connected affine algebraic group of dimension greater than one defined over a field of characteristic $p > 0$. Then the following holds:*

i) The group G is two-dimensional or uni-minimal if and only if every proper connected algebraic subgroup of G is uni-minimal.

ii) If G is uni-minimal, then either G is a chain or G contains a non-commutative algebraic subgroup. Moreover, if $p > 2$, if the group G has dimension at least three and is not a chain, then G contains a non-commutative two-dimensional algebraic subgroup J of exponent p.

Proof. If G is two-dimensional or a uni-minimal connected algebraic group, then every proper connected algebraic subgroup is uni-minimal as well. Conversely, assume that every proper connected algebraic subgroup of a connected algebraic group G is uni-minimal, but G is not. The group G must be solvable, since otherwise a Borel subgroup would be not uni-minimal. If a maximal torus T of G has dimension greater than one, then $G = T$ and G has dimension two. If a maximal torus T of G has dimension one, then the unipotent radical of G has dimension one, as well, and G has dimension two. Let now G be unipotent of dimension greater than two. Let H_1, H_2 be two different minimal connected algebraic subgroups of G. Let M_1 and M_2 be two maximal connected algebraic subgroups of G containing H_1 and H_2, respectively. As G has dimension greater than two and is not a chain, the group $M_1 \cap M_2$ contains a proper connected algebraic subgroup, hence $H_1 = H_2$, a contradiction.

Let G be a minimal counter-example to the assertion *ii)*. The group G is unipotent by Proposition 5.2.6 and it has dimension greater than two. By Theorems 5.2.27 and 5.2.30 there exists a maximal connected subgroup M of G such that M is not a chain. By minimality of G, the subgroup M contains a non-commutative algebraic subgroup H, which is a contradiction.

Now we prove the last claim of *ii)* by induction on the dimension of G. If $\dim G = 3$, then the assertion follows by Theorem 5.2.30. Let $\dim G > 3$. According to Theorem 5.2.30 not every maximal connected subgroup of G is a chain. Let M be a maximal connected subgroup of G which is not a chain. By induction, M contains a non-commutative two-dimensional subgroup of exponent p and we are done. $\qquad\qquad\square$

By Corollary 4.2.10 the unipotent non-commutative two-dimensional chains, defined over a field of characteristic $p > 2$, having exponent p cannot be properly contained in any other chain.

5.2.32 Theorem. *Let G be an affine algebraic group over a field of characteristic $p > 2$ such that G is not a chain. Let J be a connected two-dimensional subgroup of G of exponent p. If a maximal subgroup H_{n-1} of G is a chain, then either J is the unique connected two-dimensional subgroup of G of exponent p, or $\dim G = 3$ and G has exponent p.*

Proof. Since the nilpotent connected algebraic subgroup H_{n-1} is contained in a Borel subgroup of G, we can conclude that G is solvable. If the chain

H_{n-1} were a one-dimensional torus, then G had dimension two and would not contain the subgroup J. Hence H_{n-1} is a unipotent chain.

Let G be not unipotent, then H_{n-1} is the unipotent radical of G and it contains the subgroup J. Then $H_{n-1} = J$ by Corollary 4.2.10 and G has dimension three. Thus J is the unique connected two-dimensional subgroup of G of exponent p.

Let now G be unipotent. If dim $G = 3$ and J is not the unique connected two-dimensional subgroup of G of exponent p, then $G = JL$, where L is another connected two-dimensional subgroup of G of exponent p. Let K be the connected component of $J \cap L$. Since G/K is commutative, the commutator subgroup G' is one-dimensional, hence central in G. By Corollary 5.2.29, the map $x \mapsto x^p$ is a homomorphism, the kernel of which contains both J and L. Thus G has exponent p.

Assume now that G is a counter-example of dimension $n \geq 4$. Let J, L be two different connected algebraic subgroup of dimension two and exponent p. Let M be a maximal algebraic subgroup of G containing L. Since M has dimension $d \geq 3$, by Corollary 4.2.10 the subgroup M is not a chain. Let H_{n-2} be the maximal connected algebraic subgroup of the chain H_{n-1}. Clearly H_{n-2} is a maximal connected algebraic subgroup of M. Again by Corollary 4.2.10 the centre of H_{n-1} has dimension at least two, hence the subgroup H_{n-2} has exponent at least p^2. Therefore M does not have exponent p, and by induction L is the unique connected two-dimensional algebraic subgroup of exponent p contained in M. Thus L is characteristic in M, and it follows that L is normal in G.

Consider now the product $N = JL$. If $n = 4$, then $N \neq G$, because G contains the three-dimensional chain H_{n-1}. If $n > 4$, then $N \neq G$ for dimensional reasons. Thus N is contained in a maximal connected algebraic subgroup of G, but we have shown above that a maximal connected subgroup of G containing L has a unique two-dimensional subgroup of exponent p, a contradiction. □

The following proposition shows the normality of the subgroup J in Proposition 5.2.31.

5.2.33 Proposition. *Let G be an affine algebraic group of dimension greater than one defined over a field of characteristic $p > 2$ and having a unique minimal algebraic subgroup. Then either G is a chain or G contains a (noncommutative) two-dimensional normal algebraic subgroup J of exponent p.*

Proof. By Proposition 5.2.6 the group G is unipotent. Let $1 < H_1 < H_2 < \cdots < H_n = G$ be a series of normal connected algebraic subgroups of G, each of co-dimension one in the next (see [8], 10.6 (2), p. 138). Since G is uni-minimal, the subgroup H_2 is a two-dimensional chain. If H_2 has exponent p, put $J = H_2$. Otherwise let k be the smallest index such that H_k is not a chain. By Theorems 5.2.31 *ii)* and 5.2.32, the group H_k contains a unique connected algebraic subgroup J of dimension two and exponent p. Since J is characteristic in H_k, it is normal in G. □

The following theorem is the main result on the structure of uni-minimal affine algebraic groups.

5.2.34 Theorem. *If G is a non-commutative uni-minimal affine algebraic group of dimension greater than one defined over a field of characteristic $p > 2$ and having a unique minimal algebraic subgroup, then G is unipotent and can be written as a product of chains*

$$G = C J_1 \cdots J_r \quad (r \geq 0)$$

with $J_k \not< C J_1 \cdots J_{k-1} J_{k+1} \cdots J_r$ for all $k = 1, \ldots, r$, such that C is normal in G and for all k the chains J_k are two-dimensional, non-commutative, have exponent p and are normal in G.

If C is commutative, then the commutator subgroup G' of G is the unique one-dimensional connected algebraic subgroup of G, otherwise $G' = C'$.

Proof. According to Proposition 5.2.6 the group G is unipotent. Assume that G is a counter-example of minimal dimension, so G is not a chain. Let K be the unique one-dimensional connected algebraic subgroup of G and, according to Proposition 5.2.33, let J be a non-commutative two-dimensional normal connected algebraic subgroup of G having exponent p. Take $g \in J, g \notin {}_3J$. Then the homomorphism $\sigma_g : G \longrightarrow K$ given by $x \mapsto [g, x]$ has a maximal connected subgroup M, not containing J, in its kernel, forcing $G = MJ$. If M is commutative, then M is a chain because M is uni-minimal, and we put $M = C$. If M is not commutative, by minimality we have $M = C J_1 \cdots J_s$, where the chains J_k are two-dimensional, non-commutative, have exponent p for all $k = 1, \cdots, s$ and are normal in M. Thus $G = C J_1 \cdots J_s J$. Since J is normal in G, we find now $[C, J] \leq K$ and $[J_k, J] \leq K$, and this means that C and J_k are normal in G, because K is contained in all of them.

Moreover $G' = [M, J] \cdot M' \cdot J'$, hence $G' = M'K$. If M is a chain, then $G = MJ$ and $G' = K$ if M is commutative, $G' = M'$ otherwise. If M is not a chain, by minimality we have now $M' = C'K$. Thus $G' = C'K$. If C is commutative, then $G' = K$, otherwise $G' = C'$ is not a counter-example. $\square$

Using this theorem we will prove in Theorem 7.2.19 that any connected algebraic subgroup of a uni-minimal algebraic affine group is normal.

Consequences of Theorem 5.2.34 are the following

5.2.35 Corollary. *Let G be an affine algebraic group over a field of characteristic greater than two. If G is uni-minimal and uni-maximal, then G is a chain.* $\square$

5.2.36 Corollary. *Let G be an affine algebraic group of dimension n over a field of characteristic greater than two and let $1 < d < n$ be a given integer. If G has a unique connected algebraic subgroup H of dimension d which is a chain, then G is a chain as well.*

Proof. Since the chain H is the unique subgroup of G of dimension d, the group G is uni-minimal. But then G has a decomposition in a product as given in Theorem 5.2.34 with $r = 0$.   □

5.2.37 Corollary. *Let G be an algebraic affine group of dimension at least three over a field of characteristic greater than two having a unique minimal algebraic subgroup. For any two-dimensional connected algebraic subgroup H of G there exists an epimorphic image with connected kernel of G isogenous to H if and only if G is a chain.*

Proof. By Proposition 5.2.6 the group G is unipotent. We assume that G is not a chain, hence G is not commutative. Then we have $G = CJ_1 \cdots J_r$ where $r > 0$, C is a chain and the group J_r is a two-dimensional non-commutative connected algebraic subgroup. Then any proper factor group of G is a product of a chain by a vector group, and such a product is isogenous to J_r if and only if the vector group is trivial. By Corollary 4.2.10 the chain J_r cannot be isogenous to a homomorphic image of G, which is a contradiction.   □

Theorem 5.2.34 clarifies the structure of affine uni-minimal connected algebraic groups. The class of these groups is much more restricted than the class of connected algebraic groups where the commutator subgroup is the unique one-dimensional normal connected algebraic subgroup. For this class of groups we have the following

5.2.38 Proposition. *Let G be a non-simple non-commutative connected affine group. Any proper epimorphic image with connected kernel is commutative if and only if the commutator subgroup is the unique one-dimensional normal connected algebraic subgroup of G.*

Proof. First we assume that any proper epimorphic image with connected kernel is commutative. Let R_G be the solvable radical of G. Since the factor group G/R_G is semi-simple (cf. [8], p. 157), we have $G = R_G$. Let K be a one-dimensional normal connected unipotent algebraic subgroup of G (cf. [8], Theorem 10.6 (2), p. 137). Since G/K is commutative we have $K = G'$.

Conversely, we have to prove that any normal connected algebraic subgroup of G contains G'. Let N be a normal connected algebraic subgroup of G which does not contain G'. Then $G' \cap N$ is finite and the connected algebraic subgroup $[G, N]$ is contained in $G' \cap N$. It follows that N is a central normal connected algebraic subgroup of G. Hence N contains a one-dimensional normal connected algebraic subgroup of G which is different from G'. This is a contradiction.   □

5.2.39 Remark. If G is a group of maximal nilpotency class over a field of positive characteristic (see e.g. Example 3.1.4), then obviously $\dim \mathfrak{z}G = 1$ and $\dim G' = \dim G - 1$. Thus G is uni-maximal and has a unique one-dimensional normal subgroup. From this example one can see that uni-maximal groups having a unique one-dimensional *normal* subgroup do not necessarily have a unique one-dimensional subgroup.   □

Now we look at connected affine algebraic groups having a maximal connected subgroup which is a chain.

5.2.40 Theorem. *Let G be a connected affine algebraic group of dimension $n > 2$ defined over a field of characteristic $p > 2$. If G contains a maximal connected subgroup M which is a chain, then either G is a chain or M is normal unipotent and there exists a chain C of minimal dimension $d \leq 2$ such that $G = MC$. In this last case the group G is uni-minimal if and only if $d = 2$, whereas if $d = 1$ we have:*

i) *if C is a torus and $[M, C] \neq 1$, then $G' = M$ and $\mathfrak{z}°G = 1$. Moreover either M is commutative or M has dimension two and exponent p;*
ii) *if C is unipotent, $[M, C] \neq 1$ and $M' \neq 1$, then $G' = M'$;*
iii) *if C is unipotent, $[M, C] \neq 1$ and $M' = 1$, then $\mathfrak{z}°G$ is the maximal connected subgroup of M and G' is the connected one-dimensional subgroup K of M.*

Proof. If the chain M was a one-dimensional torus, then G would be two-dimensional. Hence M is a unipotent affine chain, G is solvable and M is normal. First we show that either $G = MC$ or G is a chain. If G is a counter-example of minimal dimension $n > 2$, then G is not a semi-direct product of M with a one-dimensional torus. Thus G is unipotent. Moreover, if G were uni-maximal, then G would be a chain. Therefore G is not uni-maximal.

If every maximal connected algebraic subgroup of G is a chain, then by Theorem 5.2.30 the group G has dimension three. Then $G = ML$ for a suitable maximal connected subgroup $L \neq M$ of G. Since L has dimension two, G is not a counter-example.

Thus we can assume that G has a maximal connected subgroup H which is not a chain. Put $L = (H \cap M)°$. Since L is a chain of co-dimension one in H, by minimality of G either the subgroup H has dimension two or H is a product of L by a chain C of dimension $d \leq 2$. In the former case, H is a two-dimensional vector group, so G has dimension three and again it is not a counter-example. In the latter case, since $C \not\leq M$, we have $G = MC$; hence G is not a counter-example, a contradiction.

Clearly G is uni-minimal if and only if we cannot find a one-dimensional subgroup C such that $G = MC$. Let now $d = 1$. If C is a torus, the assertion follows from Theorem 4.3.10. Now let C be unipotent, let $[M, C] \neq 1$ and denote by K the unique one-dimensional subgroup of M. As a two-dimensional vector group, the product $S = KC$ has exponent p and is contained in the fiber of 1 of the morphism $\sigma : G \longrightarrow G$, $x \mapsto x^p$, forcing G^p to have dimension at most $n - 2$. By Corollary 4.3.3, the only case where M^p is not the maximal connected algebraic subgroup of M is when M is a two-dimensional non-commutative chain of exponent p.

In this case, G is three-dimensional and G' cannot be two-dimensional, otherwise $G' = M$ would be the unique maximal connected algebraic subgroup of G and G would be a chain. Moreover if $\mathfrak{z}G$ has dimension two, then $C < \mathfrak{z}G$, because otherwise $G = C \cdot {}_3G$ would be commutative. But C is not central because $[M, C] \neq 1$. Hence ${}_3°G$ has dimension one and coincides with ${}_3°M$.

Thus we may assume that $\dim M^p = n - 2$. Hence $\dim G^p = n - 2$ and $(\sigma^{-1}(1))° = S = KC$. Thus S is a characteristic subgroup of G and $[M, C] \leq M \cap S \leq K$. From $G = MC$ it follows that $G' = M'$ if $M' \neq 1$, whereas $G' = K$ if $M' = 1$.

Since $[M, C] \leq K$, the map $\sigma_c : M \longrightarrow K$, $m \mapsto [m, c]$, is for any $c \in C$ an algebraic homomorphism, the kernel of which has co-dimension at most one in M. It follows that the maximal connected algebraic subgroup N of M centralises C. If M is not commutative, then the claim follows from the fact that $[M, C] = K \leq M'$. Finally, if M is commutative, then ${}_3°G$ is the maximal connected algebraic subgroup of M. Namely, on the one hand ${}_3°G$ has dimension at least $n - 2$, because ${}_3°G \geq N$, on the other hand ${}_3°G$ cannot be a maximal connected algebraic subgroup of G, otherwise again $G = C \cdot {}_3°G$ would be commutative. $\qquad \square$

5.2.41 Remark. About case *ii)* in the above Theorem, one may wonder whether ${}_3°G$ is always contained in ${}_3°M$ (in such a case one has ${}_3°G = {}_3°M$, because the maximal subgroup of M centralises C). This is not true, and we want to give an example of a unipotent group G, which is the semi-direct product of a non-commutative chain M of co-dimension one and a one-dimensional unipotent subgroup C such that $[M, C] \neq 1$ and ${}_3G \not\leq M$. In such a case, Proposition 3.2.12 applies and we have $\dim {}_3G = \dim {}_3M + 1$.

In order to give such an example, let M be the n-dimensional chain $\mathfrak{K}_n$ defined in Remark 4.1.6 and let C be the one-dimensional additive group.

After computing the commutator $[\mathbf{y}, \mathbf{x}]$ for elements $\mathbf{x} = (x_0, x_1, x_2, \cdots x_{n-1}) \in M$ and $\mathbf{y} = (0, y_1, y_2, \cdots, y_{n-1})$ in the maximal connected algebraic subgroup H of M

$$[\mathbf{y}, \mathbf{x}] = (0, 0, \cdots, 0, x_0^{p^2} y_1 - x_0^p y_1^p),$$

we define a homomorphism $\sigma : C \longrightarrow \mathsf{Aut}(M)$, $t \mapsto \sigma_t$, putting

$$\sigma_t(x_0, x_1, \cdots, x_{n-1}) = (x_0, x_1, \cdots, x_{n-1}) \cdot [(0, t, 0, \cdots, 0), (x_0, x_1, \cdots, x_{n-1})] =$$

$$(x_0, x_1, \cdots, x_{n-1} + x_0^{p^2} t - x_0^p t^p).$$

The semi-direct product $G = M \rtimes C$ defined by σ gives now an example where ${}_3G = \{(x_0, x_1, \cdots, x_{n-1}) \cdot t \in M \rtimes C : x_0 = 0, x_1 = -t\}$ is not contained in M. In fact, let $\mathbf{x}_0 \cdot t_0 \in {}_3G$, with $\mathbf{x}_0 = (a_0, a_1, \cdots, a_{n-1})$ and let $\mathbf{x} \cdot t \in G$. Hence

$$\mathbf{x}_0 \, \sigma_{t_0}(\mathbf{x}) = \mathbf{x} \, \sigma_t(\mathbf{x}_0). \tag{5.1}$$

For $\mathbf{x} = 1$ we find $\mathbf{x}_0 = \sigma_t(\mathbf{x}_0)$, that is $a_0 = 0$. Moreover, by 5.1 we have $\mathbf{x}_0 \, \sigma_{t_0}(\mathbf{x}) = \mathbf{x} \, \mathbf{x}_0$, that is $\sigma_{t_0}(\mathbf{x}) = \mathbf{x} \cdot [\mathbf{x}, \mathbf{x}_0]$. As by definition we have $\sigma_{t_0}(\mathbf{x}) = \mathbf{x} \cdot [(0, t_0, 0, \cdots, 0), \mathbf{x}]$, it follows that $a_1 = -t_0$. $\qquad \square$

At the end of this section we want to mention that finite groups having a unique maximal subgroup as well as finite groups having a unique minimal subgroup are well-known. In contrast to the richness of examples of uni-maximal algebraic groups (see Remark 5.2.16), a finite group G having only one maximal subgroup is necessarily cyclic and has prime-power order (see [46], Satz 3.16, p. 274).

Finite groups having a unique minimal subgroup have a similar structure to the one of uni-minimal algebraic groups given in Theorem 5.2.34. Clearly such a group G is a p-group and any commutative subgroup of G is cyclic. By Satz 7.6, p. 304 in [46], the finite group G is either cyclic or a non-commutative 2-group. Collecting information from Satz 13.10, p. 357, Satz 13.8, p. 355, Satz 14.9, p. 90 and p. 93 in [46], we obtain the following

5.2.42 Remark. A finite non-commutative 2-group G has a unique minimal subgroup if and only if G is either isomorphic to a generalised quaternion group or to a direct product with amalgamated center of the quaternion group with a cyclic group or with a generalized quaternion group.   □

5.3 Semi-Commutative Groups and Lie Algebras

In Section 5.2 we have given characterisations of uni-minimal and uni-maximal groups. In these characterisations the condition that all proper connected algebraic subgroups or that all proper epimorphic images are commutative plays an important rôle. But Corollary 5.2.23 and Proposition 5.2.38 illustrate that for algebraic groups over fields of positive characteristic, no two of these conditions are sufficient for a concrete classification. In contrast to this we shall show in this section that for real or complex Lie groups, for formal groups and for algebraic groups over fields of characteristic zero, the assumption of commutativity of all proper connected subgroups as well as the dual condition of commutativity of all proper epimorphic images are sufficient for a classification.

5.3.1 Definition. A Lie algebra $\mathfrak{g}$ is called *semi-commutative*, if it is not commutative, but all of its proper subalgebras are commutative.

The class of semi-commutative Lie algebras has found attention in the past (see [21, 25, 30–32, 92]); the standard name for these Lie algebras is semi-abelian, a name which we do not use, in order to avoid confusion with abelian varieties. The best known examples of simple semi-commutative Lie algebras over a field k are the non-split simple Lie algebras of dimension three over k. These algebras exist if and only if a quaternion division algebra over k exists (see [25], Theorem 3.1, p. 512).

5.3.2 Remark. In [31] A. Gejn has given a method to construct simple semi-commutative Lie algebras of arbitrarily large dimension. More precisely, if the

Brauer group of a perfect field F is not a 2-group, then there is a simple semi-commutative Lie algebra over F which over its centroid has dimension greater than three ([31], Theorem 4.3, p. 323). If the field F has characteristic zero and if p is a prime divisor of the order of some element of the Brauer group of F, then a simple semi-commutative Lie algebra over F exists which has dimension $p^2 - 1$.

In [30] examples of simple semi-commutative Lie algebras over fields F of positive characteristic are given which have dimension > 3.

The Brauer group of a local field K is isomorphic to the group $\mathbb{Q}/\mathbb{Z}$ (see [65], Corollary 7.1.6, p. 322). Hence for such fields of characteristic zero and for every prime number q the existence of simple semi-commutative Lie algebras of dimension $q^2 - 1$ follows. A detailed description of this situation is given in [63].

If K is a global field, the exact sequence in [65], Corollary 8.1.17, p. 376, shows that the Brauer group $Br(\mathsf{K})$ contains a subgroup isomorphic to $\mathbb{Q}/\mathbb{Z}$, since $Br(\mathsf{K}_\mathfrak{p})$ has this property for every prime $\mathfrak{p}$ of K. Thus for every algebraic number field K and every prime number q there is a simple semi-commutative Lie algebra of dimension $q^2 - 1$. $\qquad\square$

The structure of non-simple semi-commutative Lie algebras is described implicitly in the papers [92] and [32]. We give here a short proof for Lie algebras over fields of characteristic zero.

5.3.3 Proposition. *A non-simple Lie algebra $\mathfrak{g}$ of finite dimension over a field k of characteristic zero is semi-commutative if and only if for $\mathfrak{g}$ one of the following holds:*

(1) $\mathfrak{g}$ is a semi-direct product of a commutative subalgebra $\mathfrak{a}$ with a one-dimensional subalgebra $\langle x \rangle$ such that $\operatorname{ad} x$ acts irreducibly on $\mathfrak{a}$,

(2) $\mathfrak{g}$ is the three-dimensional nilpotent non-commutative Lie algebra of dimension three.

Proof. Assume first that $\mathfrak{g}$ is semi-commutative. If $\mathfrak{g}$ were not solvable, then according to Levi's theorem (cf. [49], Chapter III, p. 91) and by cf. [49], Structure theorem, Chapter III, p. 71, the algebra $\mathfrak{g}$ would contain a simple subalgebra, contradiction.

Now we consider a solvable semi-commutative Lie algebra $\mathfrak{g}$. Then there is a commutative ideal $\mathfrak{a}$ of co-dimension 1 in $\mathfrak{g}$. For $x \in \mathfrak{g}\backslash\mathfrak{a}$ and $0 \neq a \in \mathfrak{a}$ consider the elements $a(\operatorname{ad} x)^n$. These elements generate an $(\operatorname{ad} x)$-invariant subspace $\mathfrak{b}$. If $\mathfrak{g}$ is not nilpotent, then $\mathfrak{b}$ must coincide with $\mathfrak{a}$. In the case that $\mathfrak{g}$ is nilpotent, it has nilpotency class two. Thus any two independent elements of $\mathfrak{g}$ which do not lie in the centre of $\mathfrak{g}$ generate a three-dimensional non-commutative subalgebra $\mathfrak{c}$ of $\mathfrak{g}$. But then necessarily $\mathfrak{g} = \mathfrak{c}$ (cf. [49], p. 13).

Conversely it is obvious that the Lie algebras under *(1)* and *(2)* have the property that any proper subalgebra is commutative. $\qquad\square$

The proposition 5.3.3 can be used to obtain results about groups. For an arbitrary field k we denote by $L_2(\mathsf{k})$ the linear group

$$\left\{ \begin{pmatrix} 1 & a \\ 0 & t \end{pmatrix} : t, a \in \mathsf{k}, t \neq 0 \right\}$$

and by $H_3(\mathsf{k})$ the linear group

$$\left\{ \begin{pmatrix} 1 & x & z \\ 0 & 1 & y \\ 0 & 0 & 1 \end{pmatrix} : x, y, z \in \mathsf{k} \right\}.$$

Over a field k of characteristic zero, there are only two three-dimensional affine algebraic groups having a simple Lie algebra, namely the groups $SL_2(\mathsf{k})$ and $PSL_2(\mathsf{k})$. These groups contain two-dimensional non-commutative closed subgroups. Now, the correspondence between groups and Lie algebras (see [45], Chapter V, p. 87ff.) yields the following

5.3.4 Proposition. *Let K be an algebraically closed field containing a field k of characteristic zero. A connected non-commutative affine algebraic group $G(K)$ defined over k in which every proper closed connected subgroup is commutative is isomorphic either to the group $L_2(K)$ or to the three-dimensional group $H_3(K)$ of nilpotency class two.* □

Let $\mathfrak{l}_2$ be the two-dimensional non-commutative Lie algebra over a field k of characteristic zero. There is a basis $\{X, Y\}$ of $\mathfrak{l}_2$ such that for $\mathbf{x} = \alpha X + \beta Y, \mathbf{y} = \gamma X + \delta Y \in \mathfrak{l}_2$ one has $[\mathbf{x}, \mathbf{y}] = [\alpha X + \beta Y, \gamma X + \delta Y] = (\alpha\delta - \beta\gamma)Y$. The formal Lie group $F(X, Y) = FL_2(\mathsf{k})$ is given by the group law

$$\mathbf{x} \cdot \mathbf{y} = \mathbf{x} + \mathbf{y} + \frac{1}{2}[\mathbf{x}, \mathbf{y}] + \frac{1}{12}\big[\mathbf{x}, [\mathbf{x}, \mathbf{y}]\big] + \frac{1}{12}\big[\mathbf{y}, [\mathbf{y}, \mathbf{x}]\big] + \cdots .$$

The unique three-dimensional Lie algebra of nilpotency class two over k gives rise to a formal group $FH_3(\mathsf{k})$ with the multiplication

$$(x_0, x_1, x_2) \cdot (y_0, y_1, y_2) = (x_0 + y_0, x_1 + y_1, x_2 + y_2 + \frac{1}{2}(x_0 y_1 - x_1 y_2)).$$

Over a field k of characteristic zero one calls a formal group G *simple*, if its Lie algebra is simple. With this notation, Proposition 5.3.3 gives rise to

5.3.5 Proposition. *A non-commutative non-simple formal group G over a field k of characteristic zero in which every proper formal subgroup is commutative is isomorphic either to $FL_2(\mathsf{k})$ or to $FH_3(\mathsf{k})$.* □

Let G be a Lie group over a field $K \in \{\mathbb{R}, \mathbb{C}\}$. We denote by Z the connected component of the centre of G and by $\overline{G'}$ the closure of the commutator subgroup of the abstract group G. If $\dim_K(G/Z) = 2$ and if $Z = \overline{G'}$ we call the group G *extra-special*.

5.3.6 Remark. a) All examples of extra-special real or complex Lie groups which are not simply connected can be obtained in the following way: Consider the (unique) three-dimensional connected Lie group H_3 of nilpotency class two over the field $\mathsf{K} \in \{\mathbb{R}, \mathbb{C}\}$. As a manifold, H_3 coincides with the three-dimensional affine space, and the multiplication is given by

$$(x_0, x_1, x_2)(y_0, y_1, y_2) = (x_0 + y_0, x_1 + y_1, x_2 + y_2 + f(x_0, y_0, x_1, y_1)),$$

where the map $f : \mathsf{K}^4 \longrightarrow \mathsf{K}$ is defined by $f(x_0, x_1, y_0, y_1) = x_0 y_1$.

b) Consider an analytic K-homomorphism $\phi : \mathsf{K} \longrightarrow C$, where C is some K-Lie group and the analytic closure of $\phi(\mathsf{K})$ coincides with C.
For $\mathsf{K} = \mathbb{R}$, one has the possibilities $C \cong \mathsf{K}$ or $C \cong (SO_2(\mathbb{R}))^n$ for some $n \geq 1$.

For $\mathsf{K} = \mathbb{C}$ the group C is one of the following: $\mathbb{C}^+, (\mathbb{C}^*)^n$, a complex torus or non-compact toroidal group which is the complex analytic closure of a complex analytic one-parameter group (see Remark 5.2.3). In the last case, the group C is an extension of $(\mathbb{C}^*)^m$ by a complex torus.

c) We define a multiplication on the set $\mathsf{K} \times \mathsf{K} \times C$ by

$$(x_0, x_1, \mathbf{x_2})(y_0, y_1, \mathbf{y_2}) = (x_0 + y_0, x_1 + y_1, \mathbf{x_2} + \mathbf{y_2} + \phi \circ f(x_0, y_0, x_1, y_1)),$$

for $x_0, x_1, y_0, y_1 \in \mathsf{K}$ and $\mathbf{x_2}, \mathbf{y_2} \in C$.

d) A connected real or complex Lie group is extra-special if and only if it is topologically generated by two non-commuting one-parameter subgroups.

e) If G is a extra-special Lie group with $\dim_\mathsf{K} G = r > 3$, then its Lie algebra is a direct product of the three-dimensional nilpotent non-commutative Lie algebra $\mathfrak{h}_3$ and a commutative subalgebra $\mathfrak{a}$ with $\dim_\mathsf{K} \mathfrak{a} = r - 3$.   □

5.3.7 Theorem. *Let G be a connected real or complex Lie group which does not have a commutative direct factor $\neq 1$. Then every proper closed connected subgroup of G is commutative if and only if G belongs to one the following classes:*

(1) One of the real Lie groups $SO_3(\mathbb{R})$ or $\mathsf{Spin}_3(\mathbb{R})$,
(2) The connected component of the real Lie group $L_2(\mathbb{R})$,
(3) A covering of the connected component of the group $E_2(\mathbb{R})$ of motions of the real Euclidean plan,
(4) The semi-direct product of the group $\mathbb{R}^2$ with a one-parameter group

$$\Xi_d = \left\{ \begin{pmatrix} d^t \cos t & d^t \sin t \\ -d^t \sin t & d^t \cos t \end{pmatrix} : t \in \mathbb{R} \right\},$$

for $d > 1$,
(5) A covering of the complex Lie group $L_2(\mathbb{C})$,
(6) An extra-special real or complex Lie group.

Proof. Let $\mathsf{K} \in \{\mathbb{R}, \mathbb{C}\}$. If G is a semi-simple Lie group over K, then G is even simple. From the knowledge of the simple real and complex Lie groups (cf. [94]) one sees that G is a real Lie group which is isomorphic either to $SO_3(\mathbb{R})$ or $\mathsf{Spin}_3(\mathbb{R})$. If G is not solvable, we denote by S the non-trivial solvable radical of G. It follows that G/S is a simple compact real Lie group of dimension three. The subgroup S of G is commutative. If S contains a maximal torus $T \neq 1$, then T is an almost direct factor of G, a contradiction. It follows from [44], Corollary VII.1, p. 25, that S has a simple compact complement in G. Thus $S = 1$ and we are in case (1). For the rest of the proof we may assume that G is solvable.

First we assume that G has a trivial connected component of the centre. Since every real torus respectively every complex torus and every subgroup isomorphic to $(\mathbb{C}^*)^n$ of the topological commutator subgroup $\overline{G'}$ of G is central in G, the group $\overline{G'}$ is a vector group. It follows from Lie's theorem (and its real equivalent) (see [27], p. 69) that $\overline{G'}$ contains a normal subgroup M of G such that $\dim M = 1$ for the complex case and $\dim M \in \{1, 2\}$ for the real case. There is a one-dimensional connected closed subgroup E of G acting non-trivially on M. If $\mathsf{K} = \mathbb{C}$, then E is isomorphic to the multiplicative group of the field $\mathbb{C}$. Because of $G = ME$ we are in case (5). In the real case, the group E is isomorphic to the additive group of the field $\mathbb{R}$ or it is a one-dimensional torus. This just covers the cases (2), (3) and (4).

Now we consider the case that G has a centre Z of positive dimension. For the complex case we first assume that the maximal toroidal subgroup X of G is trivial. Let G be a counter-example of minimal dimension to the assertion that G is nilpotent. Let E be a closed one-parameter subgroup lying in the centre of G. Since G/E is not nilpotent, too, it follows that G/E is one of the groups described under (2)–(5). In case (2) as well as in case (5), the group G/E has dimension two over K. Thus G is a three-dimensional Lie group over K which has a one-dimensional centre. It follows from [49], p. 12, that G is the direct product of its centre and a two-dimensional non-commutative Lie group. This contradiction shows that $\dim_{\mathsf{K}}(G/E) = 3$ and that we are in case (3) or (4). In these cases, the pre-image S in G of the commutator subgroup of G/E is a three-dimensional commutative group. The group G is a semi-direct product of S with a one-dimensional subgroup D which acts semi-simply on S. Hence E has a D-invariant complement C in S and CD is a proper closed connected subgroup of G which is not commutative. This contradiction shows that G is nilpotent.

If $\mathsf{K} = \mathbb{C}$ and the non-nilpotent group G contains a non-trivial toroidal subgroup X, then X is the connected component of the center of G and G/X is a covering of the complex Lie group $L_2(\mathbb{C})$. The group G/X is a semi-direct product of the normal subgroup $V/X \cong \mathbb{C}^+$ with a subgroup $C/X \cong \mathbb{C}^*$ and C, V as proper subgroups of G are commutative. By [1], Theorem 1.1.5, p. 6, there are uniquely determined subgroups $V_1 \leq V$, $C_1 \leq C$ such that $V = X \times V_1$, $C = X \times C_1$. Hence the group $V_1 C_1$ is a proper non-commutative subgroup of G, a contradiction.

We now claim that G is nilpotent of class two. A counter-example of minimal dimension has nilpotency class at least 3. First we assume that in the Lie algebra $\mathfrak{g}$ of G there is a one-dimensional central subalgebra $\mathfrak{e}$ which corresponds to a closed one-parameter subgroup E of G. Since G is a minimal counter-example, according to remark 5.3.6 the factor algebra $\mathfrak{g}/\mathfrak{e}$ is the direct product of a three-dimensional nilpotent non-commutative Lie algebra $\mathfrak{h}/\mathfrak{e}$ and a commutative Lie algebra. Since $\mathfrak{g}$ has nilpotency class three, the same holds for the algebra $\mathfrak{h}$. Then $\mathfrak{h}$ has a basis $\{e_1, e_2, e_3, e_4\}$ with the only non-trivial relations $[e_1, e_2] = e_3$ and $[e_1, e_3] = e_4$ (see [61], p. 162). But then the subalgebra $\mathfrak{h}_0$ of $\mathfrak{h}$ generated by e_1, e_3 is not commutative and has dimension three. Thus the closure H of the exponential image of $\mathfrak{h}_0$ is a proper non-commutative subgroup of G.

If $\mathsf{K} = \mathbb{C}$ and G is a minimal counter-example of nilpotency class three which does not contain a closed normal subgroup of dimension one, we choose a minimal closed connected subgroup T of the center of G. Then T is a complex torus (see [1], Proposition 1.1.17(2), p. 15). Then G/T is extra-special by induction. As above we consider the Lie algebra $\mathfrak{g}$ of G and its subalgebra $\mathfrak{t}$ corresponding to T. Then again $\mathfrak{g}/\mathfrak{t}$ is the direct product of a three-dimensional nilpotent non-commutative Lie algebra $\mathfrak{h}/\mathfrak{t}$ and a commutative Lie algebra. As above in $\mathfrak{h}$ we find a 4-dimensional subalgebra $\mathfrak{h}_1$ with a basis $\{e_1, e_2, e_3, e_4\}$ such that $e_1, e_2, e_3 \notin \mathfrak{t}$, $e_4 \in \mathfrak{t}$ and the only non-trivial relations are $[e_1, e_2] = e_3$ and $[e_1, e_3] = e_4$. The analytic closure of the exponential image of the subalgebra generated by e_1, e_3 is a proper non-commutative subgroup of G, a contradiction. Hence G has nilpotency class two.

Let $\mathfrak{g}$ be the Lie algebra of G. Since the nilpotency class of the algebra $\mathfrak{g}$ is two, there are elements a and b in $\mathfrak{g}$ which generate a three-dimensional non-commutative subalgebra $\mathfrak{c}$ of nilpotency class two. The closure C in G of the exponential image $\exp \mathfrak{c}$ of $\mathfrak{c}$ is not commutative, thus $G = C$. As mentioned in Remark 5.3.6, this means that G is an extra-special group. □

The simplest Lie algebra of dimension $2n + 1$, $n \geq 1$, of nilpotency class two is the so called *Heisenberg algebra* H_n. This algebra is defined by $[e_{2i+1}, e_{2i+2}] = e_{2n+1}$ for $0 \leq i \leq n - 1$ and the other products are zero.

The Heisenberg algebras are characterised by the following property (see [37], Proposition 31, p. 635):

5.3.8 Proposition. *Every Lie algebra having a one-dimensional centre which coincides with the commutator subalgebra is isomorphic to a Heisenberg algebra.* □

Therefore a Heisenberg algebra is given as soon as one defines a symplectic non degenerate form on an even-dimensional k-vector space as the following corollary shows (cf. [44], Section p XI, p. 42–49).

5.3.9 Corollary. *Let $\mathfrak{g}$ be a nilpotent Lie algebra over the field k such that the commutator algebra $\mathfrak{g}'$ is one-dimensional. Let $\{e_1, e_2, \cdots, e_n\}$ be a basis* *of $\mathfrak{g}$ such that $\mathfrak{g}' = \langle e_n \rangle$, let $\mathfrak{h}$ be the subspace generated by $\{e_1, e_2, \cdots, e_{n-1}\}$ and let $B : \mathfrak{h} \times \mathfrak{h} \longrightarrow \mathsf{k}$ be the alternating bilinear form defined by*

$$[e_i, e_j] = B(e_i, e_j) \cdot e_n.$$

The algebra $\mathfrak{g}$ is a Heisenberg algebra if and only if B is non-degenerate. This is the only case where the centre of $\mathfrak{g}$ is one-dimensional. $\square$

A connected algebraic group over a field of characteristic zero or a connected Lie group such that its Lie algebra is a Heisenberg algebra is called a *Heisenberg group*. In a non-simply connected Heisenberg Lie group G, the center is isomorphic to the real torus $SO_2(\mathbb{R})$ if G is a real Lie group and to the group $\mathbb{C}^*$ if G is a complex Lie group.

We now turn our attention to Lie algebras for which every proper epimorphic image is commutative.

5.3.10 Proposition. *Let $\mathfrak{g}$ be a non-commutative non-simple finite-dimensional Lie algebra over an arbitrary field. Every proper epimorphic image of $\mathfrak{g}$ is commutative if and only if for $\mathfrak{g}$ one of the following holds:*

(1) $\mathfrak{g}$ is an extension of a commutative subalgebra $\mathfrak{a}$ by a commutative Lie algebra $\mathfrak{b}$ acting faithfully and irreducibly on $\mathfrak{a}$,
(2) $\mathfrak{g}$ is a Heisenberg algebra.

Proof. Since for the solvable radical $\mathfrak{s}$ of $\mathfrak{g}$ the factor algebra $\mathfrak{g}/\mathfrak{s}$ is semi-simple (cf. [49], p. 24), the algebra $\mathfrak{g}$ is solvable.

Let $\mathfrak{n}$ be a minimal ideal of $\mathfrak{g}$. Since $\mathfrak{g}/\mathfrak{n}$ is commutative, the commutator subalgebra $\mathfrak{g}'$ of $\mathfrak{g}$ coincides with $\mathfrak{n}$. Hence $\mathfrak{g}'$ is the only minimal ideal of $\mathfrak{g}$.

If the centre of $\mathfrak{g}$ is not trivial, then it is equal to $\mathfrak{g}'$ and one has $\dim(\mathfrak{g}') = 1$. This means that $\mathfrak{g}$ is a Heisenberg algebra and we are in the case (2).

If, however, the centre of $\mathfrak{g}$ is trivial, then the Lie algebra $\mathfrak{b} = \mathfrak{g}/\mathfrak{g}'$ acts irreducibly on the commutative Lie algebra $\mathfrak{a} = \mathfrak{g}'$. For an arbitrary element $x \in \mathfrak{g} \backslash \mathfrak{a}$ denote by $\mathfrak{c}_\mathfrak{a}(x)$ the centraliser of the subalgebra $\langle x \rangle$ in $\mathfrak{g}$.

Assume that $\mathfrak{c}_\mathfrak{a}(x) \neq 0$. Since $\mathfrak{b}$ acts irreducibly on $\mathfrak{a}$, it follows that $[\mathfrak{c}_\mathfrak{a}(x), \mathfrak{b}]$ spans all of $\mathfrak{a}$. But then $\mathfrak{c}_\mathfrak{a}(x) = \mathfrak{a}$ because

$$[[c, b], x] = -[[b, x], c] - [[x, c], b] = 0$$

for all $b \in \mathfrak{b}, c \in \mathfrak{b}$. This contradicts the fact that the centre of $\mathfrak{g}$ is trivial. Hence, in particular, the algebra $\mathfrak{b}$ acts faithfully on $\mathfrak{a}$. $\square$

5.3.11 Proposition. *Let K be an algebraically closed field containing the field k of characteristic zero and let $G(K)$ be a non-commutative non-simple connected affine algebraic group over k. Every epimorphic image of $G(K)$ of dimension smaller than $\dim G(K)$ is commutative if and only if $G(K)$ is either isomorphic to the group $L_2(K)$ or to a Heisenberg group.*

Proof. We apply the correspondence between the closed connected subgroups of the algebraic group G and the subalgebras of the Lie algebra of G (see [45], Chapter V, p. 87ff.). If the centre of G is not trivial, our assertion follows immediately from Proposition 5.3.10. Otherwise, the group G is a semi-direct product of a minimal connected normal subgroup A with a connected commutative group B. Then A has dimension one (see [45], 19.3 Theorem, p. 123) and B is isomorphic to $\mathbf{G}_m$. □

5.3.12 Proposition. *Let G be a non-simple connected real or complex Lie group which is not commutative. Then every Lie factor group of smaller dimension is commutative if and only if G belongs one the following classes:*

(1) the connected component of the real Lie group $L_2(\mathbb{R})$,
(2) a covering of the connected component of the group $E_2(\mathbb{R})$ of motions of the real Euclidean plane,
(3) the semi-direct product of the group $\mathbb{R}^2$ with a one-parameter group

$$\Xi_d = \left\{ \begin{pmatrix} d^t\cos t & d^t\sin t \\ -d^t\sin t & d^t\cos t \end{pmatrix} : t \in \mathbb{R} \right\},$$

for $d > 1$,
(4) a covering of the complex Lie group $L_2(\mathbb{C})$,
(5) an extra-special real or complex Lie group such that the connected component Z of the centre of G is a minimal connected subgroup of G, i.e. for $\mathsf{K} = \mathbb{R}: Z \cong \mathbb{R}^+, SO_2(\mathbb{R})$ and for $\mathsf{K} = \mathbb{C}: Z \cong \mathbb{C}^+, \mathbb{C}^*$ or Z is a simple complex torus.

Proof. Since the factor group G/S, where S denotes the solvable radical of G, has to be commutative if G/S in non-trivial, the group G is solvable. The topological commutator subgroup $\overline{G'}$ is characteristically simple, hence commutative. Furthermore $\overline{G'}$ is a minimal closed connected normal subgroup of G. Thus the commutative factor group $B = G/\overline{G'}$ acts either trivially or irreducibly on $\overline{G'}$.

In the first case, $\overline{G'}$ is contained in the centre of G. Furthermore it is a minimal closed connected subgroup of G, because otherwise we would find a proper closed connected normal subgroup $M \neq G'$ of G. But then the group G/M would not be commutative. This contradiction shows that we are in case (5).

In the second case, the group $\overline{G'}$ is a vector group. It follows from Lie's theorem (and its real equivalent) (see [27], p. 69) that one has $\dim \overline{G'} = 1$ for the complex case and $\dim \overline{G'} \in \{1, 2\}$ for the real case. The centraliser C of $\overline{G'}$ in G is discrete, otherwise one would find a closed connected normal subgroup of G intersecting $\overline{G'}$ trivially. Thus $G/\overline{G'}$ has dimension one, and one of the assertions (1), (2), (3) or (4) holds (cf. [84], p. 255). □

5.3.13 Remark. In contrast to the semi-commutative Lie algebras, the structure of non-commutative finite groups in which every proper subgroup is abelian is well known (cf. [57]). The order of a group G with this property is divisible by at most two prime numbers.

A representation of finite semi-commutative groups which shows the analogy with our theorems in this section is given in [73], § 6, p. 241 (see also [74], p. 720–735):

If G is a p-group, then either G is a quaternion group of order 8 or G belongs to one of the following classes:

i) Let $\mathbb{Z}_{p^n}$ be the ring of residues modulo p^n. Define on the set $\mathbb{Z}_{p^u} \times \mathbb{Z}_{p^v}$ the multiplication

$$(i,a) \cdot (k,b) = (i+k, a+b+(p^{v-1}i) \cdot b),$$

where for $p = 2$ one has $(u,v) \neq (1,1)$.

ii) Define on the set $\mathbb{Z}_{p^u} \times \mathbb{Z}_{p^v} \times \mathbb{Z}_p$ the multiplication

$$(i,j,a) \cdot (k,l,b) = (i+k, j+l, a+b+il),$$

where one reduces i and l modulo p in the last entry, and for $p = 2$ one has $(u,v) \neq (1,1)$.

In both cases the commutator subgroup G' has order p and G has nilpotency class two.

If G is not a p-group, then one can assume that $|G| = p^u q^v$, where p divides $q^v - 1$, and G is isomorphic to the following group defined on the set $\mathbb{Z}_{p^u} \times \mathsf{GF}(q^v)$, where $\mathsf{GF}(q^v)$ is the finite field of order q^v, by the following multiplication:

$$(i,a) \cdot (k,b) = (i+k, a+r^i b),$$

where r is an element of order p in the multiplicative group of $\mathsf{GF}(q^v)$.   □

5.3.14 Remark. In this work we considered only connected algebraic groups and Lie groups. In these classes, the existence of the solvable radical yields that every non-simple group for which any proper epimorphic image is commutative must be solvable. For finite groups the situation is more involved, because the major part of finite simple groups has a non-trivial outer automorphism group, which must be solvable (cf. [10]). Thus any extension of a finite simple group by a commutative group of outer automorphisms has only commutative epimorphic images.

Under the assumption that in the finite group G having only commutative proper epimorphic images t here exists a non-trivial normal solvable subgroup N, the group G is either a central extension of a cyclic group of prime order p by a commutative group or the centre of G is trivial and G is an extension of the commutator subgroup G', which is an elementary abelian p-group, by a commutative group acting irreducibly on G'.   □

5.4 Aligned Groups and Aligned Lie Algebras

We call a connected real or complex Lie group or an algebraic group *aligned*, if every two closed connected subgroups of the same dimension are isogenous. Here two Lie groups are called isogenous if one of them is a finite covering of the other one. Similarly, we call a Lie algebra *aligned* if all subalgebras of equal dimension are isomorphic.

In Section 5.2 we have seen that the class of aligned algebraic groups is so rich that no characterisation seems to be possible. Now we will show that for aligned real Lie groups and for aligned affine algebraic groups over a field of characteristic zero a detailed classification is possible. For complex Lie groups, the complex tori are serious obstacles for such a classification. For algebraic groups in positive characteristic other difficulties arise and one needs further hypotheses to get a classification (see Section 5.7).

As usual, a Lie algebra $\mathfrak{g}$ over a field of characteristic zero is called *reductive* if the solvable radical of $\mathfrak{g}$ coincides with the centre of $\mathfrak{g}$. According to Levi's theorem ([49], p. 91 and [49], Theorem 4, p. 72), any reducible Lie algebra $\mathfrak{g}$ over a field of characteristic zero is the direct product of its centre and some simple Lie algebras.

A classification of all aligned Lie algebras over an arbitrary field of characteristic zero is puzzling, even if for reductive algebras. If the reductive Lie algebra $\mathfrak{g}$ is the direct product of its centre and some simple semi-commutative Lie algebras $\mathfrak{a}_1, \cdots, \mathfrak{a}_r$, then $\mathfrak{g}$ is aligned if and only if the dimensions of the maximal commutative subalgebras of $\mathfrak{g}$ are smaller than the minimum of the dimensions of the subalgebras $\mathfrak{a}_i$. Therefore we restrict our attention for this question to solvable Lie algebras. To treat this case we need the following

5.4.1 Lemma. *Let $\mathfrak{g}$ be a Lie algebra over a field k of characteristic zero such that its commutator subalgebra $\mathfrak{a}$ coincides with the nilradical of $\mathfrak{g}$ and is a commutative minimal ideal of $\mathfrak{g}$. If for all $g, h \in \mathfrak{g}\backslash\mathfrak{a}$ the subalgebras $\mathfrak{a} + \langle g \rangle$ and $\mathfrak{a} + \langle h \rangle$ are isomorphic, then $\dim(\mathfrak{g}/\mathfrak{a}) = 1$.*

Proof. Assume that the assertion of the lemma is not true and put $\mathcal{M} = \{\mathrm{ad}\, g : g \in \mathfrak{g}\backslash\mathfrak{a}\}$. No element of $\mathcal{M}$ is nilpotent, otherwise the nilradical of $\mathfrak{g}$ would be greater than $\mathfrak{a}$. Given $g, h \in \mathfrak{g}\backslash\mathfrak{a}$, the fact that the algebras $\mathfrak{a} + \langle g \rangle$ and $\mathfrak{a} + \langle h \rangle$ are isomorphic implies that there is linear map $T \in GL(\mathfrak{a})$ and an element $\mu \in \mathsf{k}$ such that $T(\mathrm{ad}\, g)T^{-1} = \mu(\mathrm{ad}\, h)$. Let $\alpha_1, \cdots, \alpha_r$ be the eigenvalues of $(\mathrm{ad}\, g)$ and let $\beta_1, \cdots, \beta_r$ be the eigenvalues of $(\mathrm{ad}\, h)$ taken in the algebraic closure $\overline{\mathsf{k}}$ of k. Then $\beta_i = \mu\alpha_i$ for $i = 1, \cdots, r$. Since $(\mathrm{ad}\, g)$ and $(\mathrm{ad}\, h)$ commute, these maps can be brought simultaneously into upper triangular form (cf. [97]). Thus all eigenvalues of $\mu(\mathrm{ad}\, g) - (\mathrm{ad}\, h)$ are zero, i.e. this mapping is nilpotent, a contradiction. $\qquad\square$

5.4.2 Theorem. *A non-commutative solvable Lie algebra $\mathfrak{g}$ over a field k of characteristic zero is aligned if and only if $\mathfrak{g}$ is semi-commutative or has the following structure:*

(S) $\mathfrak{g}$ *a semi-direct product of a three-dimensional nilpotent non-commutative subalgebra $\mathfrak{h}$ and a one-dimensional subalgebra which acts irreducibly on $\mathfrak{h}/\mathfrak{c}$, where $\mathfrak{c}$ denotes the centre of $\mathfrak{h}$.*

Proof. Assume that $\mathfrak{g}$ is a counter-example of minimal dimension to the theorem. First we assume that the centre of $\mathfrak{g}$ is trivial and that $\mathfrak{g}$ has dimension $d > 2$. We consider a maximal ideal $\mathfrak{m}$ of $\mathfrak{g}$. Since $\mathfrak{g}$ is solvable, one sees that $\mathfrak{g} = \mathfrak{m} + \langle x \rangle$ for every $x \in \mathfrak{g} \backslash \mathfrak{m}$.

If $\mathfrak{m}$ is commutative, then $\mathrm{ad}(x)$ acts irreducibly on $\mathfrak{m}$. Otherwise $\mathfrak{g}$ would contain a non-commutative subalgebra of a dimension smaller than the dimension of $\mathfrak{m}$. Hence in this case $\mathfrak{g}$ is semi-commutative.

If $\mathfrak{m}$ is not commutative, but its centre $\mathfrak{z}$ is not trivial, then the subalgebra $\mathfrak{z} + \langle x \rangle$ is not commutative. From Proposition 5.3.3 we know the structure of the three-dimensional nilpotent Lie algebra $\mathfrak{m}$. If an inner derivation of $\mathfrak{g}$ acted non-trivially on the one-dimensional centre $\mathfrak{z}$ of $\mathfrak{m}$, then $\mathfrak{m}$ would contain a two-dimensional non-commutative subalgebra, a contradiction. Hence it follows with Proposition 5.3.3 that $\mathfrak{m} = \mathfrak{a} + \langle b \rangle$, where $\mathfrak{a} = \mathfrak{m}'$ is commutative and $\mathrm{ad}\, b$ acts irreducibly on $\mathfrak{a}$. Since $\mathrm{ad}\, b$ is not nilpotent, the same is true for every element of $\mathfrak{g}$ which is not contained in $\mathfrak{a}$. Hence $\mathfrak{a}$ is the nilradical of $\mathfrak{g}$. Since in every solvable Lie algebra the commutator subalgebra lies in the nilradical and $\mathfrak{a}$ is a minimal ideal of $\mathfrak{g}$, we see that $\mathfrak{a} = \mathfrak{g}'$. Furthermore $\mathfrak{g} = \mathfrak{a} + \langle b, x \rangle$ for every $x \in \mathfrak{g} \backslash \mathfrak{m}$. But this is not possible because of Lemma 5.4.1.

Now suppose that $\mathfrak{g}$ is nilpotent. Since $\mathfrak{g}$ is not commutative, there is a maximal non-commutative subalgebra $\mathfrak{h}$ of $\mathfrak{g}$. As $\mathfrak{h}$ is semi-commutative, it follows from Proposition 5.3.3 that $\mathfrak{h}$ has dimension three. Thus $\mathfrak{g}$ has dimension four. Hence by [61], p. 162, either $\mathfrak{g}$ is the direct product of the three-dimensional non-commutative Lie algebra with a one-dimensional subalgebra, or $\mathfrak{g}$ has a basis $\{e_1, e_2, e_3, e_4\}$ such that $[e_1, e_2] = e_3$, $[e_1, e_3] = e_4$ and all other products of basis elements are zero. In the first case, it is obvious that $\mathfrak{g}$ contains commutative and non-commutative subalgebras of dimension three. In the second case, the three-dimensional subalgebra with the basis $\{e_1, e_3, e_4\}$ is not commutative, whereas $\{e_2, e_3, e_4\}$ spans a three-dimensional commutative algebra.

Proceeding further, we suppose that $\mathfrak{g}$ is not nilpotent but that the centre $\mathfrak{z}$ of $\mathfrak{g}$ has dimension at least two. Let $\mathfrak{h}$ be a maximal subalgebra of $\mathfrak{g}$ containing $\mathfrak{z}$. Then the algebra $\mathfrak{h}$ is commutative, because of the minimality of $\mathfrak{g}$, hence $\mathfrak{g}$ is semi-commutative, a contradiction.

Thus we know that $\mathfrak{g}$ is not nilpotent and has a one-dimensional centre $\mathfrak{z}$. Let $\mathfrak{m}$ be a maximal subalgebra of $\mathfrak{g}$ containing $\mathfrak{z}$. Since $\mathfrak{g}$ is not semi-commutative, the algebra $\mathfrak{m}$ is not commutative. If $\mathfrak{m}$ is nilpotent, then the algebra $\mathfrak{m}$ has dimension three, because of the minimality of $\mathfrak{g}$, and one sees that $\mathfrak{g} = \mathfrak{m} + \langle x \rangle$ for any $x \in \mathfrak{g} \backslash \mathfrak{m}$. We now claim that $\langle x \rangle$ acts irreducibly on $\mathfrak{m}/\mathfrak{z}$. Otherwise, the algebra $\mathfrak{g}/\mathfrak{z}$ would be a three-dimensional solvable Lie algebra that is not nilpotent. Then $\mathfrak{g}/\mathfrak{z}$ would contain a one-dimensional ideal $\mathfrak{b}/\mathfrak{z}$ on which $\langle x \rangle$ acts irreducibly (cf. [49], p. 12). But then the three-dimensional Lie algebra $\mathfrak{h} = \mathfrak{b} + \langle x \rangle$ would not be nilpotent and would have the non-trivial centre $\mathfrak{z}$. Thus $\mathfrak{h}$ would be the direct product of $\mathfrak{z}$ and a two-dimensional non-commutative Lie algebra (cf. [49], p. 12). This contradicts the fact that $\mathfrak{g}$ contains two-dimensional commutative subalgebras. So $\langle x \rangle$ acts irreducibly on $\mathfrak{m}/\mathfrak{z}$, and $\mathfrak{g}$ is not a counter-example, a contradiction.

Finally we have to handle the case where $\mathfrak{m}$ has the structure described under (S). The nilradical $\mathfrak{n}$ of $\mathfrak{m}$ has dimension three and is not commutative. We consider the action of the two-dimensional algebra $\mathfrak{l} = \mathfrak{g}/\mathfrak{n}$ on the two-dimensional commutative Lie algebra $\mathfrak{n}/\mathfrak{z}$. If $\mathfrak{l}$ were not commutative, one would have $\mathfrak{l}' = \mathfrak{m}/\mathfrak{z}$, a contradiction to the fact $\mathfrak{m}$ is not nilpotent. Hence $\mathfrak{l}$ is commutative and acts irreducibly on $\mathfrak{n}/\mathfrak{z}$. Since $\mathfrak{m}/\mathfrak{z}$ is not nilpotent, it follows that $\mathfrak{n}/\mathfrak{z}$ is the nilradical of $\mathfrak{g}/\mathfrak{z}$. Otherwise there would be a nilpotent subalgebra of $\mathfrak{g}$ having the same dimension as $\mathfrak{m}$. Since $\mathfrak{l}$ has dimension two, we get a contradiction to Lemma 5.4.1. □

5.4.3 Remark. According to the classification of 4-dimensional solvable Lie algebras over a field k of characteristic zero in [62], p. 121, there is a unique 4-dimensional non-nilpotent Lie algebra having a one-dimensional centre and a three-dimensional non-commutative commutator subalgebra, the algebra denoted there by $g_{4,9}$ and having the parameter $p = 0$. This algebra has a basis $\{e_1, e_2, e_3, e_4\}$ such that

$$[e_2, e_3] = e_1 \,, \ [e_2, e_4] = -e_3 \,, \ [e_3, e_4] = e_2$$

and such that all other products the elements of this basis are 0. This algebra satisfies condition (S) if and only if $\mathrm{ad}\, e_4$ has no eigenvalue $\neq 0$ in k, i.e. if -1 is not a square in k.

5.4.4 Theorem. *A non-commutative connected affine algebraic group G over a field of characteristic zero is aligned if and only if G is unipotent and has dimension three.*

Proof. Since G is aligned and not commutative, the group G is unipotent. Hence the Lie algebra $\mathfrak{g}$ of G is unipotent. Using the correspondence between affine algebraic groups G over (the algebraic closure of) a field of characteristic zero and their Lie algebras (cf. [45], Theorem, p. 87), one sees that $\mathfrak{g}$ is aligned. The assertion now follows from Theorem 5.4.2 and Proposition 5.3.3. □

5.4.5 Corollary. *Let G be a connected non-affine algebraic group over a field k such that G is not commutative and that it is not isogenous to the direct product of two proper subgroups. Then G is aligned if and only if the following holds:*

(a) The group $D(G)$ is aligned and contains no subgroup of dimension two or three,

(b) the group $L(G)$ is the three-dimensional unipotent group of nilpotency class two.

Moreover, the field k has characteristic zero and $D(G)$ is a non-splitting extension of a one-dimensional vector group by an abelian variety.

Proof. Since G is not commutative, the maximal connected affine subgroup $L(G)$ of G is not trivial. The non-commutative group $L(G)$ contains a Borel subgroups with a non-trivial unipotent radical (cf. [45], Chapter 28 and p. 123). Thus $L(G)$ is unipotent and every one-dimensional connected algebraic subgroup of $L(G)$ is isomorphic to $\mathbf{G}_a$. It follows from Proposition 1.3.2 *iii)* that the characteristic of the field $\mathbf{k}$ is zero because G is not isogenous to the direct product of two proper subgroups. Since $L(G)$ is aligned by Theorem 5.4.4, the validity of *(b)* follows. In $L(G)$ one finds subgroups of dimension two and three. Thus $D(G)$ cannot contain connected algebraic subgroups of these dimensions. $\square$

5.4.6 Lemma. *Let G be a non-commutative connected complex Lie group G and let X be the maximal toroidal subgroup of G. If all closed one-parameter subgroups of G/X are isomorphic, then each of them is isomorphic to $\mathbb{C}^+$ and G is nilpotent.*

Proof. If G were not solvable, then G/X would contain a closed simple complex Lie subgroup D (see [55], Corollary 4.18, p. 113). But the group D contains closed one-parameter subgroups isomorphic to $\mathbb{C}^+$ and closed one-parameter subgroups isomorphic to $\mathbb{C}^*$. Hence G is solvable.

If G/X is solvable but not nilpotent, then the Lie algebra of G contains a two-dimensional non-commutative subalgebra $\mathfrak{s}$ ([49], Exercise 4, p. 54). Hence G contains a complex analytic subgroup H which is a semi-direct product of a one-dimensional normal subgroup N with the analytic subgroup $E = \exp(\mathbb{C}x)$ satisfying $H' = N$. Since the analytic group E acts semi-simply on N, which, as an analytic group, is isomorphic to $\mathbb{C}^+$, it follows that E is isomorphic to $\mathbb{C}^*$. Assume that N is not closed in G/X. As E acts transitively on the set $N \setminus \{1\}$, the topological closure of N has to be a real torus, a contradiction. If E is not closed in G/X, then the topological closure of E is an at least two-dimensional real torus. But this group does not act transitively on $\mathbb{R}^2$. Hence the non-commutative group G/X is nilpotent. Since X lies in the center of G, the group G is nilpotent, too.

Assume now that all closed complex one-parameter subgroups of G/X are isomorphic to the group $\mathbb{C}^*$. Let $\mathfrak{g}$ be the Lie algebra of the non-commutative group G/X. We consider the central series $\mathcal{C}^{i+1}\mathfrak{g} = [\mathfrak{g}, \mathcal{C}^i\mathfrak{g}]$, where $\mathcal{C}^0\mathfrak{g} = \mathfrak{g}$. Choose n minimal such that $\mathcal{C}^{n+1}\mathfrak{g} = 0$. Then there are elements $x \in \mathcal{C}^{n-1}\mathfrak{g} \setminus \mathcal{C}^n\mathfrak{g}$ and $0 \neq y \in \mathfrak{g}$ such that $0 \neq z = [x, y] \in \mathcal{C}^n\mathfrak{g}$. So $\mathfrak{h} = \mathbb{C}x + \mathbb{C}y + \mathbb{C}z$ is a non-commutative nilpotent Lie algebra. The closure H of $\exp \mathfrak{h}$ contains the closed one-parameter subgroup $\exp(\mathbb{C}x)$ which is isomorphic to $\mathbb{C}^+$, a contradiction. $\square$

5.4.7 Lemma. *Every connected complex Lie group G which is not a complex torus contains a closed complex non-compact one-dimensional subgroup.*

Proof. Assume that G does not contain a closed complex one-dimensional subgroup and let X be the maximal toroidal subgroup of G. It follows from [1],

Proposition 1.1.17(2), p. 15, that X is a torus. As $G \neq X$, in the factor group G/X there is closed one-dimensional subgroup E/X. Since X is the maximal toroidal subgroup of the commutative group E, it follows from [1], Proposition 1.1.5, p. 5, that $E = X \times E_1$ with a one-dimensional closed subgroup E_1. □

5.4.8 Theorem. *A non-commutative connected real Lie group G is aligned if and only if it belongs to one of the following types:*

(1) the three-dimensional simply connected non-commutative nilpotent real Lie group,

(2) a compact Lie group which is locally isomorphic to one of the following groups: $SO_3(\mathbb{R})$, $SO_2(\mathbb{R}) \times SO_3(\mathbb{R})$, $SO_3(\mathbb{R}) \times SO_3(\mathbb{R})$, $SU_3(\mathbb{C}, 0)$, $SO_5(\mathbb{R})$ and the 14-dimensional exceptional Lie group G_2,

(3) the universal covering group of the group $PSL_2(\mathbb{R})$,

(4) the group $L_2(\mathbb{R})$,

(5) the universal covering of the connected component of the group $E_2(\mathbb{R})$ of motions of the real Euclidean plane,

(6) the semi-direct product of the group $\mathbb{R}^2$ with a one-parameter group

$$\Xi_d = \left\{ \begin{pmatrix} d^t \cos t & d^t \sin t \\ -d^t \sin t & d^t \cos t \end{pmatrix} : t \in \mathbb{R} \right\},$$

for $d > 1$,

(7) the non-splitting central extension of the group $\mathbb{R}$ by the universal covering of the connected component of the group $E_2(\mathbb{R})$.

Proof. Assume first that G is a connected nilpotent Lie group which is not commutative. A nilpotent aligned real Lie group which is not simply connected is a (real) torus. Hence G is simply connected. As a consequence, the exponential map of G is bijective, and assertion *(1)* follows from Theorem 5.4.2.

Next we consider the case that all closed one-parameter subgroups of G are compact. Then G is compact (see [58], Theorem p. 188). Since G is not commutative, it contains a simple subgroup S. Such a group S always contains a three-dimensional subgroup locally isomorphic to $SO_3(\mathbb{R})$ (see [49], Theorem 17, p. 100). So every maximal torus of G has dimension at most two. The compact Lie groups of rank ≤ 2 are locally isomorphic to the following groups: $SO_2(\mathbb{R})$, $SO_2(\mathbb{R}) \times SO_2(\mathbb{R})$, $SO_3(\mathbb{R})$, $SO_2(\mathbb{R}) \times SO_3(\mathbb{R})$, $SO_3(\mathbb{R}) \times SO_3(\mathbb{R})$, $SU_3(\mathbb{C}, 0)$, $SO_5(\mathbb{R})$ and the 14-dimensional exceptional Lie group G_2 (cf. [94]). Since all these groups have different dimensions, all non-commutative groups in this list are aligned.

Now we treat the case that all one-parameter subgroups of G are isomorphic to $\mathbb{R}$. First we remark that in this case the group G is simply connected. If G is semi-simple, then G is the universal covering group of the group $PSL_2(\mathbb{R})$ (see [94]). If G is neither solvable nor semi-simple, then we consider the solvable non-trivial radical R of G. Since G is simply connected, it contains a closed semi-simple subgroup H (see [68], p. 19). The group H

must be isomorphic to the universal covering of the $PSL_2(\mathbb{R})$. If $\dim R = 1$, then G is isomorphic to the direct product $H \times \mathbb{R}$. But then G contains two-dimensional commutative subgroups as well as the non-commutative two-dimensional Borel subgroup of H. If $\dim R > 1$ let N be a minimal connected characteristic subgroup of G contained in R. If N is isomorphic to $\mathbb{R}$, then we get a contradiction as before. If $\dim N > 1$, then the commutative group N contains a two-dimensional commutative connected subgroup and we have again a contradiction.

From now on we only have to consider the case that G is a simply connected solvable real Lie group of dimension greater than two, that is not nilpotent. Let A be a minimal normal subgroup of G. Then A has dimension $d \leq 2$. First we assume that in G the connected component Z of the centre is trivial. If $\dim A = 1$, then the centraliser C of A in G has co-dimension one in G; hence G contains a two-dimensional closed commutative subgroup. For a one-parameter subgroup E of G not contained in C one gets the two-dimensional non-commutative subgroup AE, a contradiction. So, if $\dim A = 2$, then AE is a three-dimensional semi-commutative Lie group. If $\dim G = 3$, then we are done. Assume that $\dim G > 3$. If the dimension of the centraliser C of A is greater than two, then G contains a three-dimensional commutative subgroup, a contradiction. If, however, $C = A$, then $\dim G = 4$ and G/C is isomorphic to the group $\mathbb{R} \times SO_2(\mathbb{R})$ in its natural action on the two-dimensional vector space A. But this is not possible in an aligned group.

Now assume that Z is not trivial. If the exponential map of G is not bijective, we assume that G is a counter-example of minimal dimension. According to [20], Theoreme 3, p. 120, the group G has closed connected subgroups X, N such that N is normal in X and X/N is isomorphic to the universal covering of the group $E_2(\mathbb{R})$. If M is a maximal closed connected subgroup of G containing X, then because of the minimality of G one has $M = X$. So G has dimension five. The commutator subgroup X' of X has dimension three and is nilpotent. If G' had dimension four, then being aligned and nilpotent it would be commutative, a contradiction. Thus $G' = X'$. The commutative group Γ induced by G/G' on G' has dimension one, otherwise it could not act trivially on Z. Hence there is a one-dimensional subgroup E of G centralising G'. The group $G'E$ is not aligned, a contradiction.

If the exponential map of G is bijective, then we can apply Theorem 5.4.2. Since the algebra under (S) has an epimorphism onto the Lie algebra of the group of motions of the Euclidean plane, according to [20], Theoreme 3, p. 120, the Lie algebra of G is semi-commutative. Hence the group G is found under (5) or (6) as a three-dimensional simply connected solvable non-nilpotent semi-commutative Lie group.   □

5.4.9 Proposition. *Let G be a connected complex Lie group and assume that G/X is not commutative, where X is the maximal toroidal subgroup of G. Then G is aligned if and only if it is the direct product of the three-dimensional* *simply connected non-commutative nilpotent complex Lie group and the torus X, which is aligned and does not contain a non-trivial closed analytic subgroup of dimension $d \leq 3$.*

Proof. Since G is not commutative, the group G/X is not trivial. Then the group G contains closed one-parameter subgroups according to Lemma 5.4.7. By Lemma 5.4.6 all closed analytic one-dimensional subgroups are isomorphic to $\mathbb{C}^+$. Hence it follows from Remark 5.3.6 b) that X is a complex torus, which does not contain a closed one-dimensional subgroup.

Since any two subgroups of G of equal dimension containing X are isomorphic, it follows that the group G/X is aligned. By Lemma 5.4.6 this group is simply connected. Since the condition (S) of Theorem 5.4.2 is never satisfied in an aligned group, it follows from Theorem 5.3.3 that the Lie algebra of G/X is either commutative or the three-dimensional non-commutative Lie algebra $\mathfrak{h}_3$.

As G/X is not commutative, the Lie algebra of G is the direct product $\mathfrak{h}_3 \times \mathfrak{a}$, where $\mathfrak{a}$ is a commutative ideal. In this situation $\mathfrak{a}$ is the Lie algebra of X. By [1], Proposition 1.1.5, p. 6, the centre of X is $\exp \mathfrak{h}' \times X$. It follows that G is the direct product of X with the three-dimensional simply connected non-commutative nilpotent complex Lie group H. This direct product is aligned if and only if the complex torus X is aligned and does not contain a non-trivial closed analytic subgroup of dimension $d \leq 3$. $\square$

In the situation of Proposition 5.4.9, if G is not commutative but the group G/X is commutative, then G is nilpotent of class two and X is a complex torus (see Lemma 5.4.7). We give an example which shows that there are plenty of aligned complex Lie groups.

5.4.10 Example. Let V, W be complex vector spaces with a basis $\{e_1, e_2, e_3\}$ respectively $\{e_{12}, e_{23}, e_{31}\}$. We define an alternating bilinear map $\Gamma : V \times V \longrightarrow W$ putting

$$\Gamma(e_1, e_2) = e_{12}, \; \Gamma(e_2, e_3) = e_{23}, \; \Gamma(e_3, e_1) = e_{31}.$$

Let A_{12}, A_{23}, A_{31} be simple complex abelian varieties of dimension d_{12}, d_{23}, d_{31}, respectively. For $(i, j) \in I = \{(1, 2), (2, 3), (3, 1)\}$ we choose vectors a_{ij} in the Lie algebra $\mathfrak{a}_{ij}$ of A_{ij}. The maps α_{ij} defined by $\alpha_{ij}(e_{ij}) = a_{ij}$ and $\alpha_{ij}(e_{kl}) = 0$ for $i, j \neq k, l$ give us a linear maps $\alpha_{ij} : W \longrightarrow \mathfrak{a}_{ij}$. We define $\gamma_{ij} = \exp_{A_{ij}} \circ \alpha_{ij}$. Then mappings $\gamma_{ij} : \mathbb{C}e_{ij} \longrightarrow A_{ij}$ are injective complex homomorphisms. Since the varieties A_{ij} are simple, the image of γ_{ij} is dense in A_{ij}. We define a central extension G_1 of W by V by the commutator rule

$$[(v_1, w_1), (v_2, w_2)] = (0, \Gamma(v_1, v_2)).$$

Putting $A = A_{12} \times A_{23} \times A_{31}$, we have the analytic embedding $(\gamma_{12}, \gamma_{23}, \gamma_{31})$ of G_1 in a central extension G of A by V.

We choose the abelian varieties A_{ij} in such a way that for the dimensions d_{ij} one has

$$d_{12} > 1, \, d_{23} > d_{12} + 2, \, d_{31} > d_{12} + d_{23}.$$

Then A has non-trivial proper subgroups only of dimensions d_{ij} and $d_{ij} + d_{kl}$ for $(i,j), (k,l) \in I$ and $(i,j) \neq (k,l)$. If d is the dimension of a non-trivial subgroup of A, then by construction the group G has proper subgroups of dimension $d + 1$ and $d + 2$. Finally all closed one-dimensional analytic subgroups, which exist by Lemma 5.4.7, are isomorphic to $\mathbb{C}^+$. Since we have chosen the parameters in such a way that all these subgroups are isomorphic if they have equal dimension, the group G is aligned. $\qquad\square$

5.5 Co-Aligned Groups and Co-Aligned Lie Algebras

We call a connected real or complex Lie group or an algebraic group *co-aligned* if for every two homomorphisms, with connected kernels, the images are isogenous if they have the same dimension. Similarly a Lie algebra is called *co-aligned* if all epimorphic images of equal dimension are isomorphic.

In contrast to the results in Section 5.4, it turns out that even for a classification of Lie groups, algebraic groups over a field of characteristic zero or Lie algebras that are co-aligned, additional conditions are necessary in order to get a classification. For algebraic groups in positive characteristic a classification is obtained under restricted hypothesis (see Section 5.7).

5.5.1 Proposition. *Let $\mathfrak{g}$ be a nilpotent Lie algebra over an algebraically closed field k. Let the commutator subalgebra $\mathfrak{g}'$ have dimension greater than one and let $\mathfrak{n}_1 \neq \mathfrak{n}_2$ be one-dimensional ideals of $\mathfrak{g}$ such that $\mathfrak{g}/\mathfrak{n}_i$ is a $2n + 1$-dimensional Heisenberg algebra H_n, $i = 1, 2$. Then the centre $\mathfrak{z}$ of $\mathfrak{g}$ has dimension two, $\mathfrak{g}' = \mathfrak{z}$ and $\mathfrak{g}$ is isomorphic to the Lie algebra defined by*

$$[e_i^k, e_{n_k+i}^k] = l_1 + \lambda_k \cdot l_2 \quad \text{for } i = 1, \cdots, n_k$$

$$[e_i^k, e_{n_k+i+1}^k] = l_2 \qquad \text{for } i = 1, \cdots, n_k - 1$$

where $\{l_1, l_2\}$ is a basis of $\mathfrak{z} = \mathfrak{g}'$ and the vectors e_i^k give, for $k = 1, \cdots, s$ and $i = 1, \cdots, n_k$, a basis of a complement $\mathfrak{h}$ of $\mathfrak{z}$.

Proof. Let d be the dimension of $\mathfrak{z}$. Since $\mathfrak{n}_1$ and $\mathfrak{n}_2$ are central, the centre of $\mathfrak{g}/\mathfrak{n}_i$ has dimension $t \geq d - 1$. But $\mathfrak{g}/\mathfrak{n}_i$ is a Heisenberg algebra, so $t = 1$ and this forces $d = 2$ and $\mathfrak{z} = \mathfrak{n}_1 \oplus \mathfrak{n}_2$. Hence $\mathfrak{g}' \leq \mathfrak{n}$. Since $\dim \mathfrak{g}' > 1$ and $\dim \mathfrak{z} = 2$, we have $\mathfrak{g}' = \mathfrak{z}$. The commutator map induces on the complement $\mathfrak{h} = \langle e_1, \cdots, e_{2n} \rangle$ of $\mathfrak{z}$ two non-degenerate symplectic bilinear forms $B_i : \mathfrak{h} \times \mathfrak{h} \longrightarrow k$ by putting

$$[h_1, h_2] = B_1(h_1, h_2) \cdot l_1 + B_2(h_1, h_2) \cdot l_2.$$

Changing the basis of $\mathfrak{h}$ we can simultaneously reduce B_1 and B_2 to the canonical forms represented by matrices

$$\left(\begin{array}{c|c} & I_n \\ \hline -I_n & \end{array}\right), \left(\begin{array}{c|c} & J \\ \hline -J^t & \end{array}\right),$$

where $J = J_1(\lambda_1) \oplus \cdots \oplus J_s(\lambda_s)$ is a Jordan-block matrix, each block

$$J_k(\lambda_k) = \begin{pmatrix} \lambda_k & 1 & 0 & \cdots & 0 \\ 0 & \lambda_k & 1 & \cdots & 0 \\ 0 & 0 & \lambda_k & \cdots & 0 \\ & & & \vdots & \\ 0 & 0 & 0 & \cdots & \lambda_k \end{pmatrix}$$

corresponding to a (B_1, B_2)-indecomposable n_k-dimensional subspace $\mathfrak{h}_k \leq \mathfrak{h}$ (see [85] and [23]). Collecting the vectors of the basis of $\mathfrak{h}$ in subsets $\mathcal{B}_k = \{e_1^k, \cdots, e_{n_k}^k; e_{n_k+1}^k, \cdots, e_{2n_k}^k\}$ which are a basis of the indecomposable subspace $\mathfrak{h}_k$, the relations defining $\mathfrak{g}$ are precisely

$$[e_i^k, e_{n_k+i}^k] = l_1 + \lambda_k \cdot l_2 \text{ for } i = 1, \cdots n_k$$

$$[e_i^k, e_{n_k+i+1}^k] = l_2 \qquad \text{for } i = 1, \cdots n_k - 1;$$

that is our assertion.   □

In the previous proposition there appear Lie algebras over an algebraically closed field k with nilpotency class two and a two-dimensional centre. These algebras are classified in [29] provided that the characteristic of k is greater than 2.

5.5.2 Proposition. *Let $\mathfrak{g}$ be a non-commutative co-aligned Lie algebra of nilpotency class two over a field k. Then the centre $\mathfrak{z}$ of $\mathfrak{g}$ coincides with the commutator subalgebra $\mathfrak{g}'$ of $\mathfrak{g}$, and for any ideal $\mathfrak{n}$ of $\mathfrak{g}$ we have either $\mathfrak{n} \leq \mathfrak{g}'$ or $\mathfrak{g}' \leq \mathfrak{n}$.*

Proof. Let $\mathfrak{g}' \cap \mathfrak{n} \neq \mathfrak{g}', \mathfrak{n}$. If $\dim \mathfrak{g}' > \dim \mathfrak{n}$, then $\mathfrak{g}'$ contains an ideal $\mathfrak{m}$ with $\dim \mathfrak{m} = \dim \mathfrak{n}$. Then the factor algebras $\mathfrak{g}/\mathfrak{m}$ and $\mathfrak{g}/\mathfrak{n}$ have commutator subalgebras of different dimension, a contradiction. If $\dim \mathfrak{g}' \leq \dim \mathfrak{n}$ we take an ideal $\mathfrak{m} \geq \mathfrak{g}'$ with $\dim \mathfrak{m} = \dim \mathfrak{n}$. The factor algebra $\mathfrak{g}/\mathfrak{m}$ is commutative, whereas the algebra $\mathfrak{g}/\mathfrak{n}$ is not commutative, again a contradiction.

Since $\mathfrak{g}$ has nilpotency class two, we have $\mathfrak{g}' \subseteq \mathfrak{z}$. Assume that $\mathfrak{z} = \mathfrak{g}' \oplus \mathfrak{c}$ with an ideal $\mathfrak{c} \neq 0$. If $\dim \mathfrak{g}' \leq \dim \mathfrak{c}$, then in $\mathfrak{c}$ there is an ideal $\mathfrak{c}_1$ having the same dimension as $\mathfrak{g}'$. Then $\mathfrak{g}/\mathfrak{c}_1$ is not commutative, whereas $\mathfrak{g}/\mathfrak{g}'$ is commutative. If $\dim \mathfrak{g}' > \dim \mathfrak{c}$ then $\mathfrak{g}'$ contains an ideal $\mathfrak{m}$ with $\dim \mathfrak{m} = \dim \mathfrak{c}$. In this case the factor algebras $\mathfrak{g}/\mathfrak{m}$ and $\mathfrak{g}/\mathfrak{n}$ have commutator algebras of different dimensions, a contradiction.   □

Using Lie's Theorem ([49], Theorem 14, p. 52) and the first part of the proof of the previous proposition we get the following

5.5.3 Corollary. *Let $\mathfrak{g}$ a co-aligned Lie algebra over an algebraically closed field of characteristic zero such that the commutator subalgebra $\mathfrak{g}'$ is commutative. Then for any ideal $\mathfrak{n}$ of $\mathfrak{g}$ we have either $\mathfrak{n} \leq \mathfrak{g}'$ or $\mathfrak{g}' \leq \mathfrak{n}$.*

5.5.4 Proposition. *Let $\mathfrak{g}$ be a non-commutative nilpotent Lie algebra of nilpotency class two over an algebraically closed field. Then $\mathfrak{g}$ is co-aligned if and only if $\mathfrak{g}$ is a Heisenberg algebra.*

Proof. It is clear that two epimorphic images of the same dimension of a Heisenberg algebra are isomorphic. Conversely, assume that every two epimorphic images of $\mathfrak{g}$ of the same dimension are isomorphic and that $\mathfrak{g}$ is a minimal counter-example to the assertion, with $\dim \mathfrak{g} = n$. Since the three-dimensional Heisenberg algebra H_1 is the unique three-dimensional nilpotent non-commutative Lie algebra, we infer that $n \geq 4$. As the nilpotency class of $\mathfrak{g}$ is two, we have either $\dim \mathfrak{z} = 1$ and $\mathfrak{g}/\mathfrak{z}$ is commutative, or $\dim \mathfrak{z} > 1$.

If $\dim \mathfrak{z} = 1$, then the centre is the commutator subalgebra of $\mathfrak{g}$, and by Proposition 5.3.8 the algebra $\mathfrak{g}$ is a Heisenberg algebra, a contradiction. Therefore $\mathfrak{z}$ must have dimension greater than one.

By proposition 5.5.2 one has $\mathfrak{g}' = \mathfrak{z}$. Since $\mathfrak{g}$ is a counter-example of minimal dimension, it has two one-dimensional ideals $\mathfrak{n}_i$, $i = 1, 2$, such that $\mathfrak{g}/\mathfrak{n}_i$ is a Heisenberg algebra. Now apply Proposition 5.5.1 and find a basis $\{l_1, l_2, e_i^k : k = 1, \cdots, s \text{ and } i = 1, \cdots, n_k\}$ such that

$$[e_i^k, e_{n_k+i}^k] = l_1 + \lambda_k \cdot l_2 \text{ for } i = 1, \cdots n_k$$

$$[e_i^k, e_{n_k+i+1}^k] = l_2 \qquad \text{for } i = 1, \cdots n_k - 1$$

where $\{l_1, l_2\}$ is a basis of $\mathfrak{z} = \mathfrak{g}'$ and the vectors e_i^k give, for $k = 1, \cdots, s$ and $i = 1, \cdots, n_k$, a basis of a complement $\mathfrak{h}$ of $\mathfrak{z}$. Since the factor algebra $\tilde{\mathfrak{g}} = \mathfrak{g}/\langle l_1 + \lambda_1 \cdot l_2 \rangle$ has a two dimensional centre $\tilde{\mathfrak{z}} = \langle \tilde{l}_2, \tilde{e}_{n_1+1}^1 \rangle$, it cannot be isomorphic to a Heisenberg algebra, our final contradiction.   □

Using the one-to-one correspondence between subalgebras and subgroups for formal groups, respectively between subalgebras and connected algebraic subgroups for connected affine algebraic groups over a field of characteristic zero ([39], Theorem 14.2.3, p. 81, [45] 13.1 Theorem, p. 87), the transfer of the assertion of Proposition 5.5.4 to formal groups, as well as to connected affine algebraic groups over an algebraically closed field of characteristic zero, is immediate.

5.5.5 Remark. The following examples show that there are many co-aligned Lie algebras over fields of characteristic zero having nilpotency class greater than two and having big automorphisms groups.

1) A large class of co-aligned Lie algebras of arbitrary nilpotency class can be obtained by taking a filiform Lie algebra $\mathfrak{g}$ and considering the *direct sum with amalgamated commutator subalgebra* $\mathfrak{g}'$ of an arbitrary number of algebras, each isomorphic to $\mathfrak{g}$. These algebras do not exhaust the class of algebras such that any two factor algebras of the same dimension are isomorphic, as the following point 2) shows.

2) Let $\mathfrak{g}_5$ be the 5-dimensional Lie algebra of nilpotency class three over an algebraically closed field k of characteristic zero defined on the basis $\{e_1, \cdots, e_5\}$ by the non-trivial relations

$$[e_1, e_2] = e_3 \ [e_1, e_3] = e_4, \ [e_2, e_3] = e_5.$$

This algebra has a three-dimensional commutator algebra $\mathfrak{g}_5' = \langle e_3, e_4, e_5 \rangle$ (which is therefore the unique three-dimensional ideal of $\mathfrak{g}_5$) and a two-dimensional centre $\mathfrak{z} = \langle e_4, e_5 \rangle$.

If $\mathfrak{n}$ is a one-dimensional ideal, then the factor algebra $\mathfrak{g}_5/\mathfrak{n}$ is a 4-dimensional Lie algebra of nilpotency class three, hence it is a filiform Lie algebra (cf. Remark 3.1.3). But according to [61], p. 162, all such filiform Lie algebras are isomorphic. If $\mathfrak{n}$ is a two-dimensional ideal, then the factor Lie algebra $\mathfrak{g}_5/\mathfrak{n}$ is a three-dimensional non-commutative nilpotent Lie algebra, and all such Lie algebras are isomorphic. Since $\mathfrak{g}_5'$ is the unique three-dimensional ideal, the Lie algebra $\mathfrak{g}_5$ is co-aligned.

Now we look at the 6-dimensional Lie algebra $\mathfrak{g}_6$ over k defined by the following non-trivial relations

$$\begin{aligned}
&[e_1, e_2] = e_3, \ [e_2, e_3] = e_5, \\
&[e_1, e_3] = e_4, \ [e_2, e_5] = e_6, \\
&[e_1, e_4] = e_6.
\end{aligned}$$

One sees that the centre $\mathfrak{z}$ of $\mathfrak{g}_6$ is one-dimensional and it is generated by e_6. Moreover $\mathfrak{g}_6/\mathfrak{z}$ is isomorphic to the above algebra $\mathfrak{g}_5$, and $\mathfrak{g}_6$ has nilpotency class four.

Finally we look at the Lie algebra $\mathfrak{g}_7$ over k defined by the following non-trivial relations

$$\begin{aligned}
&[e_1, e_4] = e_3, \ [e_2, e_6] = e_3, \\
&[e_1, e_5] = e_4, \ [e_2, e_7] = e_3 + e_4, \\
&[e_1, e_6] = e_5, \ [e_5, e_7] = e_3, \\
&[e_1, e_7] = e_6, \ [e_6, e_7] = e_2 + e_4.
\end{aligned}$$

The center $\mathfrak{z}$ of $\mathfrak{g}_7$ is $\langle e_3 \rangle$ and the Lie algebra $\mathfrak{g}_7/\mathfrak{z}$ is isomorphic to a Lie algebra $\mathfrak{h}$ with a basis $\{e_1', e_2', e_4', e_5', e_6', e_7'\}$ with the non-trivial relations

$$\begin{aligned}
&[e_1', e_5'] = e_4', &&[e_1', e_6'] = e_5', \\
&[e_1', e_7'] = e_6', &&[e_2', e_7'] = e_4', \\
&[e_6', e_7'] = e_2' + e_4'.
\end{aligned}$$

Now the linear mapping $f : \mathfrak{g}_6 \longrightarrow \mathfrak{h}$ determined by $f(e_1) = e_1' - e_6', f(e_2) = e_7', f(e_3) = -e_2' - e_4' + e_6', f(e_4) = e_5', f(e_5) = -e_2', f(e_6) = e_4'$

is an isomorphism from the Lie algebra $\mathfrak{g}_6$ to the Lie algebra $\mathfrak{h}$. Moreover, the factor Lie algebra $\mathfrak{g}_5$ is isomorphic to the factor algebra $\mathfrak{g}_7/\mathfrak{z}_2$, where $\mathfrak{z}_2$ is the second member of the ascending central series, and thus $\mathfrak{g}_7$ has nilpotency class five.

Summarizing the above discussion, we see that the Lie algebras $\mathfrak{g}_5$, $\mathfrak{g}_6$ and $\mathfrak{g}_7$ are co-aligned and have nilpotency class three, four and five, respectively.

In the next points we consider the automorphism group of the Lie algebras considered in 2) and discuss how the situation in 2) depends on the fact that the field k is algebraically closed.

3) Any of the Lie algebras $\mathfrak{g}_i$, $i = 5, 6$ allows a one-dimensional isotropic torus $T = \mathbf{G}_m$ as a group of automorphisms. If $\mathfrak{t} = \langle t \rangle$ is the Lie algebra of T, the action of t on $\mathfrak{g}_5$ (cf. [35]) is given by

$$[t, e_1] = ne_1, \qquad [t, e_2] = me_2,$$
$$[t, e_3] = (n + m)e_3, \quad [t, e_4] = (2n + m)e_4$$
$$[t, e_5] = (n + 2m)e_5$$

with $n, m \in \mathbb{Z}$, and the action of t on $\mathfrak{g}_6$ is given by

$$[t, e_1] = ne_1, \quad [t, e_2] = ne_2,$$
$$[t, e_3] = 2ne_3, \quad [t, e_4] = 3ne_4$$
$$[t, e_5] = 3ne_5 \quad [t, e_6] = 4ne_6.$$

In contrast to this, the Lie algebra $\mathfrak{g}_7$ does not allow a non-trivial action of a one-dimensional isotropic torus $T = \mathbf{G}_m$. Namely, otherwise for the Lie algebra $\mathfrak{t} = \langle t \rangle$ of T we obtain, using the fact that t is a derivation, the following relations:

$[t, e_1] = n(\sum_{i=1}^{7} a_i e_i)$ with $a_i \in$ k, $n \in \mathbb{Z}$ and $(a_1, a_7) \neq (0, 0)$,

$[t, e_7] = m(\sum_{i=1}^{7} b_i e_i)$ with $b_i \in$ k, $m \in \mathbb{Z}$ and $(b_1, b_7) \neq (0, 0)$,

$[t, e_6] = na_6 e_2 + (na_2 + na_5 + mb_4)e_3 + (na_2 + na_6 + mb_5)e_4 + mb_6 e_5 + (na_1 + mb_7)e_6,$

$[t, e_5] = -na_7 e_2 + (2na_2 + na_6 + mb_5)e_3 + (mb_6 - na_7)e_4 + (2na_1 + mb_7)e_5,$

$[t, e_4] = (mb_6 - 2na_7)e_3 + (3na_1 + mb_7)e_4,$

$[t, e_3] = (4na_1 + mb_7)e_3$

and from $[t, e_2 + e_4] = [t, e_2] + [t, e_4]$ and $[e_6, e_7] = e_2 + e_4$ we obtain

$[t, e_2] = (na_1 + 2mb_7)e_2 + (na_6 - mb_2 + 2na_7)e_3 + (na_6 - 2na_1 + mb_7)e_4 - mb_1 e_5.$

Now the relations

$0 = [t, [e_1, e_2]] = (-na_7 - 2na_1 + mb_7)e_3 - (na_7 + mb_1)e_4,$

$0 = [t, [e_4, e_7]] = -mb_1 e_3,$

$0 = [t, [e_5, e_7]] = (2na_1 + 2mb_7)e_3 - mb_1 e_4$

yield $0 = na_1 = na_7 = mb_1 = mb_7$ from which it follows that $n = m = 0$. This means that T centralises $\mathfrak{g}_7$.

Finally we consider the Lie algebras $\mathfrak{g}_i$, $i = 5, 6, 7$, over the field $\mathbb{R}$ of real numbers. The Lie algebras $\mathfrak{g}_5$ and $\mathfrak{g}_6$ allow the action of a one-dimensional real torus T; the action of the Lie algebra $\langle t \rangle$ of T is given in both cases by the following non-trivial relations:

$$[t, e_1] = e_2, \ [t, e_2] = -e_1,$$
$$[t, e_4] = e_5, \ [t, e_5] = -e_4.$$

In contrast to this, the real Lie algebra $\mathfrak{g}_7$ does not admit a non-trivial action of a real torus T. Indeed, the complexification of the semi-direct product of $\mathfrak{g}_7$ by the Lie algebra $\langle t \rangle$ of T is a solvable complex Lie algebra that is not nilpotent (cf. [49], Chapter 1, Section 8), and the complexification of T yields an isotropic torus which must centralise $\mathfrak{g}_7$.

4) For a non-algebraically closed field k there exist co-aligned non-commutative nilpotent Lie algebras of nilpotency class two defined over k which are not Heisenberg algebras. In fact, if $f(x) = a_0 + a_1 x + \cdots + x^d$ has no root in k, the companion matrix

$$A_f = \begin{pmatrix} 0 & 1 & 0 & \cdots & 0 \\ 0 & 0 & 1 & \cdots & 0 \\ 0 & 0 & 0 & \cdots & 0 \\ & & \vdots & & \\ -a_0 & -a_1 & -a_2 & \cdots & -a_{d-1} \end{pmatrix}$$

of $f(x)$ is a $d \times d$ matrix such that the pair of alternating bilinear forms (B_1, B_2), with the following matrix representation

$$\left(\begin{array}{c|c} & I_d \\ \hline -I_d & \end{array} \right), \left(\begin{array}{c|c} & A_f \\ \hline -A_f^t & \end{array} \right)$$

defines now canonically a nilpotent Lie algebra $\mathfrak{g}$ of dimension $2d + 2$, having a two-dimensional commutator algebra $\mathfrak{g}'$ which coincides with the centre $\mathfrak{z}$ of $\mathfrak{g}$. More precisely, $\mathfrak{g}$ has a basis $\{e_1, \cdots, e_d, e_{d+1}, \cdots, e_{2d}, l_1, l_2\}$ fulfilling the following relations

$$[e_i, e_{i+d}] = l_1 \qquad\qquad \text{for } i = 1, \cdots d - 1,$$

$$[e_i, e_{i+d+1}] = l_2 \qquad\qquad \text{for } i = 1, \cdots d - 1,$$

$$[e_d, e_{j+d}] = -a_{j-1} \cdot l_2 \quad\ \text{for } j = 1, \cdots d - 1,$$

$$[e_d, e_{2d}] = l_1 - a_{d-1} \cdot l_2.$$

Now we prove that every one-dimensional ideal $\mathfrak{n}$ is such that $\mathfrak{g}/\mathfrak{n}$ is a Heisenberg algebra. This is clear for $\mathfrak{n} = \langle l_2 \rangle$. Therefore we can assume that $\mathfrak{n} = \langle l_1 + \lambda \cdot l_2 \rangle$. The canonical basis $\{\tilde{e}_1, \cdots, \tilde{e}_d, \tilde{e}_{d+1}, \cdots, \tilde{e}_{2d}, \tilde{l}_2\}$ of the factor algebra $\tilde{\mathfrak{g}} = \mathfrak{g}/\mathfrak{n}$ is such that $\tilde{\mathfrak{g}}' = \langle \tilde{l}_2 \rangle$. If $\tilde{\mathfrak{h}}$ is the subspace of $\tilde{\mathfrak{g}}$ generated by $\{\tilde{e}_1, \cdots, \tilde{e}_d, \tilde{e}_{d+1}, \cdots, \tilde{e}_{2d}\}$, then the bilinear form $B :$ $\tilde{\mathfrak{h}} \times \tilde{\mathfrak{h}} \longrightarrow \mathsf{k}$ defined by

$$[\tilde{e}_i, \tilde{e}_j] = B(\tilde{e}_i, \tilde{e}_j) \cdot \tilde{l}_2$$

has the following matrix representation

$$\Lambda = \begin{pmatrix} & A_f - \lambda I_d \\ \hline -A_f^{\mathrm{t}} + \lambda I_d & \end{pmatrix}.$$

Since $|A_f - \lambda I_d| = f(\lambda)$ and $f(x)$ has no root in k, the form B is non-degenerate, and $\tilde{\mathfrak{g}}$ is a Heisenberg algebra by Corollary 5.3.9.

As a consequence, the commutator algebra $\mathfrak{g}'$ is the only two-dimensional ideal of $\mathfrak{g}$. Thus any two iso-dimensional images of $\mathfrak{g}$ are isomorphic, but in this case $\mathfrak{g}$ is not a Heisenberg algebra.

Finally we want to remark that, if $\mathsf{k} = \mathbb{R}$, a one-dimensional anisotropic torus T acts on this algebra $\mathfrak{g}$, because $f(x) = \Pi_k g_k(x)^{n_k}$ is a product of powers of irreducible polynomials $g_k(x)$ (of degree two). In this case the pseudo-Jordan form of A_f is a block matrix, each block being a $2n_k \times 2n_k$ matrix of the following form

$$J_{n_k}(\rho_k) = \begin{pmatrix} \rho_k & I & 0 & \cdots & 0 \\ 0 & \rho_k & I & \cdots & 0 \\ 0 & 0 & \rho_k & \cdots & 0 \\ & & \vdots & & \\ 0 & 0 & 0 & \cdots & \rho_k \end{pmatrix}$$

where $\rho_k = \begin{pmatrix} a_k & b_k \\ -b_k & a_k \end{pmatrix}$ has g_k as minimal polynomial. Since any one-dimensional anisotropic torus having the form of the block matrix $R_k(t)^{n_k}$, where $R_k(t) = \begin{pmatrix} \cos a_k t & -\sin a_k t \\ \sin a_k t & \cos a_k t \end{pmatrix}$ with $a_k \in \mathbb{R}$, commutes with $J_{n_k}(\rho_k)$, we can find a one-dimensional torus T acting on $\mathfrak{g}$.   □

5.5.6 Definition. Let $\mathfrak{g}$ be a nilpotent Lie algebra of nilpotency class two over a field k of characteristic zero and let $\mathfrak{z}$ be the centre of $\mathfrak{g}$. The algebra $\mathfrak{g}$ is called *regular* if the map $\mathrm{ad}(x) : \mathfrak{g} \longrightarrow \mathfrak{z}$, $\mathrm{ad}(x)(g) = [x, g]$, is surjective for all $x \notin \mathfrak{z}$ (cf. [56]).

The regular Lie algebras form a wide class. The class of the so called $H - Lie$ *algebras* over the reals, which are classified by modules over Clifford algebras (cf. [14]), consists of regular Lie algebras. In the case that $\mathfrak{g}$ is a regular Lie algebra over an algebraically closed field of characteristic zero the centre $\mathfrak{z}$ of $\mathfrak{g}$ is one-dimensional (cf. [56], p. 2426). Proposition 5.3.8 and Proposition 5.5.4

yield the fact that over an algebraically closed field of characteristic zero, the class of regular Lie algebras as well as the class of co-aligned Lie algebras of nilpotency class two coincides with the class of Heisenberg algebras. Also over arbitrary fields the regular Lie algebras and the co-aligned Lie algebras of nilpotency class two are related, as Proposition 5.5.2 and the following proposition show.

5.5.7 Proposition. *In a regular Lie algebra* $\mathfrak{g}$, *the centre* $\mathfrak{z}$ *of* $\mathfrak{g}$ *coincides with the commutator subalgebra* $\mathfrak{g}'$ *of* $\mathfrak{g}$, *and for any ideal* $\mathfrak{n}$ *of* $\mathfrak{g}$ *we have either* $\mathfrak{n} \le \mathfrak{g}'$ *or* $\mathfrak{g}' \le \mathfrak{n}$.

Proof. Let $\mathfrak{n}$ be an ideal of $\mathfrak{g}$ not contained in $\mathfrak{z}$. Since regular Lie algebras are nilpotent of class two, one has for every $x \in \mathfrak{n} \setminus \mathfrak{z}$ the inclusions

$$\mathfrak{g}' \subseteq \mathfrak{z} = \mathsf{ad}(x)\mathfrak{g} \subseteq [\mathfrak{n}, \mathfrak{g}] \subseteq \mathfrak{g}' \cap \mathfrak{n} \subseteq \mathfrak{z}.$$

$\square$

In contrast to algebraic groups over fields of characteristic zero, there are non-isomorphic three-dimensional non-commutative nilpotent real or complex Lie groups G. Let T be a topological torus of dimension one in the case of a real Lie group, and two in the case of a complex Lie group. Let $\phi : \mathsf{k} \longrightarrow T$ be a covering homomorphism, $\mathsf{k} \in \{\mathbb{R}, \mathbb{C}\}$. Then a non-simply connected real respectively complex Lie group is for instance given by the following operation:

$$(x_1, x_2, \phi(x_3))(y_1, y_2, \phi(y_3)) = (x_1 + y_1, x_2 + y_2, \phi(x_3 + y_3 + x_1 y_2)).$$

5.5.8 Proposition. *Let* G *be a connected non-affine algebraic group of nilpotency class two over an algebraically closed field* k *of characteristic zero such that* $G/D(G)$ *is not commutative. Then* G *is co-aligned if and only if it is isogenous to the direct product of a Heisenberg group* H *of dimension* $2d + 1$ *and the subgroup* $D(G)$ *and the following holds:*

(a) The group $D(G)$ *is co-aligned,*
(b) for any two subgroups D_1, D_2 *of* $D(G)$ *with* $\dim D_2 > \dim D_1$ *there are no closed subgroups* U_1, U_2 *of* H *such that* $\dim D_1 + \dim U_1 = \dim D_2 + \dim U_2$.

Proof. Since the affine algebraic group $G/D(G)$ is not commutative, it follows from Proposition 5.5.4 that $G/D(G)$ is a Heisenberg group H. In this case the commutator subgroup G' of G has dimension one and is not contained in $D(G)$. Consequently G is isogenous to the direct product $H \times D(G)$. $\square$

By giving an example we now show that without the assumptions made in Proposition 5.5.8 there is no hope to give a detailed description of all co-aligned algebraic groups which are not affine.

5.5.9 Example. Let k be a field of characteristic zero and let H be the Heisenberg group of dimension $2d + 1$ over k. We choose a co-aligned abelian variety A such that for any two subgroups A_1, A_2 of A with $\dim A_2 > \dim A_1$ there are no closed subgroups U_1, U_2 of H such that $\dim A_1 + \dim U_1 = \dim A_2 + \dim U_2$. Let D be a non-splitting central extension of a one-dimensional vector group E by A ([79], Proposition 11, p. 711). In the direct product $H \times D$ we identify E with the one-dimensional centre of H. This direct product with a amalgamated one-dimensional central subgroup is co-aligned.

For complex Lie groups we have

5.5.10 Proposition. *Let G be a connected complex Lie group of nilpotency class two such that G/X, where X is the maximal toroidal subgroup of G, is non-commutative. Then G is co-aligned if and only if it is isomorphic to the direct product of a Heisenberg group H and a co-aligned complex torus group X such that for any two subgroups X_1, X_2 of X with $\dim X_2 > \dim X_1$ there are no closed subgroups U_1, U_2 of H such that $\dim X_1 + \dim U_1 = \dim X_2 + \dim U_2$.*

Proof. If G is co-aligned, then it follows from Proposition 5.5.4 that G/X is a Heisenberg group. According to 1.1.5 in [1] the pre-image of the center of G/X is the direct product $G' \oplus X$ where G' is isomorphic either to $\mathbb{C}^+$ or $\mathbb{C}^*$. Hence G is the direct product of a Heisenberg group H with X. The toroidal group X has no closed subgroup of dimension $1 \leq n \leq \dim H$ because G is co-aligned. It follows from [1] 1.1.14, p. 12, that X is a complex torus. The converse is trivial. $\qquad\square$

Let H be an extra-special complex Lie group such that the closure of the commutator subgroup is a simple complex torus group X; then X is also the center of H (cf. Remark 5.3.6). The group G which is the direct product of n copies of H with amalgamated center X is a co-aligned group.

Let $\tilde{H}$ be a complex Heisenberg group such that its commutator subgroup $\tilde{H}'$ is isomorphic to $\mathbb{C}^*$. The group G which is the direct product of n copies of $\tilde{H}$ with amalgamated center $\tilde{H}'$ is a co-aligned group.

Let G be isomorphic to H or to $\tilde{H}$. The direct product $G \times Q$, where Q is a co-aligned complex torus, is co-aligned if and only if for any two subgroups Q_1, Q_2 of Q with $\dim Q_2 > \dim Q_1$ there is no closed subgroups U_1, U_2 of G such that $\dim Q_1 + \dim U_1 = \dim Q_2 + \dim U_2$.

5.5.11 Proposition. *Let*

$$\mathfrak{g} = \mathfrak{g}_1 \oplus \cdots \oplus \mathfrak{g}_{i_1} \oplus \mathfrak{g}_{i_1+1} \oplus \cdots \oplus \mathfrak{g}_{i_1+i_2} \oplus \cdots \oplus \mathfrak{g}_{i_1+i_2\cdots+i_t},$$

be a semi-simple Lie algebra of dimension n, where the $\mathfrak{g}_i$ are simple Lie algebras such that for $i_0 + \ldots + i_{k-1} < j \leq i_0 + \ldots + i_k$ the Lie algebras $\mathfrak{g}_j$ have the same dimension l_k for any $k = 1, \ldots, t$ and $i_0 = 0$.

The Lie algebra $\mathfrak{g}$ is co-aligned precisely if all subalgebras $\mathfrak{g}_j$ with $i_0 + \ldots + i_{k-1} < j \leq i_0 + \ldots + i_k$ are isomorphic, and $\sum_{k=1}^{t} r_k l_k = \sum_{k=1}^{t} s_k l_k$, with $i_{k-1} < r_k, s_k \leq i_k$ for any $k = 1, \ldots, t$, holds if and only if $r_k = s_k$ for any $k = 1, \ldots, t$.

Proof. Assume that for some k with $1 \le k \le t$ there are algebras $\mathfrak{g}_{j_1}, \mathfrak{g}_{j_2}$ with $i_0 + \ldots + i_{k-1} < j_1, j_2 \le i_0 + \ldots + i_k$ which are not isomorphic. Then the factor algebras $\mathfrak{g}/\mathfrak{h}_1$ and $\mathfrak{g}/\mathfrak{h}_2$, where $\mathfrak{h}_i = \mathfrak{g}_1 \oplus \cdots \oplus \mathfrak{g}_{j_i - 1} \oplus \mathfrak{g}_{j_i + 1} \oplus \cdots \oplus \mathfrak{g}_{i_1 + i_2 + \cdots + i_t}$, for $i = 1, 2$, are not isomorphic, a contradiction.

If the integer $\sum_{k=1}^{t} r_k l_k$ has a unique representation with $i_{k-1} < r_k, s_k \le i_k$ for any $k = 1, \ldots, t$, then there is only one factor algebra of dimension $n - \sum_{k=1}^{t} r_k l_k$. Conversely, if there is not a unique representation of $\sum_{k=1}^{t} r_k l_k$, then the factor algebras $\mathfrak{g}/\mathfrak{h}$ and $\mathfrak{g}/\mathfrak{l}$, where $\mathfrak{h}$ is isomorphic to a direct product of r_k copies of $\mathfrak{g}_{i_1 + \cdots + i_k}$ for any $1 \le k \le t$, whereas $\mathfrak{l}$ is isomorphic to a direct product of s_k copies of $\mathfrak{g}_{i_1 + \cdots + i_k}$ for any $1 \le k \le t$, are not isomorphic, a contradiction. $\qquad \square$

5.5.12 Proposition. *Let $\mathfrak{g}$ be a finite-dimensional non-solvable Lie algebra over a field k of characteristic zero such that the non-trivial solvable radical $\mathfrak{r}$ of $\mathfrak{g}$ is commutative and a Levi complement $\mathfrak{s}$ of $\mathfrak{g}$ acts faithfully on $\mathfrak{r}$. Then the Lie algebra $\mathfrak{g}$ is co-aligned if and only if the following conditions hold:*

  i) if $\mathfrak{m}_1$ and $\mathfrak{m}_2$ are invariant subspaces under $\mathfrak{s}$, then the linear representations of $\mathfrak{s}$ on $\mathfrak{m}_1$ and $\mathfrak{m}_2$ are equivalent;

  ii) if $\mathfrak{g}_1$ and $\mathfrak{g}_2$ are ideals of $\mathfrak{g}$ of the same dimension, then the Levi complements are isomorphic.

Proof. The Lie algebra $\mathfrak{g}$ is a semi-direct product of $\mathfrak{r}$ by $\mathfrak{s}$. The ideals of $\mathfrak{g}$ which are contained in $\mathfrak{r}$ are invariant subspaces of $\mathfrak{r}$ under $\mathfrak{s}$. Any other ideal has the form $\mathfrak{v} + \mathfrak{t}$ where $\mathfrak{v}$ is a non-trivial invariant subspace and $\mathfrak{t}$ is an ideal of $\mathfrak{s}$ such that $\mathfrak{t}$ centralises the invariant complement $\mathfrak{w}$ of $\mathfrak{v}$ in $\mathfrak{r}$ ([49], p. 79).

Now we assume that any two epimorphic images $\mathfrak{g}/\mathfrak{g}_1$ and $\mathfrak{g}/\mathfrak{g}_2$ of $\mathfrak{g}$ of the same dimension are isomorphic. If $\mathfrak{g}_i$ is contained in $\mathfrak{r}$, for $i = 1, 2$, then the invariant complement $\mathfrak{m}_i$ of $\mathfrak{g}_i$ in $\mathfrak{r}$ is an ideal of $\mathfrak{g}$. Since the factor algebras $\mathfrak{g}/\mathfrak{g}_1$ and $\mathfrak{g}/\mathfrak{g}_2$ are isomorphic, the linear representations of $\mathfrak{s}$ on $\mathfrak{m}_1$ and $\mathfrak{m}_2$ are equivalent.

Let $\mathfrak{g}_1$ and $\mathfrak{g}_2$ be non-solvable ideals of $\mathfrak{g}$ of the same dimension. The Lie algebra $\mathfrak{g}/\mathfrak{g}_i$ is a semi-direct product of a commutative ideal $\mathfrak{m}'_i$ by a semisimple algebra $\mathfrak{s}'_i \ne 0$, for $i = 1, 2$. The ideal $\mathfrak{g}_i$ is a semi-direct product of a commutative ideal $\mathfrak{m}_i$ by a semisimple algebra $\mathfrak{s}_i \ne 0$, for $i = 1, 2$. Since $\mathfrak{m}_i$ is invariant under $\mathfrak{s}$, we can choose for $\mathfrak{m}'_i$ the invariant complement of $\mathfrak{m}_i$ in $\mathfrak{r}$ and for $\mathfrak{s}'_i$ the complement of $\mathfrak{s}_i$ in $\mathfrak{s}$. Since $\mathfrak{g}/\mathfrak{g}_1$ and $\mathfrak{g}/\mathfrak{g}_2$ are isomorphic, one has $\dim \mathfrak{m}'_1 = \dim \mathfrak{m}'_2$ and $\mathfrak{s}'_1$ and $\mathfrak{s}'_2$ are isomorphic. It follows that $\mathfrak{s}_1$ is isomorphic to $\mathfrak{s}_2$.

Now we assume that the Lie algebra $\mathfrak{g}$ satisfies the conditions $i)$ and $ii)$. Let $\mathfrak{g}_1$ and $\mathfrak{g}_2$ be ideals of $\mathfrak{g}$ having the same dimension. Since by $ii)$ two Levi complements $\mathfrak{s}_i$ of $\mathfrak{g}_i$ are isomorphic they may be identified with a subalgebra $\mathfrak{t}$ of $\mathfrak{s}$. Moreover we have $\dim (\mathfrak{g}_1 \cap \mathfrak{r}) = \dim (\mathfrak{g}_2 \cap \mathfrak{r})$. Hence the factor algebra $\mathfrak{g}/\mathfrak{g}_i$ is isomorphic to the semi-direct product $\mathfrak{m}'_i + \mathfrak{u}$, where $\mathfrak{m}'_i$ is the invariant complementary subspace of $\mathfrak{g}_i \cap \mathfrak{r}$ in $\mathfrak{r}$ and $\mathfrak{u}$ is the complementary ideal of $\mathfrak{t}$ in $\mathfrak{s}$. The Lie algebra $\mathfrak{t}$ centralises the invariant subspaces $\mathfrak{m}'_i$. Therefore $\mathfrak{u}$ has on

$\mathfrak{m}'_i$ the same representation as $\mathfrak{s}$. Since $\mathfrak{m}'_1$ and $\mathfrak{m}'_2$ have the same dimension, by i) the linear representations of $\mathfrak{u}$ on $\mathfrak{m}'_i$ are equivalent and the Theorem is proved. □

5.5.13 Remark. We note that Proposition 5.5.11 and Proposition 5.5.12 can be easily transferred to affine algebraic groups over fields of characteristic zero, to real Lie groups and to linear complex Lie groups. For arbitrary algebraic groups over fields of characteristic zero and for general complex Lie groups this also brings no essential difficulties as the following claim shows.

Let G be a connected algebraic group over a field of characteristic zero or a connected complex Lie group such that the solvable radical of G is commutative. Then G is the direct product of its commutator subgroup and a central subgroup of G.

Indeed, in both cases G is the semi-direct product of its solvable radical R, which is commutative, and a closed semi-simple group S ([45], p. 184, [55], Corollary 4.18, p. 113). The centraliser of C of S in R has a complement V in R. Since VS is the commutator subgroup of G, the claim is proved.

After this claim we feel justified not to formulate in detail the combinatorial conditions which characterise the co-aligned groups in this class. □

5.5.14 Proposition. *Let $\mathfrak{g}$ be a finite-dimensional solvable non-nilpotent Lie algebra over a field k of characteristic zero such that the non-trivial nilpotent radical $\mathfrak{n}$ of $\mathfrak{g}$ is commutative and the factor algebra $\mathfrak{g}/\mathfrak{n}$ acts faithfully and completely reducibly on $\mathfrak{n}$. Then the Lie algebra $\mathfrak{g}$ is co-aligned if and only if the following conditions hold:*

i) if $\mathfrak{m}_1$ and $\mathfrak{m}_2$ are invariant subspaces of $\mathfrak{n}$ under $\mathfrak{g}/\mathfrak{n}$, then the linear representations of $\mathfrak{g}/\mathfrak{n}$ on $\mathfrak{m}_1$ and $\mathfrak{m}_2$ are equivalent;

ii) if $\mathfrak{g}_1$ and $\mathfrak{g}_2$ are ideals of $\mathfrak{g}$ of the same dimension, then $\dim(\mathfrak{g}_1 \cap \mathfrak{n}) = \dim(\mathfrak{g}_2 \cap \mathfrak{n})$.

Proof. The factor algebra $\mathfrak{g}/\mathfrak{n}$ is commutative since the commutator subalgebra of a solvable Lie algebra is nilpotent.

The Lie algebra $\mathfrak{g}$ is a semi-direct product of $\mathfrak{n}$ by a commutative subalgebra $\mathfrak{k}$. Namely, if $\mathfrak{m}$ is a subalgebra of $\mathfrak{g}$ which is a semi-direct product $\mathfrak{n} + \mathfrak{k}_1$ with commutative $\mathfrak{k}_1$ and $\mathfrak{n} \cap \mathfrak{k}_1 = 0$, then any subalgebra of $\mathfrak{g}$ having $\mathfrak{m}$ as a subalgebra of co-dimension one is a semi-direct product $\mathfrak{n} + \mathfrak{k}_2$ with commutative $\mathfrak{k}_2$ and $\mathfrak{n} \cap \mathfrak{k}_2 = 0$.

The ideals of $\mathfrak{g}$ which are contained in $\mathfrak{n}$ are the invariant subspaces of $\mathfrak{n}$ under $\mathfrak{k}$. Any other ideal $\mathfrak{l}$ has the form $(\mathfrak{l} \cap \mathfrak{n}) + \mathfrak{h}$ where $\mathfrak{h}$ is an ideal of $\mathfrak{k}$ such that $\mathfrak{h}$ centralises the invariant complement of $(\mathfrak{l} \cap \mathfrak{n})$ in $\mathfrak{n}$.

Now we assume that any two epimorphic images $\mathfrak{g}/\mathfrak{g}_1$ and $\mathfrak{g}/\mathfrak{g}_2$ of $\mathfrak{g}$ of the same dimension are isomorphic. If $\mathfrak{g}_i$ is contained in $\mathfrak{n}$, for $i = 1, 2$, then the invariant complement $\mathfrak{m}_i$ of $\mathfrak{g}_i$ in $\mathfrak{n}$ is an ideal of $\mathfrak{g}$. Since the factor algebras $\mathfrak{g}/\mathfrak{g}_1$ and $\mathfrak{g}/\mathfrak{g}_2$ are isomorphic, the linear representations of $\mathfrak{k}$ on $\mathfrak{m}_1$ and $\mathfrak{m}_2$ are equivalent.

Let $\mathfrak{g}_1$ and $\mathfrak{g}_2$ be non-nilpotent ideals of $\mathfrak{g}$ of the same dimension. The Lie algebra $\mathfrak{g}/\mathfrak{g}_i$ is a semi-direct product of a commutative ideal $\mathfrak{m}_i'$ by a commutative algebra $\mathfrak{k}_i' \neq 0$, for $i = 1, 2$. The ideal $\mathfrak{g}_i$ is a semi-direct product of $\mathfrak{g}_i \cap \mathfrak{n}$ by a commutative algebra $\mathfrak{k}_i \neq 0$, for $i = 1, 2$. Since $\mathfrak{g}_i \cap \mathfrak{n}$ is invariant under $\mathfrak{k}$, we can choose for $\mathfrak{m}_i'$ the invariant complement of $\mathfrak{g}_i \cap \mathfrak{n}$ in $\mathfrak{n}$ and for $\mathfrak{k}_i'$ a complement of $\mathfrak{k}_i$ in $\mathfrak{k}$. Since $\mathfrak{g}/\mathfrak{g}_1$ and $\mathfrak{g}/\mathfrak{g}_2$ are isomorphic, one has $\dim \mathfrak{m}_1' = \dim \mathfrak{m}_2'$ from which it follows that $\dim (\mathfrak{g}_1 \cap \mathfrak{n}) = \dim (\mathfrak{g}_2 \cap \mathfrak{n})$.

Now we assume that the Lie algebra $\mathfrak{g}$ satisfies the conditions $i)$ and $ii)$. Let $\mathfrak{g}_1 = (\mathfrak{g}_1 \cap \mathfrak{n}) + \mathfrak{l}_1$ and $\mathfrak{g}_2 = (\mathfrak{g}_2 \cap \mathfrak{n}) + \mathfrak{l}_2$ be ideals of $\mathfrak{g}$ having the same dimension. By $ii)$ we have $\dim (\mathfrak{g}_1 \cap \mathfrak{n}) = \dim (\mathfrak{g}_2 \cap \mathfrak{n})$. Since $\mathfrak{k}$ is commutative it follows that $\mathfrak{l}_1$ and $\mathfrak{l}_2$ may be identified with a subalgebra $\mathfrak{l}$ of $\mathfrak{k}$. Hence the factor algebra $\mathfrak{g}/\mathfrak{g}_i$ is isomorphic to the semi-direct product $\mathfrak{m}_i' + \mathfrak{f}$, where $\mathfrak{m}_i'$ is the invariant complementary subspace of $\mathfrak{g}_i \cap \mathfrak{n}$ in $\mathfrak{n}$ and $\mathfrak{f}$ is a complementary ideal of $\mathfrak{l}$ in $\mathfrak{k}$. The algebra $\mathfrak{l}$ centralises the invariant subspaces $\mathfrak{m}_i'$. Therefore $\mathfrak{f}$ has on $\mathfrak{m}_i'$ the same representation as $\mathfrak{k}$. Since $\mathfrak{m}_1'$ and $\mathfrak{m}_2'$ have the same dimension, by $i)$ the linear representations of $\mathfrak{l}$ on $\mathfrak{m}_i'$ are equivalent and the theorem is proved.   □

5.5.15 Lemma. *Let G be a co-aligned group which is solvable but not nilpotent and belongs to one of the following classes:*

(1) connected affine algebraic groups over an algebraically closed field k of characteristic zero,

(2) connected complex linear Lie groups.

Then G is a semi-direct product of the unipotent radical U and a one-dimensional torus T isomorphic to k^*, respectively $\mathbb{C}^*$, acting non-trivially on U.

Proof. By [49], Theorem 14, p. 52, there is a normal subgroup N of G having co-dimension one in the unipotent radical U of G. The factor group G/N is co-aligned, solvable but not nilpotent and has the one-dimensional unipotent radical U/N ([55], Proposition 3.15, p. 96). With $d = \dim G/U$ the group $(G/N)/(U/N)$ is isomorphic to $(\mathsf{k}^*)^d$, where in the case (2) one puts $\mathsf{k} = \mathbb{C}$ and observes [55], Proposition 1.22, p. 33. If d were greater than one, then G/N would contain a one-dimensional central subgroup C/N isomorphic to k^*. But then the factor groups G/U and G/C would not be isomorphic, a contradiction.

Since G is solvable but not nilpotent, the group G is a semi-direct product of U with a group isomorphic to k^* acting non-trivially on U.   □

5.5.16 Theorem. *Let G be a solvable non-nilpotent connected group of dimension $n > 2$ in one of the following classes:*

(1) connected affine algebraic groups over an algebraically closed field k of characteristic zero,

(2) connected linear complex Lie groups.

Let the unipotent radical U of G be commutative. Then G is co-aligned if and only if the following holds:

i) for $n = 3$ the group G is isomorphic to the group defined by the following operation

$$(t, x_1, x_2)(s, y_1, y_2) = (ts, s^n x_1 + y_1, s^{\varepsilon n} x_2 + y_2),$$

with a positive integer n, $\varepsilon = \pm 1$, $t, s \in \mathsf{k}^*$ and $x_i, y_i \in \mathsf{k}$;
ii) for $n > 3$ the group G is a semi-direct product of the vector group U with a one-dimensional torus T such that every one-dimensional closed connected subgroup of U is normalized but not centralised by T.

Proof. First we assume that every two epimorphic images of G of the same dimension are isogenous. From Lemma 5.5.15 it follows that G is a semi-direct product of U with a group $T \cong \mathsf{k}^*$, where in the case (2) one puts $\mathsf{k} = \mathbb{C}$, and that T does not centralise U.

Let $n = 3$. Then U is two-dimensional and according to [49], p. 12, the group G has either two or infinitely many one-dimensional normal subgroups. In the latter case we have $\epsilon = 1$. In the former case the assertion follows by Proposition 5.5.14 (i).

Now we treat the case where $n > 3$. Let $U = M_1 \times \cdots \times M_{n-1}$, where M_i are one-dimensional invariant subspaces of U under T. By Proposition 5.5.14 it follows that the actions of T on any two invariant subspaces of U are equivalent. This holds in particular for $M_2 \times M_3 \times \cdots \times M_{n-1}$ and $M_1 \times M_3 \times \cdots \times M_{n-1}$, which yields that any subspace of U is invariant under T.

Conversely, since U is commutative, two epimorphic images of G of the same dimension are isomorphic, since G satisfies the conditions of Proposition 5.5.14. $\qquad\square$

For a connected not necessarily affine algebraic group over an algebraically closed field of characteristic zero, we put $F(G) = D(G)$, whereas for a connected complex Lie group we denote by $F(G)$ the greatest toroidal subgroup of G. With this notation we have

5.5.17 Corollary. *Let G be a solvable non-nilpotent connected group of dimension $n > 2$ in one of the following classes:*

(1) connected non-affine algebraic groups over an algebraically closed field k of characteristic zero,
(2) connected complex Lie groups.

Let the unipotent radical of $G/F(G)$ be commutative. Then G is co-aligned if and only if the following holds:

(a) G_1 satisfies the condition i) or ii) of Theorem 5.5.16, where G_1 denotes the maximal connected affine subgroup, respectively the maximal connected linear complex Lie subgroup of G,

(b) $F(G)$ is an abelian variety if G is an algebraic group, respectively a complex torus if G is a complex Lie group, and G is the direct product of G_1 and $F(G)$,

(c) $F(G)$ is co-aligned, has no closed connected subgroup of co-dimension m in $F(G)$ with $1 \leq m \leq \dim G_1$ and has no closed connected subgroup having the same dimension as a closed connected proper diagonal subgroup of $G_1 \times F(G)$.

Proof. Assume first that G is co-aligned. Then $G/F(G)$ has the structure given in Theorem 5.5.16. In $G/F(G)$ and hence in G we find a one-dimensional closed subgroup T isomorphic to k^* respectively to $\mathbb{C}^*$ acting on the unipotent radical U^* of $G/F(G)$ such that the centraliser of T in U^* is trivial. Then the centraliser of T in the pre-image of U^* in G is the group $F(G)$. It follows that (a), (b), (c) hold.

The converse implication is obvious.   □

5.5.18 Lemma. *Let $\mathfrak{g}$ be a nilpotent Lie algebra of class two over an algebraically closed field k of characteristic zero and let Γ be a group of automorphisms of $\mathfrak{g}$ acting such that every subspace of $\mathfrak{g}/\mathfrak{g}'$ is invariant under Γ. Then every subspace of $\mathfrak{g}'$ is invariant under Γ, too.*

Proof. Consider a Γ-invariant complement $\mathfrak{c}$ of $\mathfrak{g}'$. Then there is a basis

$$\{e_{1,1}, \cdots e_{1,r_1}, e_{2,1}, \cdots e_{2,r_2}\}$$

of $\mathfrak{g}$ for which $\{e_{1,1}, \cdots e_{1,r_1}\}$ is a basis of $\mathfrak{c}$, $\{e_{2,1}, \cdots e_{2,r_2}\}$ is a basis of $\mathfrak{g}'$ and $e_{2,k} = [e_{1,i}, e_{1,j}]$ for every $1 \leq k \leq r_2$ and suitable $1 \leq i,j \leq r_1$. For $\gamma \in \Gamma$ there is $c \in \mathsf{k}$ such that $e_{1,i}^{\gamma} = ce_{1,i}$ for all $1 \leq i \leq r_1$. But then

$$e_{2,k}^{\gamma} = [e_{1,i}, e_{1,j}]^{\gamma} = [e_{1,i}^{\gamma}, e_{1,j}^{\gamma}] = c^2[e_{1,i}, e_{1,j}] = c^2 e_{2,k}$$

for every $1 \leq k \leq r_2$ and suitable $1 \leq i,j \leq r_1$.   □

5.5.19 Proposition. *Let $\mathfrak{g}$ be a nilpotent Lie algebra of class two over an algebraically closed field k of characteristic zero and let Γ be a group of automorphisms of $\mathfrak{g}$ acting such that every subspace of $\mathfrak{g}/\mathfrak{g}'$ is invariant under Γ. Then every ideal $\mathfrak{h}$ of $\mathfrak{g}$ with $\mathfrak{h} \cap \mathfrak{g}' \neq 0$ is invariant under Γ.*

Proof. Assuming that the proposition is wrong, we consider a counter-example $\mathfrak{g}$ of minimal dimension. Let $\mathfrak{h}$ be an ideal of $\mathfrak{g}$ which is not Γ-invariant. Then by Proposition 5.5.18 the ideal $\mathfrak{h}$ is not contained in $\mathfrak{g}'$ and $\mathfrak{h}_1 = \mathfrak{h} \cap \mathfrak{g}'$ is Γ-invariant. Since $\mathfrak{g}$ is a minimal counter-example, we see that the algebra $\mathfrak{h}/\mathfrak{h}_1$ is Γ-invariant in $\mathfrak{g}/\mathfrak{h}_1$.   □

5.5.20 Theorem. *Let $\mathfrak{g}$ be a solvable, but not nilpotent Lie algebra of dimension $n \geq 4$ over an algebraically closed field k of characteristic zero such that the unipotent radical $\mathfrak{u}$ of $\mathfrak{g}$ has nilpotency class two. The Lie algebra $\mathfrak{g}$ is co-aligned if and only if the following holds:*

i) *for* $n = 4$ *the ideal* $\mathfrak{u}$ *is a Heisenberg algebra with a basis* $\{e_1, e_2, e_3\}$ *such that* $[e_1, e_2] = e_3$. *The ideal* $\mathfrak{u}$ *has a one-dimensional complementary subspace* $\mathfrak{t} = \langle t \rangle$ *in* $\mathfrak{g}$, *acting on* $\mathfrak{u}$ *as follows:* $[t, e_1] = m e_1$, $[t, e_2] = \epsilon m e_2$ *and* $[t, e_3] = (1 + \epsilon) m e_3$ *for some* $m \in \mathbb{N}$, *where* $\epsilon = \pm 1$.

ii) *for* $n > 4$ *the ideal* $\mathfrak{u}$ *is a Heisenberg algebra, the ideal* $\mathfrak{u}$ *has a one-dimensional complementary subspace* $\mathfrak{t}$, *acting on* $\mathfrak{u}$ *such that every non-trivial subspace of* $\mathfrak{u}$ *is normalized but not centralised.*

Proof. First we assume that $\mathfrak{g}$ is co-aligned. In analogy to the proof of Lemma 5.5.15 one shows that $\mathfrak{u}$ has co-dimension one in $\mathfrak{g}$. Let $\mathfrak{t} = \langle t \rangle$ be a one-dimensional subalgebra of $\mathfrak{g}$ which is complementary to $\mathfrak{u}$. If $n = 4$, then $\mathfrak{u}$ is the Heisenberg algebra of dimension three and we can take a basis $\{e_1, e_2, e_3\}$ such that $[e_1, e_2] = e_3$ and $\langle e_1, e_2 \rangle$ is invariant under t, because t acts semisimply. By Proposition 5.5.14 the assertion follows, because the action of t on $\langle e_1, e_2 \rangle$ is given by $[t, e_1] = m e_1$, $[t, e_2] = \epsilon m e_2$, with $m \in \mathbb{N}$ and $\epsilon = \pm 1$.

Now we treat the case where $n > 4$. Analogously to the proof of Theorem 5.5.16, it follows that under the action of t every non-trivial subspace of $\mathfrak{u}/\mathfrak{z}\mathfrak{u}$ is normalized but not centralised. Thus by Lemma 5.5.18 every subspace of $\mathfrak{u}'$ is invariant under t. It follows from Proposition 5.5.19 that every ideal $\mathfrak{n}$ of $\mathfrak{u}$ satisfying $\mathfrak{u} \cap \mathfrak{u}' \neq 0$ is an ideal in $\mathfrak{g}$.

Assume that $\mathfrak{u}'$ is properly contained in the center $\mathfrak{z}\mathfrak{u}$ of $\mathfrak{u}$. Then $\mathfrak{z}\mathfrak{u} = \mathfrak{u}' \oplus \mathfrak{c}$ with a $\mathfrak{t}$-invariant subspace $\mathfrak{c}$ of $\mathfrak{z}\mathfrak{u}$. Let $\mathfrak{f}$ be a one-dimensional subspace of $\mathfrak{u}'$ and let $\mathfrak{e}$ be a one-dimensional subspace of $\mathfrak{c}$. Then $\mathfrak{e}$ and $\mathfrak{f}$ are ideals by Lemma 5.5.18. But the commutator algebras of $(\mathfrak{g}/\mathfrak{e})'$ and $(\mathfrak{g}/\mathfrak{f})'$ have different dimensions, a contradiction to the fact that $\mathfrak{g}$ is co-aligned. Hence $\mathfrak{z}\mathfrak{u} = \mathfrak{u}'$ which implies that every ideal $\neq 0$ of $\mathfrak{u}$ intersects $\mathfrak{u}'$ non-trivially. Thus every ideal of $\mathfrak{u}$ is an ideal of $\mathfrak{g}$. Let $\mathfrak{n}_1, \mathfrak{n}_2$ be ideals of $\mathfrak{u}$. Since $\mathfrak{u}/\mathfrak{n}_i$ is the unipotent radical of $\mathfrak{g}/\mathfrak{n}_i$, it follows that $\mathfrak{u}$ is co-aligned. Hence $\mathfrak{u}$ is a Heisenberg algebra by Proposition 5.5.4.

Choose $e_1, e_2 \in \mathfrak{u}$ such that $[e_1, e_2]$ spans the centre of $\mathfrak{u}$ and let T be the one-parameter subgroup corresponding to $\mathfrak{t}$. Since every subspace of the vector space spanned by an element $a_1 e_1 + a_2 e_2$ is invariant modulo $\mathfrak{z}\mathfrak{u}$ under $Ad\ T$, for every $t \in T$ there is a scalar c_t such that $(Ad\ t) e_i = c_t e_i$ for $i = 1, 2$. Hence $(Ad\ t)[e_1, e_2] = c_t^2 [e_1, e_2]$. This means that T operates non-trivially on $\mathfrak{z}\mathfrak{u}$.

Conversely, with the help of Proposition 5.5.12 one sees that every Lie algebra satisfying the conditions of the theorem is co-aligned.   □

5.5.21 Theorem. *Let G be a solvable, but non-nilpotent connected group of dimension $n > 3$ in one of the following classes:*

(1) connected affine algebraic groups over an algebraically closed field k *of characteristic zero,*

(2) connected complex linear Lie groups.

Let the unipotent radical U of G be of nilpotency class two. Then G is co-aligned if and only if the following holds:

i) for $n = 4$ the group G is isomorphic to the group defined by the following operation

$$(t, x_1, x_2, x_3)(s, y_1, y_2, y_3) = (ts, s^m x_1 + y_1, s^{\varepsilon m} x_2 + y_2, s^{(1+\varepsilon)m} x_3 + y_3 + s^m x_1 y_2),$$

with $\varepsilon = \pm 1$, $0 < m \in \mathbb{Z}$, $t, s \in \mathsf{k}^*$ and $x_i, y_i \in \mathsf{k}^+$, where $\mathsf{k} = \mathbb{C}$ for complex Lie groups.

ii) for $n > 4$ the unipotent radical U is a Heisenberg group and has a one-dimensional complementary subgroup T, acting on U such that every non-trivial connected subgroup of U is normalized but not centralised.

Proof. The assertion follows from Theorem 5.5.20 and Lemma 5.5.15, using for algebraic groups the correspondence between subalgebras and connected algebraic subgroups ([45], Theorem 13.1, p. 87) and for complex Lie groups [40], Theorem 3.2, p. 80 and [55], Theorem 1.15, p. 23, since the exponential map is bijective on the unipotent radical. □

As before, we use the notation $F(G)$ for the subgroup $D(G)$ of a connected non-affine algebraic group G over an algebraically closed field of characteristic zero and for the greatest toroidal subgroup of a connected complex Lie group G.

5.5.22 Corollary. *Let G be a solvable, but non-nilpotent connected group of dimension $n > 2$ in one of the following classes:*

(1) connected non-affine algebraic groups over an algebraically closed field k of characteristic zero,

(2) connected complex Lie groups.

If the unipotent radical of $G/F(G)$ is nilpotent of class two, then G is co-aligned if and only if $G = H \times F(G)$ where:

i) H is a connected affine algebraic group, respectively a linear complex Lie group isomorphic to a group in Theorem 5.5.21,

ii) $F(G)$ is a co-aligned abelian variety if G is an algebraic group, respectively a co-aligned complex torus if G is a complex Lie group,

iii) $F(G)$ has no closed connected subgroup of co-dimension m in $F(G)$ with $1 \leq m \leq \dim H$ and has no closed connected subgroup having the same dimension as a closed connected proper diagonal subgroup of $H \times F(G)$.

Proof. We first assume that G is co-aligned. Let U^* be the unipotent radical of $G/F(G)$ and let U be the pre-image of U^*.

According to Theorem 5.5.21 the group $U/F(G)$ is a Heisenberg group of dimension $2d + 1$ with $d \geq 1$ and G itself is a semi-direct product of U with a one-dimensional torus T. Let $\mathfrak{g}$ be the Lie algebra of G. Then in $\mathfrak{g}$ there exist elements e_i, e_{d+i} with $1 \leq i \leq d$ such that every subalgebra generated by e_i is normalized but not centralised by T and e_i does not commute with e_{d+i} for any i. Since for any i the elements e_i and e_{d+i} generate modulo the subalgebra corresponding to $F(G)$ a three-dimensional Heisenberg algebra, the subalgebra generated by $[e_i, e_{d+i}]$ is normalized but not centralised by $Ad(T)$. It follows that any non-trivial connected subgroup of the group H corresponding to the Lie algebra generated by $e_1, \cdots, e_{2d}$ is normalized but not centralised by T. Since $F(G)$ is contained in the center of G, one has $H \cap F(G) = 1$. As H is Heisenberg group of dimension $2d + 1$, we have that G is the direct product of H and $F(G)$. Moreover $F(G)$ must be a co-aligned abelian variety if G is an algebraic group, respectively a co-aligned complex torus if G is a complex Lie group.

Conversely, one easily checks that the groups described in the claim are co-aligned.   □

Let G be a formal group, a connected affine algebraic group or a connected real, respectively linear complex Lie group. We recall that G is called *filiform* if the Lie algebra of G is filiform.

There are even solvable but not nilpotent affine algebraic groups G over an algebraically closed field of characteristic zero, respectively connected linear complex Lie groups which are co-aligned and which have a unipotent radical of nilpotency class $\dim G - 2$. Namely in [35] (see also [2], Proposition 1 and Theorem 2) there are determined all filiform Lie algebras $\mathfrak{u}$ over an algebraically closed field of characteristic zero which admit a non-trivial action of an algebraic, respectively linear complex torus group T. If $\mathfrak{u}$ admits only a one-dimensional torus T, then $\mathfrak{u}$ belongs to one of three families of filiform Lie algebras $A_n^k(\lambda_1, \cdots, \lambda_{t-1})$, $B_n^k(\lambda_1, \cdots, \lambda_{t-1})$, and $C_n(\lambda_1, \cdots, \lambda_t)$ described in [2], Theorem 2; the family $A_n^k(\lambda_1, \cdots, \lambda_{t-1})$ contains for any dimension ≥ 5 infinitely many non-isomorphic algebras, whereas the families $B_n^k(\lambda_1, \cdots, \lambda_{t-1})$, and $C_n(\lambda_1, \cdots, \lambda_t)$ contain for any even dimension ≥ 6 infinitely many non-isomorphic Lie algebras.

If $\mathfrak{u}$ has dimension n and admits a two-dimensional torus group T, then we may choose for the Lie algebra $\mathfrak{u}$ a basis $\{e_1, \ldots, e_n\}$ such that the multiplication is given either by

i) $[e_1, e_i] = e_{i+1}, \quad 2 \leq i \leq n - 1$

or $n = 2p$ and the multiplication is given by

ii) $[e_1, e_i] = e_{i+1}, \quad 2 \leq i \leq n - 2$
$[e_i, e_{n-i+1}] = (-1)^i e_n, \quad i = 2, \ldots, p,$

(cf. [2], Proposition 1, p. 187), and for the Lie algebra $\mathfrak{t}$ of T we may choose a basis $\{t_1, t_2\}$ such that $\mathfrak{t}$ acts on $\mathfrak{u}$ in the case *i)* as

$[t_1, e_i] = ie_i, \quad 1 \leq i \leq n$
$[t_2, e_i] = e_i, \quad 2 \leq i \leq n$

and in the case *ii)* as

$[t_1, e_1] = e_1$
$[t_1, e_i] = (i - 2)e_i, \quad 3 \leq i \leq n - 1$

$$[t_1, e_n] = (n-3)e_n, \; [t_2, e_n] = 2e_n$$
$$[t_2, e_i] = e_i, \qquad 2 \le i \le n-1,$$

where in all cases all products which are not mentioned are zero (cf. [35], p. 121).

It follows by Lemma 5.5.15 that G is co-aligned only if a one-dimensional subalgebra $\langle at_1 + bt_2 \rangle$ of $\mathfrak{t}$ acts on $\mathfrak{u}$. If a were zero, then in case i), as well as in case ii), the factor algebra $\mathfrak{g}/\langle e_2, \dots, e_n \rangle$ would be the Lie algebra of a direct product of a one-dimensional vector group with an algebraic, respectively complex one-dimensional linear torus. Hence in these cases, G is co-aligned if and only if the Lie algebra of T has the form $\langle t_1 + bt_2 \rangle$.

5.5.23 Proposition. *Let G be a connected non-commutative solvable real Lie group such that the commutator subgroup G' is commutative and the factor group G/G' has a non-trivial compact subgroup. Then the group G is co-aligned if and only if one of the following conditions holds:*

> *i) the Lie group G is a semi-direct product of a $2d$-dimensional vector group $\mathbb{R}^{2d}$ by a one-dimensional torus such that any irreducible subspace of $\mathbb{R}^{2d}$ has dimension two;*
>
> *ii) the Lie group G is a semi-direct product of the vector group $\mathbb{R}^4$ by a two-dimensional torus such that any irreducible subspace of $\mathbb{R}^4$ has dimension two;*
>
> *iii) the Lie group G is a direct product of a one-dimensional torus and a semi-direct product of a $2d$-dimensional vector group $\mathbb{R}^{2d}$ by a one-dimensional torus such that any irreducible subspace of $\mathbb{R}^{2d}$ has dimension two.*

Proof. If the group G has the structure given in the proposition and the torus T has dimension one, then any normal subgroup of G is contained in $\mathbb{R}^{2d}$. The cases ii) and iii) are obvious.

Now we assume that any two epimorphic images of G of the same dimension are isomorphic. Since G/G' is a commutative Lie group such that the maximal compact subgroup is not trivial and connected, it follows that G/G' is a torus T. Moreover $G = G'T$ is a semi-direct product of G' by T. Let $\langle t_1, \dots, t_s \rangle$ be the Lie algebra of T. The group T acts on the commutative group G' which is isomorphic to $\mathbb{R}^{2d}$. If the connected component C of the centraliser of G' in T has dimension at least two, then we choose a two-dimensional torus W in C. Let Z be a two-dimensional invariant subspace of $\mathbb{R}^{2d}$. Then G/W and G/Z are non-isogenous factor groups of the same dimension, which is a contradiction. It follows that C has dimension at most one. Hence G is a direct product of C and a semi-direct product of a $2d$-dimensional vector group $\mathbb{R}^{2d}$ by a torus K such that any irreducible subspace of $\mathbb{R}^{2d}$ has dimension two.

Now we suppose that the torus group K has dimension greater than two. Then K contains a two-dimensional subgroup K_2 such that K_2 fixes element-wise a non-trivial subspace R of $\mathbb{R}^{2d}$ and acts on a non-trivial complement S of R in $\mathbb{R}^{2d}$ with 0-dimensional centraliser. The group $K_2 S$ is a normal subgroup of the semi-direct product $K\mathbb{R}^{2d}$ and the factor group $K\mathbb{R}^{2d}/K_2S$ is isomorphic to K_1R where K_1 is the unique complement of K_2 in K which fixes S elementwise. If $\dim R = 2a$ and $\dim S = 2b$, then $a + b = d$. Moreover $\dim K_2S = 2 + 2b$ and $2 + 2b \leq 2d$. In $\mathbb{R}^{2d}$ there exists an invariant subspace W of K of dimension $2 + 2b$. Since the factor groups $K\mathbb{R}^{2d}/K_2S$ and $K\mathbb{R}^{2d}/W$ are not isomorphic, it follows that $\dim K \leq 2$.

If C has dimension one and K has dimension two, then the direct product $C \cdot (K_1R)$ has dimension $2 + 2a \leq 2d$. Since in $\mathbb{R}^{2d}$ there is an invariant subspace of dimension $2 + 2a$, we obtain a contradiction.

If C has dimension one and K has dimension one, too, then we are in case $iii)$. Now we suppose that K is a maximal torus of G and has dimension two. Then G is isogenous to the direct product $(K_1R) \times (K_2S)$. Clearly $\dim R = \dim S$. If $\dim R > 2$, then let R_1 and S_1 be invariant subspaces of R, respectively of S, of co-dimension two in R, respectively in S. Let R_2 be an invariant subspace of R of co-dimension four. Then the factor groups $G/(R_1S_1)$ and $G/(R_2S)$ are not isogenous. This contradiction proves the assertion. $\square$

Let H be a Lie group such that the Lie algebra $\mathfrak{h} = \langle e_0, e_1, \ldots, e_{2n} \rangle$ of H satisfies the relations $[e_{2i+1}, e_{2i+2}] = e_0$ for $i = 0, \ldots, n - 1$ and all the other products are zero. The group $H(m_0, \ldots, m_n)$ is a semi-direct product of the group H by a one-dimensional torus group T such that $T = \{e^{it} : t \in \mathbb{R}\}$ acts on $\mathfrak{h}$ in the following way:

$$e_{2i+1} \mapsto \cos(m_i t)e_{2i+1} - \sin(m_i t)e_{2i+2}$$

$$e_{2i+2} \mapsto \sin(m_i t)e_{2i+1} + \cos(m_i t)e_{2i+2}$$

for $i = 0, \ldots, n-1$, with natural numbers m_i, and $Ad e^{it}(e_0) = e_0$ for all $t \in T$.

5.5.24 Lemma. *Let G be a nilpotent real Lie group of nilpotency class two such that G' is two-dimensional. Let T be a one-dimensional torus group of automorphisms of G such that any irreducible subspace of G/G' has dimension two. Then T centralises G'.*

Proof. Let $\mathfrak{g} = \langle e_1, \ldots, e_n \rangle$ be the Lie algebra of G such that e_{n-1}, e_n generate the commutator subalgebra $\mathfrak{g}'$ of $\mathfrak{g}$ and such that $[e_1, e_i] = e_{n-1}$ for some i. Let $\mathfrak{t} = \langle t \rangle$ be the Lie algebra of T. Since T acts completely reducibly on the Lie algebra $\mathfrak{g}$, we may choose a basis of $\mathfrak{g}$ such that $\langle e_{2i+1}, e_{2i+2} \rangle$ are invariant subspaces under T for any $i = 0, \ldots, n - 2$. Then $[e_{2i+1}, e_{2i+2}] = 0$ since otherwise T would leave invariant a one-dimensional subgroup of G'. Moreover, we may assume that $[e_1, e_3] = e_{n-1}$ and we may choose the vectors e_3, e_4 and e_n in such a way that for the derivations induced by $\mathfrak{t}$ on $\mathfrak{g}$ we have

$$[t, e_1] = -e_2, \quad [t, e_2] = e_1, \quad [t, e_3] = -e_4, \quad [t, e_4] = e_3,$$

$$[t, e_{n-1}] = -e_n, \quad [t, e_n] = e_{n-1}.$$

These relations yield

$$-e_n = [t, e_{n-1}] = [t, [e_1, e_3]] = [[t, e_1], e_3] + [e_1, [t, e_3]] = -[e_2, e_3] - [e_1, e_4]$$

and with this we obtain

$$e_{n-1} = [t, e_n] = [t, [e_2, e_3]] + [t, [e_1, e_4]] =$$
$$[[t, e_2], e_3] + [e_2, [t, e_3]] + [[t, e_1], e_4] + [e_1, [t, e_4]] =$$
$$2[e_1, e_3] - 2[e_2, e_4] = 2e_{n-1} - 2[e_2, e_4]$$

or $[e_2, e_4] = \frac{1}{2} e_{n-1}$.

This is impossible since

$$-\frac{1}{2} e_n = [t, \frac{1}{2} e_{n-1}] = [t, [e_2, e_4]] = [[t, e_2], e_4] + [e_2, [t, e_4]] =$$
$$[e_1, e_4] + [e_2, e_3] = e_n.$$ $\qquad\square$

5.5.25 Lemma. *Let G be a connected solvable real Lie group such that the commutator subgroup G' is a filiform Lie group and the factor group G/G' is isomorphic to $SO_2(\mathbb{R})$. Then G is isogenous to a semi-direct product of the three-dimensional non-commutative nilpotent Lie group G' by a one-dimensional torus T which acts irreducibly on G'/G'', where G'' is the commutator subgroup of G'.*

Proof. The group G is a product of G' by a one-dimensional torus T. Since G' is filiform, we choose a basis $\{e_1, e_2, \ldots, e_n\}$ of the Lie algebra $\mathfrak{g}'$ of G' in such a way that $[e_1, e_i] = e_{i+1}$ holds for $i = 2, \ldots, n-1$. The group T must act irreducibly on G'/G''. Namely, if it is not the case, then for the one-dimensional Lie algebra $\mathfrak{t} = \langle t \rangle$ of T we would have $[t, e_1] = [t, e_2] = 0$ and $[t, e_i] = [t, [e_1, e_{i-1}]] = 0$ if $[t, e_{i-1}] = 0$. The Lie algebra $\mathfrak{g}'$ contains a chain of characteristic subalgebras $\mathfrak{z}^{n-i+1} = \langle e_i, \ldots, e_n \rangle$ such that $\mathfrak{z}^{n-2} = \mathfrak{g}''$ and $\mathfrak{z}^{n-i+1}/\mathfrak{z}^{n-i+2}$ is a one-dimensional Lie algebra. Since the group T acts completely reducibly on $\mathfrak{g}'$, the action of $\mathfrak{t}$ on $\mathfrak{g}'$ may be given by $[t, e_1] = -e_2$, $[t, e_2] = e_1$ and $[t, e_i] = 0$ for $i = 3, \ldots, n$. Since $[e_2, e_3]$ is an element of $\mathfrak{z}^{n-3}$, we have $0 = [t, [e_2, e_3]] = [[t, e_2], e_3] + [e_2, [t, e_3]] = [e_1, e_3] = e_4$ and hence $[e_1, e_i] = 0$ for $i = 3, \ldots, n$. $\qquad\square$

5.5.26 Lemma. *Let $\mathfrak{g} = \langle e_1, \ldots, e_n \rangle$, with $n \geq 5$ odd, be a nilpotent real Lie algebra such that $[e_{2i+1}, e_{2i+2}] = e_{n-2}$ for all $0 \leq i \leq (n-5)/2$, the subalgebra $\mathfrak{n} = \langle e_{n-1}, e_n \rangle$ is an ideal of $\mathfrak{g}$ and the commutator subalgebra $\mathfrak{g}' = \langle e_{n-2}, e_{n-1}, e_n \rangle$. Let T be a one-dimensional torus group of automorphisms of $\mathfrak{g}$ such that any subspace $\mathbb{R}e_{2i+1} + \mathbb{R}e_{2i+2}$ as well as $\mathfrak{n}$ is invariant under T. If T acts irreducibly on $\mathfrak{n}$ as well as on one of the subspaces $\mathbb{R}e_{2i+1} + \mathbb{R}e_{2i+2}$, then $\langle e_{n-2} \rangle$ is an ideal of $\mathfrak{g}$ invariant under T.*

Proof. If $\mathfrak{g}'$ were not commutative, then we could choose the basis $\mathcal{B} = \{e_1, \ldots, e_n\}$ of $\mathfrak{g}$ such that $[e_{n-2}, e_{n-1}] = e_n$, $[e_{n-2}, e_n] = [e_{n-1}, e_n] = 0$, and $\langle e_n \rangle \subset \mathfrak{n}$ would be a characteristic subalgebra of $\mathfrak{g}$ and hence invariant under T. It follows that $\mathfrak{g}'$ is commutative. Since $\mathfrak{n}$ is contained in $\mathfrak{g}'$ there is $1 \leq i \leq n-3$ with $[e_i, x] \neq 0$ for suitable $x \in \mathfrak{n}$. Therefore we may choose a basis $\mathcal{B}$ such that $[e_1, e_{n-1}] = e_n$ and $[e_1, e_n]$

$= [e_{n-1}, e_n] = 0$ (cf. [61], p. 162). Since for the Lie algebra $\mathfrak{t} = \langle t \rangle$ of T one has $[t, e_1] = -e_2$, $[t, e_2] = e_1$, $[t, e_{n-1}] = -e_n$ and $[t, e_n] = e_{n-1}$, we obtain $0 = [t, [e_1, e_n]] = [[t, e_1], e_n] + [e_1, [t, e_n]] = -[e_2, e_n] + [e_1, e_{n-1}]$ and hence $[e_2, e_n] = [e_1, e_{n-1}] = e_n$, which is a contradiction to the fact that $\mathfrak{g}$ is nilpotent.   □

5.5.27 Proposition. *Let G be a connected solvable real Lie group such that the commutator subgroup G' is not commutative and the factor group G/G' has a non-trivial compact subgroup of dimension ≥ 2. Then the group G is co-aligned if and only if G is the direct product of two groups $H(m_0)$ and $H(p_0)$ with amalgamated isomorphic centres.*

Proof. If the group G has the structure given in the proposition, then G'' is the only one-dimensional connected normal subgroup of G. The factor group G/G'' is a semi-direct product of $\mathbb{R}^4$ by a two-dimensional torus group T. Furthermore, the group G/G'' is a direct product of $T_1\mathbb{R}^2$ and $T_2\mathbb{R}^2$ with one-dimensional tori T_i for $i = 1, 2$ and any invariant subspace of T_i has dimension two.

Now we assume that any two epimorphic images of G of the same dimension are isogenous. Since G/G' is a commutative Lie group such that the maximal compact subgroup has dimension ≥ 2, it follows that G/G' is a torus T of dimension ≥ 2. It follows from Proposition 5.5.23 that the torus T has dimension two and that G/G'' is of one of the following types: A semi-direct product of the vector group $\mathbb{R}^4$ by a two-dimensional torus such that any irreducible subspace of $\mathbb{R}^4$ has dimension two, or it is a direct product of a one-dimensional torus with a semi-direct product of a $2d$-dimensional vector group $\mathbb{R}^{2d}$ by a one-dimensional torus such that any irreducible subspace of $\mathbb{R}^{2d}$ has dimension two.

Let G be a counter-example to the assertion of minimal dimension. Then there exists in G'' a minimal connected subgroup N normal in G such that G/N has a commutative commutator subgroup or has the structure given in the claim. As G is solvable, the subgroup N has dimension at most two (cf. [49], p. 52).

If G'/N is commutative, then $G'' = N$. Suppose that $\dim N = 1$. As T acts completely reducibly on the Lie algebra $\mathfrak{g}'$ of G', it follows that G/N is not a direct product of a one-dimensional torus C and a semi-direct product of a $2d$-dimensional vector group $\mathbb{R}^{2d}$ by a one-dimensional torus, since otherwise G/N and G/C would be not isogenous. If G/N had the structure given in $ii)$ of Proposition 5.5.23, then N would be central in G and G would not be a counter-example since G' would be the direct product of two three-dimensional Heisenberg groups with amalgamated center.

Let $\dim N = 2$. If G/N has the structure given in $iii)$ of Proposition 5.5.23, then there exists a one-dimensional torus K acting faithfully and completely reducibly on the commutator subgroup G'. Hence there exists in G' a sequence of normal subgroups M_i of G such that $M_1 = N$ and M_i has co-dimension two in M_{i+1}. The subalgebra $\mathfrak{m}_{i+1}$ corresponding to

M_{i+1} is generated by the subalgebra $\mathfrak{m}_i$ and by two one-dimensional subalgebras generated by e_1^{i+1} and e_2^{i+1}. If $\mathfrak{m}_i$ is commutative and $[e_1^{i+1}, e_2^{i+1}] = 0$, then $\mathfrak{m}_{i+1}$ is commutative. Since $\mathfrak{g}'$ is not commutative, there exists a minimal non-commutative subalgebra $\mathfrak{m}_{i+1}$ such that e_1^{i+1}, e_2^{i+1} generate the Heisenberg algebra. As this algebra is invariant under the algebra $\langle k \rangle$ corresponding to K we have $[k, e_1^{i+1}] = -e_2^{i+1}$ and $[k, e_2^{i+1}] = -e_1^{i+1}$. This yields $[k, [e_1^{i+1}, e_2^{i+1}]] = [[k, e_1^{i+1}], e_2^{i+1}] + [e_1^{i+1}, [k, e_2^{i+1}]] = 0$, which is a contradiction since the whole torus T would centralise N.

Now we assume that the group G/N has the structure given in $ii)$ of Proposition 5.5.23 and let $\{e_1, \ldots, e_6\}$ be a basis of $\mathfrak{g}'$ such that $\langle e_5, e_6 \rangle$ is the Lie algebra of G''. Since the maximal torus T acts completely reducibly on $\mathfrak{g}'$, we may assume that the subspaces generated by e_1, e_2 and e_3, e_4, as well as e_5, e_6, are invariant under T. Since T induces on N the group $SO_2(\mathbb{R})$, we may moreover suppose that $[e_1, e_2] = [e_3, e_4] = 0$ and $[e_1, e_3] = e_5$. Let T_i be the one-dimensional subtorus of T which fixes e_{2i+1} and e_{2i+2} elementwise and induces on the subspace $\langle e_{2i+3}, e_{2i+4} \rangle$ the group $SO_2(\mathbb{R})$, where $i = 0, 1$. Since T does not centralise N, we may assume that for the Lie algebra $\mathfrak{t} = \langle t \rangle$ of T_0 we have

$$[t, e_1] = 0, \quad [t, e_2] = 0, \quad [t, e_3] = -e_4, \quad [t, e_4] = e_3,$$

$$[t, e_5] = -e_6, \quad [t, e_6] = e_5.$$

These relations yield

$$e_6 = [t, e_5] = [t, [e_1, e_3]] = [[t, e_1], e_3] + [e_1, [t, e_3]] = [e_1, -e_4]$$

from which it follows that

$$-e_5 = [t, -e_6] = [t, [e_1, e_4]] = [[t, e_1], e_4] + [e_1, [t, e_4]] = [e_1, e_3] = e_5,$$

which is a contradiction.

Now we treat the case that G'/N is not commutative. Let N be one-dimensional. Then the Lie algebra $\mathfrak{g}'$ has dimension six and has a basis $\{e_1, \ldots, e_6\}$ such that e_6 generates the Lie algebra $\mathfrak{n}$ of N, $[e_{2i+1}, e_{2i+2}] = e_5 + \lambda_i e_6$ for $i = 0, 1$ and all the other products are zero. Clearly, the Lie algebra $\mathfrak{n}$ is a central subalgebra of the Lie algebra $\mathfrak{g}$ of G. If $[e_j, e_5] = 0$ for any j, then also the subalgebra $\langle e_5 \rangle$ is a central subalgebra of the Lie algebra $\mathfrak{g}$ of G. Since $\mathfrak{n}$ is not a direct factor of $\mathfrak{g}'$, there exists $\lambda_{i_0} \neq 0$. But then the factor algebras $\mathfrak{g}'/\mathfrak{n}$ and $\mathfrak{g}'/\langle e_5 + \lambda_{i_0} e_6 \rangle$ are not isomorphic. If e_5 is not central, then we may assume that $[e_1, e_5] = e_6$. In this case we consider the subalgebra $\mathfrak{m} = \langle e_1, e_2, e_5, e_6 \rangle$. The torus T induces on the subspace $\mathbb{R}e_1 + \mathbb{R}e_2$ an automorphism group isomorphic to $SO_2(\mathbb{R})$ and centralises e_5 as well as e_6. Let $\langle t \rangle$ be the one-dimensional subalgebra corresponding to a one-dimensional subtorus of T which induces the same automorphism group as T. Since we have

$$[t, e_1] = -e_2, [t, e_2] = e_1, [t, e_5] = [t, e_6] = 0$$

we obtain

$$0 = [t, e_6] = [t, [e_1, e_5]] = [[t, e_1], e_5] + [e_1, [t, e_5]] = -[e_2, e_5]$$

which yields

$$0 = [t, [e_2, e_5]] = [[t, e_2], e_5] + [e_2, [t, e_5]] = [e_1, e_5]$$

a contradiction.

Now let dim $N = 2$. Then the Lie algebra $\mathfrak{g}'$ of G' has a basis $\{e_1, \ldots, e_7\}$ such that $[e_{2i+1}, e_{2i+2}] = e_5$ for $i = 0, 1$ and $\mathfrak{n} = \langle e_6, e_7 \rangle$ is the Lie algebra of N. Since the commutator subalgebra of $\mathfrak{g}'/\mathfrak{n}$ is $\langle e_5, e_6, e_7 \rangle/\mathfrak{n}$, it follows that $\mathfrak{g}'' = \langle e_5, e_6, e_7 \rangle$. If $\mathfrak{g}''$ is not commutative, then the commutator subalgebra of $\mathfrak{g}''$ is one-dimensional and the group G contains a one-dimensional normal subgroup. Hence the Lie algebra $\mathfrak{g}''$ is commutative. It follows that $\langle e_{2n-1} \rangle$ is an ideal of $\mathfrak{g}'$. As the torus T is two-dimensional, it contains a one-dimensional torus T_1 which acts irreducibly on the subspace $\mathbb{R}e_1 + \mathbb{R}e_2$ as well as on $\langle e_6, e_7 \rangle$. Now according to Lemma 5.5.26 the subalgebra $\langle e_5 \rangle$ would be invariant under T_1. In T there is a one-dimensional complement T_2 of T_1 which centralises the subalgebra $\langle e_1, e_2 \rangle$. This yields that G contains a one-dimensional normal subgroup which corresponds to the Lie algebra $\langle e_5 \rangle$. □

In contrast to Proposition 5.5.27, the class of co-aligned real Lie groups G for which G' is not commutative and for which the factor G/G' has a one-dimensional torus as maximal subgroup is difficult to describe. The simplest examples are the groups $H(m_0, \cdots, m_k)$. But the examples in Remark 5.5.5 (3) show that in this class one finds groups having a centre of dimension $d > 1$ and groups of nilpotency class $c > 2$.

5.6 Aligned and Co-Aligned Commutative Groups

Since any connected commutative affine algebraic G over a field of characteristic zero is isomorphic to a direct product $\mathbf{G}_m^r \times \mathbf{G}_a^s$, the aligned as well as co-aligned commutative groups G in this category are either vector groups or tori. For real Lie groups we have an analogous situation. Any aligned as well as co-aligned real Lie group is isomorphic either to $\mathbb{R}^n$ or to $SO_2(\mathbb{R})^n$. The situation changes drastically if one considers non-affine connected algebraic groups which have no non-trivial affine epimorphic image, or complex toroidal Lie groups.

For abelian varieties it is easy to decide which of them are aligned or co-aligned. Since every abelian variety defined over a field k is isogenous to a direct product of simple abelian varieties over k which are uniquely determined up to k-isogenies (cf. [54], Corollary p. 30), the classification of aligned or co-aligned abelian varieties can be refined in the following way:

5.6.1 Definition. A connected algebraic k-group G is called k-*aligned* if any two k-subgroups of the same dimension are k-isogenous.

A connected algebraic k-group G of dimension n is called k-*co-aligned* if any two k-epimorphic images of the same dimension $m < n$ are k-isogenous.

5.6.2 Proposition. *Let*

$$A = A_1 \times \cdots \times A_{i_1} \times A_{i_1+1} \times \cdots \times A_{i_1+i_2} \times \cdots \times A_{i_1+i_2+\cdots+i_t},$$

be an abelian variety defined over a field k *of dimension* n, *where the* A_i *are simple abelian varieties over* k *such that for* $i_0 + \cdots + i_{k-1} < j \le i_0 + \cdots + i_k$ *the abelian varieties* A_j *have the same dimension* l_k *for any* $k = 1, \ldots, t$ *and* $i_0 = 0$.

The abelian variety A *is* k-*co-aligned precisely if all varieties* A_j *with* $i_0 + \cdots + i_{k-1} < j \le i_0 + \cdots + i_k$ *are* k-*isogenous and* $\sum_{k=1}^{t} r_k l_k = \sum_{k=1}^{t} s_k l_k$, *with* $i_{k-1} + 1 \le r_k, s_k \le i_k$ *for any* $k = 1, \ldots, t$, *holds if and only if* $r_k = s_k$ *for any* $k = 1, \ldots, t$.

Proof. Assume that for some k with $1 \le k \le t$ there are varieties A_{j_1}, A_{j_2} with $i_0 + \cdots + i_{k-1} < j_1, j_2 \le i_0 + \cdots + i_k$ which are not k-isogenous. Then the factor groups A/B_1 and A/B_2, where $B_i = A_1 \times \cdots \times A_{j_i-1} \times A_{j_i+1} \times \cdots \times A_{i_1+i_2+\cdots+i_t}$, for $i = 1, 2$, are not k-isogenous, a contradiction.

If the integer $\sum_{k=1}^{t} r_k l_k$ has a unique representation with $i_{k-1} < r_k, s_k \le i_k$ for any $k = 1, \ldots, t$, then there is only one factor group of dimension $n - \sum_{k=1}^{t} r_k l_k$. Conversely, if there is not a unique representation of $\sum_{k=1}^{t} r_k l_k$, then the factor groups A/B and A/C, where B is k-isogenous to a direct product of r_k copies of $A_{i_1+\cdots+i_k}$ for any $1 \le k \le t$, whereas C is k-isogenous to a direct product of s_k copies of $A_{i_1+\cdots+i_k}$ for any $1 \le k \le t$, are not k-isogenous, a contradiction. □

If an abelian k-variety is k-isogenous to a direct product $B \times C$, then the connected subgroup B is defined over k if and only if the connected subgroup C is defined over k (cf. [54], Theorem 6, p. 28).

If an abelian k-variety A is k-isogenous to the direct products of k-subgroups $B_1 \times C_1$ and $B_2 \times C_2$, then B_1 and B_2 are k-isogenous if and only if C_1 and C_2 are k-isogenous.

These considerations yield

5.6.3 Corollary. *An abelian variety* A *defined over a field* k *is* k-*aligned if and only if it is* k-*co-aligned.* □

5.6.4 Corollary. *Let* G *be a non-affine connected algebraic group over a perfect field such that its maximal affine connected algebraic subgroup* L *has dimension one and such that* G *is not isogenous to a direct product of two proper connected algebraic subgroups. Then* G *is co-aligned if and only if the factor group* G/L *satisfies the conditions of Proposition 5.6.2.*

Proof. The assertion follows immediately since L is contained in any connected algebraic subgroup of G. □

5.6.5 Proposition. *Let G be a non-affine connected commutative algebraic group which is a non-split extension of an affine connected algebraic group L by an abelian variety A. If G is co-aligned, then $G = D(G)$ and L is a vector group or a torus.*

Proof. The group $L \cap D(G)$ has positive dimension. Hence it can contain a one-dimensional vector group only if the characteristic of k is zero (cf. Theorem 1.3.2). If there exists a one-dimensional connected algebraic subgroup S of G which is not contained in $D(G)$, then the factor group G/S is not isogenous to a factor group G/M, where M is a one-dimensional connected algebraic subgroup of $L \cap D(G)$. Hence $G = D(G)$. If now L is not a vector group or a torus, then in L there is a vector group U and a one-dimensional torus T such that the factor groups G/U and G/T have the same dimension but they are not isogenous.   □

A determination of aligned and co-aligned groups for groups $G = D(G)$ seems to be difficult.

Let G be an extension of a vector group, respectively of a torus, by a simple abelian variety A such that G is not decomposable into a direct product of two non-trivial proper connected subgroups. Then G is aligned, since any proper connected algebraic subgroup is affine. Moreover if dim $A = 1$ and G is a non-split extension of a vector group by A, then according to Proposition 1.3.4 the group G is a two-dimensional chain.

If $G = D(G)$ is such that the factor group G/L, where L is the maximal connected affine subgroup of G, is a simple abelian variety A, then for the decision under which circumstances the group G is co-aligned it is necessary to know explicitly the factor systems belonging to this extension. For instance if G is co-aligned and L is a vector group or a torus of dimension n, then any factor group G/M, where M is a connected affine subgroup, is a non-split extension of L/M by A. In particular if $L = Q \times Q$ with Q isomorphic to $\mathbf{G}_a$ or $\mathbf{G}_m$, then the group G is isogenous to a group defined on $A \times Q \times Q$ by the product

$$(a, x_1, x_2) \cdot (b, y_1, y_2) = (a \cdot b, \; x_1 \cdot y_1 \cdot \gamma_1(a, b), \; x_2 \cdot y_2 \cdot \gamma_2(a, b))$$

for suitable factor systems $\gamma_i : A \times A \longrightarrow Q$ (cf. [89], Proposition 6, p. 170). For the factor group G/H, where $H = \{0, q, q) : q \in Q\}$, we have $(a, x_1, x_2)H = (a, 1, x_2 x_1^{-1})$ and

$$(a, x_1, x_2)H \cdot (b, y_1, y_2)H = (a \cdot b, \; x_1 \cdot y_1 \cdot \gamma_1(a, b), \; x_2 \cdot y_2 \cdot \gamma_2(a, b))H =$$

$$(a \cdot b, \; 1, \; x_2 \cdot y_2 \cdot x_1^{-1} \cdot y_1^{-1} \cdot \gamma_2(a, b)\gamma_1^{-1}(a, b))H$$

for any $(a, x_1, x_2)H, (b, y_1, y_2)H \in G/H$. If G is co-aligned, then we must have $\gamma_1 \neq \gamma_2$ since otherwise G/H is isogenous to the direct product $A \times Q$.

For abelian varieties, Poincaré's irreducibility theorem gives a unique decomposition up to isogeny as a product of simple abelian varieties. Although

for complex tori this does not hold, as the so-called Shafarevich extensions
show (see [7], Proposition 6.1, p. 22 and Corollary 6.3, p. 23), the property
to be co-aligned can be treated also for complex tori. If G is a complex
torus, then there exists a unique abelian variety G_a and a unique epimorphism $\pi : G \longrightarrow G_a$ such that for any homomorphism $f : G \longrightarrow H$ into an
abelian variety H there exists a unique homomorphism $g : G_a \longrightarrow H$ such that
$f = g \circ \pi$ (cf. [7], Theorem 6.2, p. 56 and Universal Property 6.3, p. 57). Hence
if G is co-aligned, then G_a must satisfy the conditions of Proposition 5.6.2.

Let G be a complex toroidal group of dimension $n \geq 2$. Then G contains
in the maximal compact real subgroup of G a unique maximal complex torus
T such that G/T is torus-free i.e. does not contain any complex Lie torus $\neq 1$
(cf. [1], p. 23). According to [59], Theorem 4, there exist torus-free toroidal
groups for any dimension $n \geq 2$. If G is a two-dimensional torus-free group,
then G contains no proper connected closed complex subgroup (cf. [1], p. 6).

5.7 Characterisations of Chains by Factors

An affine chain of dimension n has only $n - 1$ proper algebraic connected subgroups. Now we characterise the chains within the class of connected algebraic
groups having only finitely many connected algebraic subgroups.

5.7.1 Theorem. *A connected affine algebraic group G has only finitely many
connected algebraic subgroups if and only if G is a chain or the direct product
$U \times T$ where T is a one-dimensional torus and U is a unipotent chain.*

Proof. As a direct consequence of the hypothesis we have that G is not semi-
simple and that any epimorphic image of G with connected kernel has only
finitely many connected algebraic subgroups. Therefore, if we denote by R_G
the solvable radical of G, the factor group G/R_G has to be trivial, which
forces G to be solvable. Let T be a maximal torus of G and U the unipotent
radical of G. If $G = T$, then G is clearly one-dimensional. If $G = U$ and
$\dim G \leq 2$, then G is a chain, whereas if $G = U$ and $\dim G > 2$, then by
induction any connected subgroup of G, as well as any epimorphic image of
G, is a chain, and by Theorem 5.1.1 the group G is a chain as well. Let now
$G = TU$ with T and U non-trivial. By the above arguments we have that U
is a chain and $\dim T = 1$. Moreover, since the maximal tori of the solvable
group G are conjugate in G, the torus T has to be normal in G. $\square$

Since an algebraic group G has only finitely many connected algebraic
subgroups if and only if the maximal connected affine algebraic group $L(G)$,
as well as the group $D(G)$, as in the decomposition of G in Theorem 1.3.2,
has only finitely many algebraic subgroups, we mention the following

5.7.2 Proposition. *Let G be a connected algebraic group such that G has no
non-trivial affine epimorphic image. Then G has only finitely many connected*

algebraic subgroups if and only if $L(G)$ is a connected subgroup of $\mathbf{G}_m \times \mathbf{G}_a$ and the factor group $G/L(G)$ is isogenous to a direct product $A_1 \times \cdots \times A_n$ of simple abelian varieties such that A_i is not isogenous to A_j for $i \neq j$.

Proof. Clearly the group G has only finitely many connected algebraic subgroups if and only if $L(G)$, as well as the factor group $G/L(G)$, has only finitely many algebraic subgroups. For the group $L(G)$ the assertion follows by Theorem 5.7.1 and by *iii*) of Theorem 1.3.2, whereas for the factor group $G/L(G)$ the assertion is a direct consequence of [64], Corollary 1, p. 174.   □

If we weaken the notion of a chain and assume only that in an algebraic group any two connected algebraic subgroups of equal dimension are isogenous, this condition is not sufficient to obtain a classification, at least not for algebraic groups over fields of positive characteristic. If, however, one adds conditions on the factor groups which are dual to the chosen restrictions on subgroups, essentially only the chains are left.

5.7.3 Proposition. *Let G be a non-commutative connected unipotent algebraic group, defined over a perfect field of positive characteristic p, such that for every normal connected algebraic k-subgroup H of G there exists a connected algebraic subgroup H^* such that G/H is isogenous to H^*. Then either G is a chain or G is not uni-maximal nor uni-minimal and contains a two-dimensional connected non-commutative algebraic subgroup, which can be taken to be defined over k if k is infinite.*

Proof. Let G be uni-maximal. We want to prove by induction on k that G has a unique connected algebraic subgroup of co-dimension k, forcing G to be a chain. If $k = 1$ the assertion is true, since G is uni-maximal. Assume now that G contains a unique connected algebraic subgroup T_i of co-dimension $i \leq k$. Hence G/T_k is a chain. Let H be a normal connected algebraic k-subgroup of G of dimension k. Since G/H has dimension $n-k$ and T_k is the unique connected algebraic subgroup of co-dimension k, the group T_k is isogenous to the uni-maximal factor group G/H, from which it follows that T_k is uni-maximal. Therefore G has a unique connected algebraic subgroup of co-dimension $k+1$ and the assertion follows.

Assume now that G is not uni-maximal. Let $\Phi(G)$ be the Frattini subgroup of G. The group $G/\Phi(G)$ is a vector k-group of dimension $d > 1$, otherwise the group G would be uni-maximal. Hence G, containing an algebraic subgroup isogenous to $G/\Phi(G)$, is not uni-minimal. Let T be a maximal connected k-subgroup of $\Phi(G)$, normal in G, and let L be an algebraic subgroup of G isogenous to G/T. Since $T \leq \Phi(G)$ we have $\Phi(G/T) = \Phi(G)/T$, from which it follows that $\Phi(L)$ is one-dimensional. But from this fact it follows that L has to be non-commutative, otherwise L should be a two-dimensional commutative chain. Since G/T is isogenous to L and T is maximal in $\Phi(G)$, the group $\Phi(G)$ is maximal in G and this forces G to be uni-maximal, a contradiction. Let now x, y be two non-commuting elements of L (respectively of the dense subgroup $L(k)$, if k is infinite). Then according to Proposition 5.2.19, there is a two-dimensional algebraic connected subgroup (respectively k-subgroup) S of L such that $\Phi(L)$ is a maximal connected algebraic subgroup of S and $x, y \in S$.   □

5.7.4 Corollary. *Let G be a non-commutative connected unipotent algebraic group defined over a perfect field of positive characteristic. Assume that the following two conditions are satisfied:*

(1) For any normal connected algebraic k-subgroup H of G there exists a connected algebraic subgroup H^* such that G/H is isogenous to H^*.
(2) Every two connected algebraic subgroups of the same dimension are isogenous.

Then G is chain.

Proof. Let G be a counter-example of minimal dimension. By the above Proposition 5.7.3, condition (1) forces G to be not uni-minimal and to contain a non-commutative connected algebraic k-subgroup of dimension two. Since by condition (2) any two algebraic connected subgroups of dimension two must be isogenous and G contains also a two-dimensional vector group, we get a contradiction.   □

5.7.5 Theorem. *Let G be a connected algebraic group defined over a perfect field k. Assume that the following two conditions are satisfied:*

(1) For any normal connected algebraic k-subgroup H of G there exists a connected algebraic subgroup H^* such that G/H is isogenous to H^*.
(2) Every two connected algebraic subgroups of the same dimension are isogenous.

Then G belongs to one of the following classes:

(a) char(k) $= 0$ and G is a vector group, a torus or the three-dimensional unipotent group of nilpotency class two,
(b) char(k) > 0 and G is a vector group, a torus or an affine chain,
(c) G is an abelian variety such that any two subvarieties A, B of equal dimension are isogenous,
(d) G is isogenous to the direct product $L(G) \times D(G)$, where $L(G)$ falls under the cases (a) or (b), the abelian $D(G)$ falls under (c) and every simple factor of $D(G)$ has a dimension greater than $\dim L(G)$.

Proof. Assume first that G is unipotent. If k has positive characteristic, the assertion follows by Corollary 5.7.4. If the characteristic of k is zero, our assertion is proved using Theorem 5.4.4.

If G is affine, but not unipotent, then G contains a one-dimensional torus. Thus every one-dimensional connected algebraic subgroup of G is a torus and it follows that G itself, together with a Borel subgroup, is a torus.

If G is an abelian variety the assertion follows by [64], Chapter 4, Corollary 1, p. 174.

Assume now that $G = L(G)D(G)$, with $L(G) \neq 1 \neq D(G)$. By (1) in G we find a subgroup A isogenous to the abelian variety $G/L(G)$. By Theorem 1.3.2 iv) we have $G = L(G)A$, with $(L(G) \cap A)^\circ = 1$. It follows that $D(G) = A$ and that G is isogenous to the direct product $L(G) \times D(G)$.

Let H be a normal subgroup of $L(G)$. Then H is normal in G, since $D(G)$ is a central subgroup of G. Thus according to condition (2) there is a connected algebraic subgroup H^* of G which is isogenous to G/H. For dimensional reasons we find $D(G) = D(H^*)$ and we have $H^* = L(H^*)D(G) = (H^* \cap L(G))D(G)$. It follows that the conditions (1) and (2) hold in $L(G)$. In the same way one sees that (1) and (2) are satisfied in $D(G)$.

The nilpotent group $L(G)$ contains a connected algebraic subgroup of any given dimension $d \leq \dim L(G)$. From condition (1) it follows that every simple factor of the abelian variety $D(G)$ has a dimension greater than $\dim L(G)$. $\square$

5.7.6 Proposition. *Let G be a non-commutative connected unipotent algebraic group defined over a perfect field of positive characteristic. If for any connected normal k-subgroup H of G there exists a normal connected algebraic subgroup H^* such that G/H^* is isogenous to H, then either G is a chain or G is not uni-minimal, nor uni-maximal.*

Proof. Let G be uni-minimal and let H_k be a k-dimensional connected algebraic subgroup of G. We want to prove by induction on k that G has a unique k-dimensional connected algebraic subgroup, that is G is a chain. For $k = 1$ the assertion is true because G is uni-minimal. We assume that there exists a unique i-dimensional connected algebraic subgroup, for any $i \leq k$. Hence H_k is a chain. Together with an $(n-k)$-dimensional connected normal algebraic k-subgroup of G, the factor group G/H_k is uni-minimal. Therefore there exists a unique connected $(k+1)$-dimensional algebraic subgroup of G and the assertion follows.

Now let G be uni-maximal and let T_h be a connected algebraic k-subgroup of co-dimension h in G. We show by induction on h that G has a unique connected algebraic k-subgroup T_h of co-dimension h, that is G is a k-chain, hence a chain by Theorem 4.1.1. For $h = 1$ the assertion is true because G is uni-maximal. Hence we assume that there is a unique connected algebraic k-subgroup T_i of co-dimension $i \leq h$. Together with an $(n-h)$-dimensional epimorphic image of G, the k-subgroup T_h is uni-maximal, hence there exists a unique connected algebraic subgroup T_{h+1} of co-dimension $(h+1)$ in G and the assertion follows. $\square$

5.7.7 Proposition. *Let G be a non-commutative connected unipotent algebraic group defined over a perfect field of positive characteristic. If for any connected k-subgroup H of G there exists a normal connected algebraic subgroup H^* such that G/H^* is isogenous to H, then either G is a chain or G is not uni-minimal, nor uni-maximal and has a non-commutative uni-maximal group as epimorphic image.*

Proof. By Proposition 5.7.6 the group G is either a chain or is not uni-minimal, nor uni-maximal. By Theorem 5.2.21 there exists a connected non-commutative algebraic k-subgroup L of G of minimal dimension. By Corollary 5.2.22, the subgroup L is uni-maximal. Then G/L^* is uni-maximal. $\qquad\square$

5.7.8 Corollary. *Let G be a non-commutative connected unipotent algebraic group defined over a perfect field of positive characteristic. Assume that the following two conditions are satisfied:*

(1) For any connected k-subgroup H of G there exists a normal connected algebraic subgroup H^* such that G/H^* is isogenous to H.

(2) Every two epimorphic images of G of the same dimension are isogenous.

Then G is chain.

Proof. Let G be a counter-example of minimal dimension. By Proposition 5.7.7 we have that G is not uni-minimal, nor uni-maximal and has a non-commutative uni-maximal group as epimorphic image of dimension $t > 1$. By condition (2), any t-dimensional epimorphic image of G is uni-maximal. Let $\Phi(G)$ be the Frattini subgroup of G and assume that $G/\Phi(G)$ has dimension s. Let $t > s$ and let H be a connected algebraic subgroup of $\Phi(G)$, normal in G and having co-dimension t in G. Since G/H is uni-maximal and $\Phi(G/H) = \Phi(G)/H$, we have that $\Phi(G)$ is maximal in G, forcing G to be uni-maximal, a contradiction. Let $t \leq s$ and let H be a connected algebraic subgroup of G, containing $\Phi(G)$ and having co-dimension t. On the one hand, the factor group G/H is uni-maximal, on the other hand it is a vector group, because $H \geq \Phi(G)$. Therefore $t = 1$, a contradiction. $\qquad\square$

5.7.9 Theorem. *Let G be a connected algebraic group defined over a perfect field k. Assume that the following two conditions are satisfied:*

(1) For any connected subgroup H of G there exists a normal connected algebraic subgroup H^* such that G/H^* is isogenous to H.

(2) Every two epimorphic images of G of the same dimension are isogenous.

Then G belongs to one of the following classes:

(a) $\mathsf{char}(\mathsf{k}) = 0$ and G is a vector group, a torus or the three-dimensional unipotent group of nilpotency class two,

(b) $\mathsf{char}(\mathsf{k}) > 0$ and G is a vector group, a torus or an affine chain,

(c) G is an abelian variety such that any two subvarieties A, B of equal dimension are isogenous,

(d) G is isogenous to the direct product $L(G) \times D(G)$, where $L(G)$ falls under the cases (a) or (b), the group $D(G)$ is an abelian variety as in case (c) and every simple factor of $D(G)$ has a dimension greater than $\dim L(G)$.

Proof. It follows from the conditions *(1)* and *(2)* that any two connected algebraic subgroups of G having the same dimension are isogenous.

First we treat the case where the group G is affine. If G is not unipotent, then G contains a one-dimensional torus. Thus every one-dimensional connected algebraic subgroup of G is a torus, and it follows that G itself is a torus. Therefore let the group G be unipotent. If k has characteristic zero, then the assertion follows from Theorem 5.4.4. If the characteristic of k is positive, the assertion follows by Corollary 5.7.8.

We consider now the case that G is non-affine. If G is an abelian variety, the assertion *(c)* follows directly by *(2)* (see [64], Chapter 4, Corollary 1, p. 174). If G is not an abelian variety, then $G = L(G)D(G)$, with $L(G) \neq 1 \neq D(G)$. By *(1)* we find a subgroup T in G such that $L(G)$ is isogenous to G/T. For dimensional reasons we now find that $L(G) \cap D(G)$ is finite, therefore $D(G)$ is an abelian variety and G is isogenous to $L(G) \times D(G)$.

Let H be a normal subgroup of $L(G)$. Then H is normal in G, since $D(G)$ is a central subgroup of G. Now according to condition *(1)* there is a connected algebraic subgroup H^* of G such that H is isogenous to G/H^*. Since H is affine, the group $D(G)$ is contained in H^* and $D(H^*) = D(G)$. Thus $H^* = L(H^*)D(G) = (H^* \cap L(G))D(G)$. It follows that the conditions *(1)* and *(2)* hold in $L(G)$. In the same way one sees that *(1)* and *(2)* are satisfied in $D(G)$. The nilpotent group $L(G)$ contains a connected algebraic subgroup of any given dimension $d \leq \dim L(G)$. It follows from conditions *(1)* and *(2)* that every simple factor of the abelian variety $D(G)$ has dimension greater than $\dim L(G)$. □

6

Three-Dimensional Affine Groups

As shown in the previous sections, three-dimensional groups which are uni-maximal or uni-minimal play a significant rôle in our theory. We now shall give a complete classification of three-dimensional connected unipotent algebraic groups defined over a field k of characteristic $p > 2$. Some of our results even hold in the case $p = 2$. A main tool is the theory of extensions, which is particularly efficient for unipotent groups defined over a perfect field, as we have seen in Remarks 2.1.4 and 2.1.5. As already mentioned in Section 4.2, we obtain at the same time a classification of the groups of k-rational points of three-dimensional connected unipotent algebraic groups, if the field k is infinite and perfect.

By Corollary 4.2.10, if $p > 2$ and the three-dimensional unipotent group G is a chain, then G' is one-dimensional, and we can refer to Theorem 4.3.1. Therefore in the present section we consider groups which are not chains.

6.1 Reduction to Subclasses

6.1.1 Proposition. *Let G be a three-dimensional unipotent group over a perfect field of characteristic greater than two. If a maximal connected algebraic subgroup M of G exists such that M is a chain, then* $\dim G' \leq 1$.

Proof. Any connected maximal subgroup of G contains G'. If the dimension of G' were two, then $G' = M$ would be the unique maximal connected subgroup of G. As G' is a chain, the group G would be a chain, which contradicts Corollary 4.2.10. □

We recall that for any two-dimensional unipotent algebraic group H, defined over a perfect field k, we can give a representation of H as a central extension of a one-dimensional unipotent group by itself

$$1 \longrightarrow \mathbf{G}_a \longrightarrow H \longrightarrow \mathbf{G}_a \longrightarrow 1.$$

As mentioned in Remark 2.1.6, in [18], II, § 3, 4.6, it is shown that $H^2(\mathbf{G}_a, \mathbf{G}_a)$ is a free left $\mathsf{End}(\mathbf{G}_a)$-module, having the following family of polynomials as a basis modulo $B^2(\mathbf{G}_a, \mathbf{G}_a)$:

$$\Phi_1(x, y) = \sum_i \frac{(p-1)!}{i!(p-i)!} x^i y^{p-i};$$

$$\eta_j(x, y) = xy^{p^j} \quad (j = 1, 2, \cdots).$$

If for any element $h \in H$ we choose a representation $h = (x_1, x_2)$, up to a coboundary, the product of two elements of H is given by

$$(x_1, x_2) \cdot (y_1, y_2) = \left(x_1 + y_1, x_2 + y_2 + f(\Phi_1(x_1, y_1)) + \sum_{i=1}^{t} g_i(x_1 y_1^{p^{n_i}}) \right),$$

for suitable factor systems $xy^{p^{n_1}}, \cdots, xy^{p^{n_t}}$, and suitable polynomials $f, g_1, \cdots, g_t \in \mathsf{End}\,\mathbf{G}_a$.

The group H is a two-dimensional vector group if and only if $f, g_1, \cdots, g_t = 0$, while if some of the polynomials $f, g_1, \cdots, g_t$ is non-zero, we get the chain which in Section 4 we denoted by $\mathfrak{C}_2(f, n_i, g_i)$. In particular, if $f \neq 0$ and $g_i = 0$ for any $i = 1, \cdots, t$, we get a commutative chain, which we denote by $\mathfrak{W}_2(f)$, isogenous to the two-dimensional Witt group $\mathfrak{W}_2$ via the isogeny $\tilde{f} : \mathfrak{W}_2 \longrightarrow \mathfrak{W}_2(f)$ which maps $(x_0, x_1) \mapsto (x_0, f(x_1))$.

6.1.2 Remark. Let G be a three-dimensional connected unipotent algebraic group which is not a chain. Observe that if $\dim G' = 1$ and G/G' is a chain, then by Theorem 3.2.10 we have that $\dim {}_3 G = 2$. In particular, since $G' \leq {}_3 G$ and G/G' is a chain, the connected component ${}_3{}^\circ G$ of ${}_3 G$ is the unique two-dimensional connected algebraic subgroup of G. It follows, as G is not a chain, that ${}_3{}^\circ G$ is isomorphic to a two-dimensional vector group. Hence, if $\dim G' = 1$, we have the following possibilities for G:

1.a) G/G' is isomorphic to a two-dimensional vector group and $\dim {}_3 G = 1$;
1.b) G/G' is isomorphic to a two-dimensional vector group and ${}_3{}^\circ G$ is isogenous to $\mathfrak{W}_2(f)$ for a given non-zero p-polynomial f;
1.c) both G/G' and ${}_3{}^\circ G$ are isomorphic to a two-dimensional vector group;
1.d) G/G' is isogenous to $\mathfrak{W}_2(f)$ for a given non-zero p-polynomial f and ${}_3{}^\circ G$ is isomorphic to a two-dimensional vector group.

Analogously, if $\dim G' = 2$, then G' is the unique two-dimensional connected algebraic subgroup of G. It follows that G' is isomorphic to a two-dimensional vector group, otherwise together with G' the group G would be a chain. Hence, if $\dim G' = 2$, we have the following possibilities for G:

2.a) both G' and ${}_3{}^\circ G$ are isomorphic to a two-dimensional vector group;
2.b) G' is isomorphic to a two-dimensional vector group and $\dim {}_3 G = 1$. $\square$

First we consider the case in which $\dim G' = 1$.

6.2 dim $G' = 1$

The following theorem is a classification of three-dimensional connected unipotent groups which are not chains over a perfect field of positive characteristic p, whereas if G is a chain we refer to Theorem 4.3.1.

6.2.1 Theorem. *Let G be a three-dimensional connected unipotent algebraic group defined over a perfect field k of positive characteristic p such that $\dim G' = 1$. If G is not a chain, then:*

1.a) If G/G' is isomorphic to a two-dimensional vector group and $\dim {}_3 G = 1$, then G is isogenous to the group defined by the following product:

$$(x_0, x_1, x_2) \cdot (y_0, y_1, y_2) =$$

$$(x_0 + y_0, x_1 + y_1, x_2 + y_2 + \beta(x_0, y_0) + \gamma(x_1, y_1) + \sigma_{y_0}(x_1)) \qquad (6.1)$$

where $\beta, \gamma \in C^2(\mathbf{G}_a, \mathbf{G}_a)$, $\sigma_y : \mathbf{G}_a \longrightarrow \mathbf{G}_a$ is a homomorphism such that $\sigma_{y_0 + z_0} = \sigma_{y_0} + \sigma_{z_0}$ and such that the identities

$$\begin{cases} \beta(y_0, \varphi(t)) - \beta(\varphi(t), y_0) = \sigma_{y_0}(\psi(t)) \\ \gamma(\psi(t), y_1) - \gamma(y_1, \psi(t)) = \sigma_{\varphi(t)}(y_1) \end{cases}$$

hold only if the p-polynomials $\varphi = \psi = 0$ are both the constant zero;

1.b) if G/G' is isomorphic to a two-dimensional vector group and $_3°G$ is isomorphic to $\mathfrak{W}_2(f)$ for a given non-zero p-polynomial f, then G is uniminimal. In particular G is isogenous to a product $\mathfrak{W}_2(f) \curlyvee H$, where H is a two-dimensional non-commutative unipotent chain. If $p > 2$, then H has exponent p;

1.c) if both G/G' and $_3°G$ are isomorphic to a two-dimensional vector group, then G is isogenous to the direct product of a two-dimensional non-commutative unipotent chain H with a one-dimensional connected algebraic subgroup;

1.d) if G/G' is isogenous to $\mathfrak{W}_2$, then $_3°G$ is the unique connected two-dimensional algebraic subgroup of G and it is a vector group. Moreover, G is isogenous to a direct product $\mathfrak{W}_2 \curlywedge H$ with amalgamated factor group, where H is a two-dimensional non-commutative unipotent chain.

Proof. Let G/G' be a two-dimensional vector group. Thus we have a central extension

$$1 \longrightarrow G' \longrightarrow G \longrightarrow \mathbf{G}_a \oplus \mathbf{G}_a \longrightarrow 1$$

and a corresponding representation of $G = G_\phi$ as the group defined on the three-dimensional affine space by the operation

$$(x_0, x_1, x_2)(y_0, y_1, y_2) = (x_0 + y_0, x_1 + y_1, x_2 + y_2 + \phi(x_0, x_1, y_0, y_1)).$$

Let T be the two-dimensional connected algebraic subgroup of G defined by $x_0 = 0$ and put $\gamma(x_1, y_1) = \phi(0, x_1, 0, y_1)$.

Since T is normal in G and G/T is one-dimensional, we can give G_ϕ also a representation as an extension of T by $\mathbf{G}_a$:

$$1 \longrightarrow T \longrightarrow G_\phi \longrightarrow \mathbf{G}_a \longrightarrow 1.$$

By Remark 2.1.15 we find that together with G_ϕ the group G is biregularly isomorphic to the group defined by the operation

$$(x_0, x_1, x_2) \cdot (y_0, y_1, y_2) =$$
$$(x_0 + y_0, x_1 + y_1, x_2 + y_2 + \beta(x_0, y_0) + \gamma(x_1, y_1) + \sigma_{y_0}(x_1)),$$

where $\delta^2 \beta = 0$ and $\sigma_{y_0} : \mathbf{G}_a \longrightarrow \mathbf{G}_a$ is a homomorphism such that $(0, x_1, x_2)^{(y_0, 0, 0)} = (0, x_1, x_2 + \sigma_{y_0}(x_1))$.

Since $(y_0, 0, 0) + (z_0, 0, 0) = (y_0 + z_0, 0, \beta(y_0, z_0))$ and $(0, 0, \beta(y_0, z_0))$ is central, we find $\sigma_{y_0 + z_0} = \sigma_{y_0} + \sigma_{z_0}$.

Furthermore, since any two-dimensional connected algebraic subgroup U of G contains $G' = \{(x_0, x_1, x_2) \in G : x_0 = x_1 = 0\}$, for U there exists a pair $(\varphi, \psi) \neq (0, 0)$ of p-polynomials such that

$$U = \{(x_0, x_1, x_2) \in G : x_0 = \varphi(t), x_1 = \psi(t) \text{ with } t \in \mathbf{G}_a\}.$$

Now it follows easily that:

1.a) The centre of G is one-dimensional if and only if for any $(\varphi, \psi) \neq (0, 0)$ the polynomial

$$\beta(\varphi(t), y_0) + \gamma(\psi(t), y_1) + \sigma_{y_0}(\psi(t)) - \beta(y_0, \varphi(t)) - \gamma(y_1, \psi(t)) - \sigma_{\varphi(t)}(y_1)$$

is not identically zero.

1.b) If $\mathfrak{z}^\circ G$ is isomorphic to $\mathfrak{W}_2(f)$ for a given non-zero p-polynomial f, then G is uni-minimal, because otherwise for a further connected one-dimensional algebraic subgroup K we would have $G = \mathfrak{z}^\circ G \cdot K$, which is not possible. The assertion follows now by Propositions 5.2.31 and 2.1.3. One can give to such a group a representation by the operation in (6.1) with $\sigma_{y_0} = 0$, $\gamma = f\Phi_1$ and $\beta \in \mathsf{C}^2(\mathbf{G}_a, \mathbf{G}_a)$ such that $\beta(x_0, y_0) \neq \beta(y_0, x_0)$.

1.c) Let $\mathfrak{z}^\circ G$ be a two-dimensional vector group. As G is not uni-maximal there exists a maximal connected algebraic subgroup $M \neq \mathfrak{z}^\circ$. Clearly, M cannot be commutative, otherwise $G = \mathfrak{z}^\circ G \cdot M$ would be commutative. Hence M is a two-dimensional non-commutative chain. The assertion follows taking a central one-dimensional subgroup K not contained in M.

For the last assertion *1.d)* we assume that G/G' is isogenous to $\mathfrak{W}_2$, hence G is uni-maximal. As mentioned in Remark 6.1.2 we have that $\mathfrak{z}^\circ G$ is a two-dimensional vector group containing G'. Let K be a connected one-dimensional algebraic subgroup of G such that $\mathfrak{z}^\circ G = K \oplus G'$. Since

$G' \cap K = 1$, the canonical homomorphism $\pi : G \longrightarrow G/G' \times G/K$ is injective, and Proposition 2.1.3 yields that G is isogenous to $\mathfrak{W}_2 \curlywedge H$, where H is a two-dimensional non-commutative chain. □

The cases 1.a), 1.b) and 1.c) of the previous Theorem 6.2.1 can be collected in the following

6.2.2 Proposition. *Let G be a three-dimensional unipotent group such that dim $G' = 1$ and G/G' is a two-dimensional vector group. Then G is isomorphic to the unipotent group defined on the affine space by the following multiplication:*

$$(x_0, x_1, x_2) \cdot (y_0, y_1, y_2) =$$

$$(x_0 + y_0, x_1 + y_1, x_2 + y_2 + \beta(x_0, y_0) + \gamma(x_1, y_1) + \sigma_{y_0}(x_1)) \qquad (6.2)$$

where $\beta, \gamma \in C^2(\mathbf{G}_a, \mathbf{G}_a)$ and $\sigma_y : \mathbf{G}_a \longrightarrow \mathbf{G}_a$ is a homomorphism such that $\sigma_{y_0 + z_0} = \sigma_{y_0} + \sigma_{z_0}$. □

The case 1.d) can be collected together with the case where G is a chain in the following

6.2.3 Proposition. *Let G be a three-dimensional unipotent group such that dim $G' = 1$ and G/G' is isogenous to a Witt group. Then G is isogenous to the unipotent group defined on the affine space by the following multiplication:*

$$(x_0, x_1, x_2) \cdot (y_0, y_1, y_2) =$$

$$(x_0 + y_0, x_1 + y_1 + \Phi_1(x_0, y_0), x_2 + y_2 + f(\Phi_2(x_0, x_1; y_0, y_1)) + \beta(x_0, y_0)) \quad (6.3)$$

where f is a p-polynomial and $\beta \in C^2(\mathbf{G}_a, \mathbf{G}_a)$ is not symmetric. □

6.2.4 Remark. Let G be as in Proposition 6.2.2. If $\sigma_y \neq 0$ and $\beta = \gamma = 0$, then the group G is isomorphic to the semi-direct product of $\mathbf{G}_a$ by $\mathbf{G}_a \oplus \mathbf{G}_a$. For instance, if $\sigma_y(x) = y \cdot x$, then G is isomorphic to the group of 3×3 unipotent matrices

$$\begin{pmatrix} 1 & x_1 & x_2 \\ & 1 & x_3 \\ & & 1 \end{pmatrix}.$$

In general, if N is the kernel of the homomorphism $\sigma : \mathbf{G}_a \longrightarrow k[\mathbf{F}]$, $y \mapsto \sigma_y$, then the central extension

$$1 \longrightarrow \mathbf{G}_a \longrightarrow G^* \longrightarrow (\mathbf{G}_a/N) \oplus \mathbf{G}_a \longrightarrow 1$$

with the factor system $\Psi^*((x_0 + N, x_1); (y_0 + N, y_1)) = \sigma_{y_0}(x_1)$ can be represented as the group consisting of the matrices

$$g(\sigma_y, x, z) = \begin{pmatrix} 1 & \sigma_y & z \\ & 1 & x \\ & & 1 \end{pmatrix},$$

where the multiplication is given by

$$g(\sigma_y, x, z)g(\sigma_{y'}, x', z') = g(\sigma_{y+y'}, x + x', z + z' + \sigma_y(x')).$$

On the other hand, if $\sigma_y = 0$, then G is isogenous to $H_1 \curlyvee H_2$, where H_1 and H_2 are the two-dimensional connected unipotent algebraic groups defined by the product

$$(x_0, x_2) \cdot (y_0, y_2) = (x_0 + y_0, x_2 + y_2 + \beta(x_0, y_0))$$

for H_1, respectively by the product

$$(x_1, x_2) \cdot (y_1, y_2) = (x_1 + y_1, x_2 + y_2 + \gamma(x_1, y_1))$$

for H_2.

It is clear that the hypothesis $\dim_3 G = 1$ forces the groups H_1, H_2 to be non-commutative. In the next Proposition we will see that this condition is also sufficient for $_3 G$ to be one-dimensional.   □

6.2.5 Proposition. *Let G be a three-dimensional connected unipotent algebraic group. If G is a product with amalgamated central subgroup of two non-commutative connected algebraic groups, then $\dim_3 G = 1$.*

Proof. By Proposition 2.1.3, the group G contains two non-commutative subgroups G_1, G_2 such that $G = G_1 G_2$ and $[G_1, G_2] = 1$. Since G_1 and G_2 are two-dimensional, the group $K = (G_1 \cap G_2)^\circ$ is one-dimensional. Moreover K is central. Let $x_i \in G_i$, $x_i \notin {}_3 G_i$ and consider the map $\sigma_{x_i} : G \longrightarrow K$ defined by $\sigma_{x_i}(g) = [g, x_i]$. Since G_i is not contained in the kernel of σ_{x_i}, these kernels are two-dimensional. Therefore G_i is the connected component of the kernel of σ_{x_j}, where $\{i, j\} = \{1, 2\}$. Since $_3°G$ is contained in both kernels, we find that $_3 G$ has dimension one.   □

6.2.6 Example. Let k be a perfect field of characteristic two. Let G be the connected unipotent algebraic group defined on the three-dimensional affine space by

$$(x_0, x_1, x_2)(y_0, y_1, y_2) = (x_0 + y_0, x_1 + y_1, x_2 + y_2 + x_0^2 y_1^4 + x_0 y_0^2 + x_1 y_1^2). \quad (6.4)$$

Then $G' = \{(x_0, x_1, x_2) \in G : x_0 = x_1 = 0\}$. Assume that there exists in G a further one-parameter subgroup K. Then $K = \{(\alpha(t), \beta(t), \gamma(t)) : t \in \mathsf{k}\}$ with 2-polynomials α, β which cannot be simultaneously trivial. Since any element in K has order two, we find by (6.4)

$$\alpha(t)^2 \beta(t)^4 + \alpha(t)^3 + \beta(t)^3 = 0,$$

that is $\beta(t)^3 = \alpha(t)^2 \beta(t)^4 + \alpha(t)^3$. If we look at the degree δ of these polynomials we see that this is not possible, because $3\delta(\beta)$ can never be equal to $\max\{3\delta(\alpha), 2\delta(\alpha) + 4\delta(\beta)\}$. Hence G is uni-minimal.

The subgroup of elements which are rational over the prime subfield k_0 of k is isomorphic to the quaternion group of order 8, because it is a non-commutative extension of a group of order 2, namely $\{(0, 0, x_2) : x_2 \in k_0\}$, by an elementary abelian group of order 4, induced by the non-symmetric factor system $x_0 y_1 + x_0 y_0 + x_1 y_1$. This shows that G cannot contain a commutative two-dimensional chain, since the set of k_0-rational points of such a group would be a cyclic group of order 4.   □

6.3 dim $G' = 2$

Now we assume that $\dim G' = 2$. We can classify the three-dimensional unipotent groups having a two-dimensional commutator subgroup only if the characteristic of the ground field is greater than two, as one can see in the proofs. By Proposition 6.1.1 in this case G' is a two-dimensional vector group.

In Theorem 6.3.3 we are forced to consider an equation of the shape $\delta^2 \beta(x, y, z) = -\sigma_z(\sum_{i=1}^{t} g_i(xy^{p^{n_i}}))$, where σ_z is a p-polynomial such that $\sigma_{z_1+z_2} = \sigma_{z_1} + \sigma_{z_2}$ and $g_1, \ldots, g_t$ are non-zero p-polynomials. Hence the next Proposition 6.3.2 is devoted to finding sufficient conditions on the p-polynomials σ_z, on the p-polynomials $g_1, \ldots, g_t$ and on the exponent $n_1, \ldots, n_t$, for the existence of a polynomial $\beta \in \bar{k}[x, y]$ satisfying the above equation.

First we recall that for any polynomial $\beta \in \bar{k}[x_1, x_2]$ the polynomial $\delta^2 \beta(x_1, x_2, x_3)$ lies in the kernel of the map $\Omega_3 : S \longrightarrow S$ defined by

$$f(x, y, z) \mapsto \sum_{\tau \in S_3} (-1)^{sign(\tau)} f\big(\tau(x), \tau(y), \tau(z)\big),$$

where S is the subspace of $\bar{k}[x, y, z]$ consisting of polynomials in which for any monomial one has $\deg x > 0, \deg y > 0, \deg z > 0$.

In general $\delta^2(\bar{k}[x, y]) \neq \ker \Omega_3$. For instance $z\Phi_1(x, y) \in \ker \Omega_3 \setminus \delta^2(\bar{k}[x, y])$ (see Proposition 4.2.3).

6.3.1 Lemma. *Let V be the $\bar{k}$-vector space of all the p-polynomials in S such that the degree of the variable x is smaller than the degree of the variable y. Then*

$$V \cap \ker \Omega_3 = \langle xy^{p^h} z : h > 0 \rangle_{\bar{k}[\mathbf{F}]} + \langle xy^{p^h} z^{p^h} : h > 0 \rangle_{\bar{k}[\mathbf{F}]} +$$

$$\langle axy^{p^h} z^{p^{h+t}} + bx^{p^t} y^{p^{h+t}} z + (a+b)xy^{p^{h+t}} z^{p^h} : h > 0, t > 0, a, b \in k \rangle_{\bar{k}[\mathbf{F}]}.$$

Proof. Since any element of V lies in the $\bar{k}[\mathbf{F}]$-module generated by the set

$$A = \left\{ xy^{p^h} z, \ xy^{p^h} z^{p^h}, \ xy^{p^h} z^{p^{h+t}}, \ x^{p^h} y^{p^{h+t}} z, \ xy^{p^{h+t}} z^{p^h} : h, t > 0 \right\},$$

and since $\Omega_3(h(f)) = h(\Omega_3(f))$ for any $h \in \bar{k}[\mathbf{F}]$ and for any $f \in S$, we can restrict to investigate the elements of A. As a simple computation shows we have that

i) $\Omega_3(xy^{p^h}z) = \Omega_3(xy^{p^h}z^{p^h}) = 0$ for any $h > 0$,

ii) $\Omega_3(xy^{p^h}z^{p^{h+t}}) \neq 0$ for any $h, t > 0$,

iii) $\Omega_3(xy^{p^h}z^{p^{h+t}}) = \Omega_3(x^{p^h}y^{p^{h+t}}z) = -\Omega_3(xy^{p^{h+t}}z^{p^h})$ for any $h, t > 0$,

and the assertion follows.   □

6.3.2 Proposition. $V \cap \delta^2(k[x,y]) = V \cap \ker\Omega_3 = \delta^2(\mathfrak{D} + \mathsf{C}^2(\mathbf{G}_a, \mathbf{G}_a))$, where

$$\mathfrak{D} = \langle x^{1+p^h}y + \tfrac{1}{2}x^{p^h}y^2 : h > 0\rangle_{k[\mathbf{F}]} + \langle \tfrac{1}{2}xy^{2p^h} : h > 0\rangle_{k[\mathbf{F}]} +$$

$$\langle (a+b)xy^{p^h+p^{h+t}} + bx^{1+p^h}y^{p^{h+t}} + bx^{p^h}y^{1+p^{h+t}} : h > 0, t > 0, a, b \in k\rangle_{k[\mathbf{F}]}.$$

Proof. As $\delta^2(h(\beta)) = h(\delta^2\beta)$ for any $h \in k[\mathbf{F}]$ and for any $\beta \in k[x,y]$, the assertion follows by the above Lemma 6.3.1 and by the following polynomial identities

i) $xy^{p^h}z = \delta^2(x^{1+p^h}y + \tfrac{1}{2}x^{p^h}y^2)$ for any $h > 0$;

ii) $xy^{p^h}z^{p^h} = \delta^2(-\tfrac{1}{2}xy^{2p^h})$ for any $h > 0$;

iii) $axy^{p^h}z^{p^{h+t}} + bx^{p^t}y^{p^{h+t}}z + (a+b)xy^{p^{h+t}}z^{p^h} =$
$$\delta^2(-(a+b)xy^{p^h+p^{h+t}} - bx^{1+p^h}y^{p^{h+t}} - bx^{p^h}y^{1+p^{h+t}})$$
for any $h, t > 0$ and for any $a, b \in k$.   □

6.3.3 Theorem. *Let G be a three-dimensional connected unipotent algebraic group defined over a perfect field k of characteristic $p > 2$ such that G' is isomorphic to a two-dimensional vector group.*

2.a) If $\dim {}_3G = 2$, *then G is isomorphic to the product $H_1 \lambda H_2$, where H_1 and H_2 are non-commutative two-dimensional groups;*

2.b) if $\dim {}_3G = 1$, *then G is isomorphic to the group defined by the following product:*

$$(x_0, x_1, x_2) \cdot (y_0, y_1, y_2) =$$

$$\left(x_0 + y_0, x_1 + y_1 + \sum_{i=1}^{t} g_i(x_0 y_0^{p^{n_i}}), x_2 + y_2 + \beta(x_0, y_0) + \sigma_{y_0}(x_1) \right),$$

where

- *$\{\sigma_y : y \in \mathbf{G}_a\}$ is a non-trivial subgroup of $k[\mathbf{F}]$ which is a homomorphic image of $\mathbf{G}_a$ and $g_1, \ldots, g_t$ are non-zero p-polynomials such that $\sigma_z(\sum_{i=1}^{t} g_i(xy^{p^{n_i}})) \in V \cap \delta^2(k[x,y])$,*
- *β is a polynomial in $\mathfrak{D} + \mathsf{C}^2(\mathbf{G}_a, \mathbf{G}_a)$ satisfying, for any $x, y, z \in \mathbf{G}_a$, the equation $\delta^2\beta(x, y, z) = -\sigma_z(\sum_{i=1}^{t} g_i(xy^{p^{n_i}}))$.*

Proof. Since G' is isomorphic to a two-dimensional vector group, we can represent G as an extension of $\mathbf{G}_a \oplus \mathbf{G}_a$ by $\mathbf{G}_a$

$$1 \longrightarrow \mathbf{G}_a \oplus \mathbf{G}_a \longrightarrow G \longrightarrow \mathbf{G}_a \longrightarrow 1.$$

Choose for any element $g \in G$ the following representation on the three-dimensional affine space:

$$g = (x_0, x_1, x_2),$$

where $G' = \{(x_0, x_1, x_2) : x_0 = 0\}$ and $Z = \{(x_0, x_1, x_2) : x_0 = x_1 = 0\}$ is a central subgroup of G.

The one-dimensional factor group G/G' acts by conjugation on G' as a unipotent group of automorphisms. Let $\{\sigma_y : y \in \mathbf{G}_a\}$ be the subgroup of $k[\mathbf{F}]$ which is a homomorphic image of $\mathbf{G}_a$ such that

$$(0, x_1, x_2)^{(y_0, 0, 0)} = (0, x_1, x_2 + \sigma_{y_0}(x_1)),$$

for any $y_0 \in \mathbf{G}_a$, $(0, x_1, x_2) \in G'$.

If $(\alpha, \beta) : \mathbf{G}_a \times \mathbf{G}_a \longrightarrow \mathbf{G}_a \oplus \mathbf{G}_a$ is the factor system associated with the above extension, the product of two elements of G is given by

$$(x_0, x_1, x_2) \cdot (y_0, y_1, y_2) =$$
$$(x_0 + y_0, x_1 + y_1 + \alpha(x_0, y_0), x_2 + y_2 + \beta(x_0, y_0) + \sigma_{y_0}(x_1)).$$

As G/Z is a two-dimensional non-commutative group, we have that

$$\alpha(x_0, y_0) = f(\Phi_1(x_0, y_0)) + \sum_{i=1}^{t} g_i(x_0 y_0^{p^{n_i}}),$$

where $t > 0$ and g_i is a non-zero p-polynomial for any $i = 1, \ldots, t$. Furthermore, the associativity of the multiplication in G forces β to fulfil the following equation:

$$\delta^2 \beta(x, y, z) = -\sigma_z(f(\Phi_1(x, y)) + \sum_{i=1}^{t} g_i(xy^{p^{n_i}})).$$

Since (as shown in Proposition 4.2.3) there is no $\beta \in k[x, y]$ satisfying the equation

$$\delta^2 \beta(x, y, z) = -\sigma_z(f(\Phi_1(x, y))),$$

and since the polynomials $\sigma_z(f(\Phi_1(x, y)))$, $\sigma_z(\sum_{i=1}^{t} g_i(xy^{p^{n_i}}))$ have no homogenous components in common, we have either $\sigma_y = 0$ for any $y \in \mathbf{G}_a$, or $\sigma_y \neq 0$ for some $y \in \mathbf{G}_a$ and $f = 0$.

Assume first that $\sigma_y = 0$ for any $y \in \mathbf{G}_a$. In this case we have that $\beta \in C^2(\mathbf{G}_a, \mathbf{G}_a)$, and that the product in G is given by

$$(x_0, x_1, x_2)) \cdot (y_0, y_1, y_2) =$$

$$(x_0 + y_0, x_1 + y_1 + f(\Phi_1(x_0, y_0)) + \sum_{i=1}^{t} g_i(x_0 y_0^{p^{n_i}}), x_2 + y_2 + \beta(x_0, y_0)).$$

This means that the group G is isogenous to the product $H_1 \curlywedge H_2$, where H_1 and H_2 are the two-dimensional connected unipotent algebraic groups defined by the product

$$(x_0, x_1) \cdot (y_0, y_1) = (x_0 + y_0, x_1 + y_1 + f(\Phi_1(x_0, y_0)) + \sum_{i=1}^{t} g_i(x_0 y_0^{p^{n_i}})),$$

for H_1, respectively by the product

$$(x_0, x_2) \cdot (y_0, y_2) = (x_0 + y_0, x_2 + y_2 + \beta(x_0, y_0))$$

for H_2.

In particular, the group H_2 is non-commutative, since otherwise $\dim G' = 1$.

Assume now that the group $\{\sigma_y : y \in \mathbf{G}_a\}$ is non-trivial. In this case the product in G is given by

$$(x_0, x_1, x_2) \cdot (y_0, y_1, y_2) =$$

$$(x_0 + y_0, x_1 + y_1 + \sum_{i=1}^{t} g_i(x_0 y_0^{p^{n_i}}), x_2 + y_2 + \beta(x_0, y_0) + \sigma_{y_0}(x_1)),$$

where the polynomial β satisfies $\delta^2 \beta(x, y, z) = -\sigma_z(\sum_{i=1}^{t} g_i(xy^{p^{n_i}}))$. This means that $\sigma_z(\sum_{i=1}^{t} g_i(xy^{p^{n_i}})) \in V \cap \delta^2(\mathsf{k}[x, y])$ and hence, by Proposition 6.3.2, that $\beta \in \mathfrak{D} + \mathsf{C}^2(\mathbf{G}_a, \mathbf{G}_a)$. □

6.4 Characterisation of Three-Dimensional Unipotent Groups

6.4.1 Proposition. *Let G be a non-commutative three-dimensional connected unipotent algebraic group defined over a field k of characteristic $p > 0$ having exponent p.*

i) If $\dim G' = 1$, then G is isogenous to the group defined on the three-dimensional affine space by the following multiplication:

$$(x_0, x_1, x_2)(y_0, y_1, y_2) =$$

$$(x_0 + y_0, x_1 + y_1, x_2 + y_2 + \beta(x_0, y_0) + \gamma(x_1, y_1) + \sigma_{y_0}(x_1)) \quad (6.5)$$

where $\beta, \gamma \in \langle \{xy^{p^i} : i > 0\} \rangle_{\mathsf{k}[\mathbf{F}]}$ and $\{\sigma_y : y \in \mathbf{G}_a\}$ is a subgroup of $\mathsf{k}[\mathbf{F}]$ which is a homomorphic image of $\mathbf{G}_a$;

ii) if $\dim G' = 2$, then the group G is isogenous either to the direct product with amalgamated factor group of two non-commutative two-dimensional connected unipotent algebraic groups having exponent p, or to the group defined on the three-dimensional affine space with multiplication

$$(x_0, x_1, x_2)(y_0, y_1, y_2) =$$

$$(x_0 + y_0, x_1 + y_1 + \alpha(x_0, y_0), x_2 + y_2 + \beta(x_0, y_0) + \gamma(x_0, y_0) + \sigma_{y_0}(x_1)) \quad (6.6)$$

where

- *$\alpha, \gamma \in \langle \{xy^{p^i} : i > 0\} \rangle_{\mathsf{k}[\mathbf{F}]}$, with $\alpha \neq 0$, and $\{\sigma_y : y \in \mathbf{G}_a\}$ is a non-trivial subgroup of $\mathsf{k}[\mathbf{F}]$ which is a homomorphic image of $\mathbf{G}_a$ such that*

- $\sigma_z(\alpha(x,y)) \in V \cap \delta^2(\mathsf{k}[x,y])$ if $p > 3$,
- $\sigma_z(\alpha(x,y)) \in \langle -xy^{p^h} z^{p^{h+t}} + x^{p^t} y^{p^{h+t}} z - xy^{p^{h+t}} z^{p^h} : h > 0, t > 0 \rangle_{\mathsf{k}[\mathbf{F}]}$ if $p = 3$,
- β is a polynomial in $\mathfrak{D}$ satisfying, for any $x, y, z \in \mathbf{G}_a$, the equation $\delta^2 \beta(x, y, z) = -\sigma_z(\alpha(x,y))$.

Proof. Since for $p = 2$ any three-dimensional connected unipotent algebraic group of exponent p is isomorphic to a three-dimensional vector group, we can assume $p > 2$.

Let $\dim G' = 1$. Since G has exponent p, the factor group G/G' is iso-morphic to a two-dimensional vector group. Then, by Theorem 6.2.1, up to isogenies the product in G is given by (6.5), where $\beta, \gamma \in \langle \{xy^{p^i} : i > 0\} \rangle_{\mathsf{k}[\mathbf{F}]}$, since G has exponent p.

Let $\dim G' = 2$. If $\dim_3 G = 2$, then by Theorem 6.3.3 the first part of asser-tion *ii)* follows. Assume therefore that $\dim_3 G = 1$. Then, by Theorem 6.3.3, up to isogenies the product in G is given by (6.6). By a simple computation we see that

$$(x_0, x_1, x_2)^p = (0, 0, \sum_{k=1}^{p-1}[\beta(x_0, kx_0) + \sigma_{kx_0}(x_1)]).$$

for any $(x_0, x_1, x_2) \in G$. In particular, since $\sigma_y + \sigma_t = \sigma_{y+t}$ for any $y, t \in \mathbf{G}_a$ and $\sum_{k=1}^{p-1} k \equiv 0\,(p)$, we have that $\sum_{k=1}^{p-1} \sigma_{kx_0}(x_1) = 0$. Furthermore, by Proposition 6.3.2 the polynomial β is a sum of polynomials of the shape

$$f(x^{1+p^n} y + \tfrac{1}{2} x^{p^n} y^2),$$

$$g(\tfrac{1}{2} xy^{2p^m}),$$

$$h((a+b)xy^{p^s + p^{s+u}} + bx^{1+p^s} y^{p^{s+u}} + bx^{p^s} y^{1+p^{s+u}}),$$

for some $f, g, h \in \mathsf{k}[\mathbf{F}]$, $a, b \in \mathsf{k}$ and $n, m, s, u > 0$. Hence, by the following polynomial identities

$$\sum_{k=1}^{p-1} f(x^{1+p^n} kx + \tfrac{1}{2} x^{p^n}(kx)^2) = f(x^{2+p^n}) \sum_{k=1}^{p-1} k + f(\tfrac{1}{2} x^{2+p^n}) \sum_{k=1}^{p-1} k^2,$$

$$\sum_{k=1}^{p-1} g(\tfrac{1}{2} x(kx)^{2p^m}) = g(\tfrac{1}{2} x^{1+2p^m}) \sum_{k=1}^{p-1} k^2,$$

$$h((a+b)x(kx)^{p^s+p^{s+u}} + bx^{1+p^s}(kx)^{p^{s+u}} + bx^{p^s}(kx)^{1+p^{s+u}}) =$$

$$h((a+b)x^{1+p^s+p^{s+u}}) \sum_{k=1}^{p-1} k^2 + h(bx^{1+p^s+p^{s+u}})(\sum_{k=1}^{p-1} k + \sum_{k=1}^{p-1} k^2),$$

and by the fact that $\sum_{k=1}^{p-1} k^2 \equiv 0\,(p)$ if and only if $p > 3$, the second part of assertion *ii)* follows. $\qquad\square$

As a corollary of Theorem 5.2.30 we find the following fact about three-dimensional unipotent groups.

6.4.2 Corollary. *Let G be a non-commutative three-dimensional connected unipotent algebraic group defined over a field k of characteristic $p > 2$. If G is not a chain, then G contains a two-dimensional connected algebraic subgroup of exponent p.* $\qquad\square$

If $\dim_3 G = 2$, then we can say a bit more:

6.4.3 Proposition. *Let G be a three-dimensional connected unipotent algebraic group defined over a field of characteristic $p > 2$, containing a two-dimensional non-commutative subgroup H and a two-dimensional commutative chain W. If $\dim_3 G = 2$, then G is uni-minimal and contains a unique two-dimensional chain J of exponent p. Moreover, the subgroup $W = {}_3{}^\circ G$ is the unique commutative two-dimensional connected algebraic subgroup of G.*

Proof. Since G is non-commutative, ${}_3{}^\circ G$ must coincide with W. Consequently, G is uni-minimal, because for a one-dimensional algebraic subgroup K, not contained in ${}_3{}^\circ G$, for dimensional reasons we would have $G = {}_3{}^\circ GK$, a contradiction. Since G is uni-minimal, every maximal connected algebraic subgroup of G is a two-dimensional chain. Furthermore, containing $H \neq W$ as a two-dimensional subgroup, G is not a chain. Hence, by Theorem 5.2.30, the group G contains a two-dimensional chain J of exponent p. Assume now that G contains a two-dimensional chain $T \neq J$ of exponent p. Then $G = JT$. In particular, since $p > 2$ and G' is central in G, for any $g = xy \in G$ (with $x \in J$, $y \in T$), it would follow that $g^p = (xy)^p = x^p y^p [x, y^{-p(p-1)/2}] = 0$ (cf. Proposition 5.2.28). Consequently, G would have exponent p, a contradiction to the fact that $W \leq G$. □

6.4.4 Remark. The hypothesis $\dim_3 G = 2$ is indispensable in the above Proposition 6.4.3, as the following example shows. Let G be the three-dimensional unipotent group defined by the following product

$$(x_0, x_1, x_2)(y_0, y_1, y_2) =$$

$$(x_0 + y_0, x_1 + y_1, x_2 + y_2 + \Phi_1(x_0, y_0) + \Phi_1(x_1, y_1) + x_0 y_0^p + y_0^p x_1).$$

The group G contains a commutative chain $W = \{(x_0, x_1, x_2) \in G : x_0 = 0\}$ and a two-dimensional non-commutative subgroup $H = \{(x_0, x_1, x_2) \in G : x_1 = 0\}$. Furthermore, by a direct computation we see that ${}_3 G = \{(x_0, x_1, x_2) \in G : x_0 = x_1 = 0\}$ and that the two-dimensional subgroup $T = \{(x_0, x_1, x_2) \in G : x_0 + x_1 = 0\}$ is a vector group. Now, if G contained a two-dimensional chain J of exponent p, then we would have $G = JT$ and, consequently, G would have exponent p, a contradiction. □

In the following theorem we characterise the three-dimensional unipotent groups which are not uni-maximal.

6.4.5 Theorem. *Let G be a three-dimensional connected unipotent algebraic group, over a field k with $\operatorname{char} \mathsf{k} = p > 0$. Then G either is the product of (at most three) chains which are normal in G, or G is uni-maximal. In particular, if G is neither commutative nor uni-maximal, the group G is a product of two non-commutative chains.*

Proof. Assume that the theorem does not hold, and that G is a counter-example.

By the classification of connected commutative unipotent algebraic groups we can assume that G is not commutative. Therefore, since G is not unimaximal, we have that $\dim G' = 1$ and that G/G' is isomorphic to a two-dimensional vector group. It follows, by Theorem 6.2.1, that the multiplication in G is given by

$$(x_0, x_1, x_2) \cdot (y_0, y_1, y_2) =$$
$$(x_0 + y_0, x_1 + y_1, x_2 + y_2 + \beta(x_0, y_0) + \gamma(x_1, y_1) + \sigma_{y_0}(x_1)). \qquad (6.7)$$

In particular, since by assumption the group G contains at least one two-dimensional commutative group, we can assume that the factor system γ is symmetric.

If $\sigma_y = 0$ for any $y \in \mathbf{G}_a$, then the factor system β is forced to be non-symmetrical by the fact that G is non-commutative by assumption. Therefore G is the product of the two-dimensional non-commutative chains $H_1 = \{(x_0, x_1, x_2) \in G : x_1 = 0\}$ and $H_2 = \{(x_0, x_1, x_2) \in G : x_0 = x_1\}$, and G is not a counter-example.

Assume therefore that the subgroup $\{\sigma_y : y \in \mathbf{G}_a\}$ of $k[\mathbf{F}]$ is not trivial. Since there exists an r_0 such that for $r > r_0$ the factor system $\beta(x, y) + \gamma(x^{p^r}, y^{p^r}) + \sigma_y(x^{p^r})$ is not symmetrical, one sees that for any $r > r_0$ the set $C_r = \{(x_0, x_1, x_2) \in G : x_1 = x_0^{p^r}\}$ is a two-dimensional non-commutative chain. It follows that the group G is a product of two two-dimensional non-commutative chains C_r, C_s for $r, s > r_0$, which again shows that G is not a counter-example. $\qquad \square$

In the next Theorem 6.4.7 we characterise the three-dimensional unipotent groups where any two two-dimensional subgroups are isogenous.

6.4.6 Lemma. *Let k be a field of characteristic $p > 0$ and let G_{Ψ_1}, G_{Ψ_2} be two-dimensional connected unipotent algebraic groups isomorphic to the central extensions of $\mathbf{G}_a$ by $\mathbf{G}_a$ corresponding to the factor system $\Psi_1 = \sum_{i \in I_1} g_i(\eta_i)$ or $\Psi_2 = \sum_{k \in I_2} h_k(\eta_k)$, respectively, with $\eta_i(x, y) = xy^{p^i}$ and p-polynomials g_i, h_k. Furthermore let $i' \in I_1, k' \in I_2$ be such that $\deg \Psi_1 = (1 + p^{i'}) \cdot \deg g_{i'}$ and $\deg \Psi_2 = (1 + p^{k'}) \cdot \deg h_{k'}$. If G_{Ψ_1} and G_{Ψ_2} are isogenous, then $i' = k'$.*

Proof. If G_{Ψ_1} and G_{Ψ_2} are isogenous, then by Proposition 2.1.13 there exist $f_1, l_1, f_2, l_2 \in k[\mathbf{F}]$ such that the factor system $f_1 \Psi_1 l_1 - f_2 \Psi_2 l_2$ is a coboundary. It follows that $\deg f_1 \Psi_1 l_1 = \deg f_2 \Psi_2 l_2$.

Since $\deg f_1 \Psi_1 l_1 = \deg f_1 \cdot \deg l_1 \cdot \deg g_{i'} \cdot (1 + p^{i'})$, as well as $\deg f_2 \Psi_2 l_2 = \deg f_2 \cdot \deg l_2 \cdot \deg h_{k'} \cdot (1 + p^{k'})$, the assertion follows from the fact that the equality $p^n(1 + p^h) = p^m(1 + p^k)$ holds if and only if $h = k$ and $n = m$. $\qquad \square$

6.4.7 Theorem. *Let G be a three-dimensional connected unipotent algebraic group defined over a field k of characteristic $p > 2$. If any two two-dimensional* *connected algebraic subgroups of G are isogenous, then G is either uni-maximal or isomorphic to the three-dimensional vector group.*

Proof. A counter-example G is a non-commutative connected unipotent group which is not uni-maximal. Hence we have that $\dim G' = 1$ and that G/G' is isomorphic to the two-dimensional vector group. Moreover, from Theorem 5.2.21 any two-dimensional connected algebraic subgroup of G is not commutative. Thus G is uni-minimal and by Theorem 5.2.30 the group G has exponent p. Therefore by Theorem 6.2.1 the group G is isomorphic to the group defined on the three-dimensional affine space by the product

$$(x_0, x_1, x_2)(y_0, y_1, y_2) =$$

$$\left(x_0 + y_0, x_1 + y_1, x_2 + y_2 + \sum_{i \in I_1} g_i(\eta_i(x_0, y_0)) + \sum_{j \in I_2} h_j(\eta_j(x_1, y_1)) + \sigma_{y_0}(x_1) \right)$$

where g_i and h_j are, for any $i \in I_1, j \in I_2$, non-zero p-polynomials, $\sigma_y : \mathbf{G}_a \longrightarrow \mathbf{G}_a$ is a homomorphism such that $\sigma_{y_0 + z_0} = \sigma_{y_0} + \sigma_{z_0}$ and $\eta_i(x_0, y_0) = x_0 y_0^{p^i}$ (see Remark 2.1.6).

We want to show that we can assume that σ_{y_0} is not the zero polynomial. In fact, $\sigma_{y_0} = 0$ if and only if the section $\tau : \mathbf{G}_a \longrightarrow G$ given by $x_0 \mapsto (x_0, 0, 0)$ satisfies the relation $(0, x_1, x_2)^{(y_0, 0, 0)} = (0, x_1, x_2 + \sigma_{y_0}(x_1)) = (0, x_1, x_2)$. If we replace τ with the section τ' given by $x_0 \mapsto (x_0, x_0, 0)$, we find

$$(0, x_1, x_2)^{(y_0, y_0, 0)} = (0, x_1, x_2 + \sum_{j \in I_2} h_j(\eta_j(-y_0, x_1)) + \sum_{j \in I_2} h_j(\eta_j(x_1, y_0))).$$

Putting $\sigma'_{y_0}(x_1) = \sum_{j \in I_2} h_j(\eta_j(-y_0, x_1)) + \sum_{j \in I_2} h_j(\eta_j(x_1, y_0))$, we find that G is birationally isomorphic to the group defined on the affine space by the product

$$(x_0, x_1, x_2)(y_0, y_1, y_2) =$$

$$\left(x_0 + y_0, x_1 + y_1, x_2 + y_2 + \beta(x_0, y_0) + \gamma(x_1, y_1) + \sigma'_{y_0}(x_1) \right),$$

where $\beta(x_0, y_0) = \sum_{i \in I_1} g_i(\eta_i(x_0, y_0)) + \sum_{j \in I_2} h_j(\eta_j(x_0, y_0))$ and $\gamma(x_1, y_1) = \sum_{j \in I_2} h_j(\eta_j(x_1, y_1))$. (Here we see concretely that the effects of changing a section for a non-central extension are not those of adding a trivial factor system, because, as mentioned in Remark 2.1.15, now we have $\delta^1(\tau - \tau') \neq \delta^1\tau - \delta^1\tau'$).

Since the subgroups $H_1 = \{(x_0, x_1, x_2) : x_1 = 0\}$ and $H_2 = \{(y_0, y_1, y_2) : y_0 = 0\}$ are isogenous, by Lemma 6.4.6 there exists $n \in I_1 \cap I_2$ such that $\deg \sum_{i \in I_1} g_i(\eta_i) = (1 + p^n) \cdot \deg g_n$ and $\deg \sum_{j \in I_2} h_j(\eta_j) = (1 + p^n) \cdot \deg h_n$. Furthermore, for any two p-polynomials φ and ψ such that $(\varphi, \psi) \neq (0, 0)$ the factor system

$$f(x, y) = \sum_{i \in I_1} g_i(\eta_i(\varphi(x), \varphi(y))) + \sum_{j \in I_2} h_j(\eta_j(\psi(x), \psi(y))) + \sigma_{\varphi(y)}(\psi(x))$$

is not trivial and the degree of any representative of the class $f(x,y) + B^2(\mathbf{G}_a, \mathbf{G}_a)$ is equal to $(1+p^n)p^v$ for a suitable non-negative integer v.

- We may assume that $\deg g_n = \deg h_n$.

In fact, if $\deg g_n = p^r$ and $\deg h_n = p^s$, with $r \neq s$, then the group G^* defined by the following product

$$(x_0, x_1, x_2)(y_0, y_1, y_2) =$$

$$(x_0+y_0, x_1+y_1, x_2+y_2+\sum_{i\in I_1} g_i(\eta_i(x_0^{p^s}, y_0^{p^s}))+\sum_{j\in I_2} h_j(\eta_j(x_1^{p^r}, y_1^{p^r}))+\sigma_{y_0^{p^s}}(x_1^{p^r})),$$

is isogenous to G via the isogeny $f : G^* \longrightarrow G$, $(x_0, x_1, x_2) \mapsto (x_0^{p^s}, x_1^{p^r}, x_2)$ and for G^* the assumption on the degrees is valid.

- if $\deg \sigma_{y_0}(x_1) = p^t(1+p^n)$, then $p^t \leq \deg g_n$.

In fact, let $\deg \sigma_{y_0}(x_1) = p^t(1+p^n)$ with $p^t > \deg g_n$ and let $ay_0^{p^t} x_1^{p^{t+n}} + by_0^{p^{t+n}} x_1^{p^t}$ be the homogeneous component of maximal degree of the polynomial $\sigma_{y_0}(x_1)$. Clearly $(a,b) \neq (0,0)$, therefore for any $s > 0$ at least one of the following relations holds:

i) $\deg \sigma_y(x^{p^s}) = p^t(1+p^{n+s}) > p^s(1+p^n) \cdot \deg g_n =$
   $\deg [\sum_{i\in I_1} g_i(\eta_i(x,y)) + \sum_{j\in I_2} h_j(\eta_j(x^{p^s}, y^{p^s}))]$
ii) $\deg \sigma_{y^{p^s}}(x) = p^t(1+p^{n+s}) > p^s(1+p^n) \cdot \deg g_n =$
   $\deg [\sum_{i\in I_1} g_i(\eta_i(x^{p^s}, y^{p^s})) + \sum_{j\in I_2} h_j(\eta_j(x,y))]$

Since each of the non-symmetrical factor systems

$$\sum_{i\in I_1} g_i(\eta_i(x,y)) + \sum_{j\in I_2} h_j(\eta_j(x^{p^s}, y^{p^s})) + \sigma_y(x^{p^s}),$$

$$\sum_{i\in I_1} g_i(\eta_i(x^{p^s}, y^{p^s})) + \sum_{j\in I_2} h_j(\eta_j(x,y)) + \sigma_{y^{p^s}}(x)$$

has degree $p^t(1+p^{n+s})$ with $s > 0$ we get in both cases a contradiction.

- Summarizing, we have that:

1) $\deg \sum_{i\in I_1} g_i(\eta_i) = (1+p^n) \cdot \deg g_n = (1+p^n) \cdot \deg h_n = \deg \sum_{j\in I_2} h_j(\eta_j)$,
2) if $\deg \sigma_{y_0}(x_1) = p^t(1+p^n)$, then $p^t \leq \deg g_n$.

Using these two relations we will show that:

- $\deg \sigma_{y_0}(x_1) \leq (1+p^n) \cdot \deg g_n$.

In fact, let $p^r = \deg g_n$ and assume that $\deg \sigma_{y_0}(x_1) = p^t(1+p^s) > p^r(1+p^n)$.

If $s \neq 0$, then $s \neq n$ by 2). Therefore, denoting by $\alpha y_0^{p^t} x_1^{p^{t+s}} + \beta y_0^{p^{t+s}} x_1^{p^t}$ the homogenous component of maximal degree of $\sigma_{y_0}(x_1)$, we have, for $a, b \in k$ with $\alpha a^{p^t} b^{p^{t+s}} \neq \beta a^{p^{t+s}} b^{p^t}$, that the polynomial

$$\sum_{i \in I_1} g_i(\eta_i(ax, ay)) + \sum_{j \in I_2} h_j(\eta_j(bx, by)) + \sigma_{ay}(bx)$$

is a non-symmetrical factor system having degree equal to $p^t(1 + p^s)$ with $s \neq n$, a contradiction. Therefore $s = 0$. Furthermore, since $2p^t > p^r(1 + p^n)$ we have $t \geq n + r$. If $t > n + r$, then for any $k > 0$ we have that

$$\deg \sigma_{y^{p^k}}(x_1) = p^t + p^{k+t} > p^{r+k}(1 + p^n) =$$

$$\deg \left[\sum_{i \in I_1} g_i(\eta_i(x^{p^k}, y^{p^k})) + \sum_{j \in I_2} h_j(\eta_j(x, y)) \right].$$

It follows that the polynomial

$$\sum_{i \in I_1} g_i(\eta_i(x^{p^k}, y^{p^k})) + \sum_{j \in I_2} h_j(\eta_j(x, y)) + \sigma_{y^{p^k}}(x)$$

is a non-symmetrical factor system having degree equal to $p^t(1 + p^k)$. In particular, if we choose $k \neq n$ we get a contradiction. Therefore $t = n + r$. Let $\tau_{y_0}(x_1)$ be the polynomial of degree $p^u + p^v$, with $u, v < r + n$, such that

$$\sigma_{y_0}(x_1) = \gamma x^{p^{r+n}} y^{p^{r+n}} + \sum_{l=1}^{r+n} (\xi_l y_0^{p^{r+n}} x_1^{p^{r+n-l}} + \delta_l y_0^{p^{r+n-l}} x_1^{p^{r+n}}) + \tau_{y_0}(x_1).$$

Furthermore, let $\alpha x^{p^r} y^{p^{r+n}}$, respectively $\beta x^{p^r} y^{p^{r+n}}$ be the monomial of maximal degree of the polynomial $\sum_{i \in I_1} g_i(\eta_i(x, y))$, respectively of $\sum_{j \in I_2} h_j(\eta_j(x, y))$, and let $a, b \in \mathsf{k} \setminus \{0\}$ be such that

$$\alpha a^{p^r + p^{n+r}} + \beta b^{p^r + p^{n+r}} + \xi_n a^{p^{n+r}} b^{p^r} + \delta_n a^{p^r} b^{p^{n+r}} = 0.$$

Now, for any p-polynomials $\varphi = a\mathbf{F}^{h+k} + c\mathbf{F}^h$, $\psi = b\mathbf{F}^{h+k} + d\mathbf{F}^h$, with h, k positive integers and $c, d \in \mathsf{k} \setminus \{0\}$, we have that

a) $\deg g_i(\eta_i(\varphi(x), \varphi(y))) = p^{h+k}(1 + p^i) \cdot \deg g_i$ for any $i \in I_1$,

b) $\deg [g_n(\eta_n(\varphi(x), \varphi(y))) - \alpha a^{p^r + p^{n+r}} x^{p^{h+k+r}} y^{p^{h+k+n+r}}] = p^{h+r}(1 + p^{k+n})$,

c) $\deg h_j(\eta_j(\psi(x), \psi(y))) = p^{h+k}(1 + p^j) \cdot \deg h_j$ for any $j \in I_2$,

d) $\deg [h_n(\eta_n(\psi(x), \psi(y))) - \beta b^{p^r + p^{n+r}} x^{p^{h+k+r}} y^{p^{h+k+n+r}}] = p^{h+r}(1 + p^{k+n})$,

e) $\deg [\gamma \cdot (\varphi(y)\psi(x))^{p^{r+n}} - \gamma \cdot (ab)^{p^{r+n}} x^{p^{h+k+r+n}} y^{p^{h+k+r+n}}] = p^{h+r+n}(1 + p^k)$,

f) $\deg [\xi_l \varphi(y)^{p^{r+n}} \psi(x)^{p^{r+n-l}} + \delta_l \varphi(y)^{p^{r+n-l}} \psi(x)^{p^{r+n}}] = p^{h+k+r+n-l}(1 + p^l)$
for any l,

g) $\deg \tau_{\varphi(y)}(\psi(x)) < p^{h+r+n}(1 + p^k)$,

from which it follows that

i) $\deg [\sum_{i \in I_1} g_i(\eta_i(\varphi(x), \varphi(y))) - \alpha a^{p^r + p^{n+r}} x^{p^{h+k+r}} y^{p^{h+k+n+r}}] = max\{p^{h+r}(1 + p^{k+n}), p^{h+k}(1 + p^i) \cdot \deg g_i : i \in I_1, i \neq n\}$,

ii) $\deg\left[\sum_{j\in I_2} h_j(\eta_j(\psi(x),\psi(y))) - \beta\, b^{p^r} {}^{+p^{n+r}} x^{p^{h+k+r}} y^{p^{h+k+n+r}}\right] =$
    $max\{p^{h+r}(1+p^{k+n}),\ p^{h+k}(1+p^j)\cdot \deg h_j : j\in I_2, j\neq n\}$,

iii) $\deg\left[\sigma_{\varphi(y)}(\psi(x)) - \gamma(ab)^{p^{r+n}} x^{p^{h+k+r+n}} y^{p^{h+k+r+n}} -\right.$
    $\left.\xi_n a^{p^{n+r}} b^{p^r} x^{p^{h+k+r}} y^{p^{h+k+n+r}} - \delta_n a^{p^r} b^{p^{n+r}} x^{p^{h+k+n+r}} y^{p^{h+k+r}}\right] =$
    $max\{p^{h+r+n}(1+p^k),\ p^{h+k+r+n-l}(1+p^l) : l\neq n\}$.

Therefore, if we take $k\neq n$ such that $p^{k+n}\neq p^i$ for any $i\in I_1$ and $p^{k+n}\neq p^j$ for any $j\in I_2$, for a suitable chosen value of c and d the degree of the non-symmetrical factor system

$$\sum_{i\in I_1} g_i(\eta_i(\varphi(x),\varphi(y))) + \sum_{j\in I_2} h_j(\eta_j(\psi(x),\psi(y))) + \sigma_{\varphi(y)}(\psi(x)) - \gamma(ab)^{p^t} x^{p^{h+k+t}} y^{p^{h+k+t}}$$

is equal to $(1+p^u)p^v$ with $u\neq n$, which is a contradiction. Therefore $\deg\sigma_{y_0}(x_1)\leq p^r(1+p^n)$ where $p^r = \deg g_n$. This inequality is crucial to obtain the desired contradiction and to prove the theorem.

• Final considerations.

Assume that $\deg\sigma_{y_0}(x_1) = p^r(1+p^n)$. Let $\xi_n y_0^{p^n} x_1^{p^{r+n}} + \delta_n y_0^{p^{r+n}} x_1^{p^n}$ be the homogeneous component of maximal degree of the polynomial $\sigma_{y_0}(x_1)$ and let $p^u + p^{u+m}$ be the degree of the polynomial $\sigma_{y_0}(x_1) - (\xi_n y_0^{p^n} x_1^{p^{r+n}} + \delta_n y_0^{p^{r+n}} x_1^{p^n})$. Taking, as before, $a, b\in k\setminus\{0\}$ such that $\alpha a^{p^r+p^{n+r}} + \beta\, b^{p^r+p^{n+r}} + \xi_n a^{p^{n+r}} b^{p^r} + \delta_n a^{p^r} b^{p^{n+r}} = 0$ and considering the p-polynomials $\varphi = a\mathbf{F}^{h+k} + c\mathbf{F}^h$, $\psi = b\mathbf{F}^{h+k} + d\mathbf{F}^h$, we get, in addition to i) and ii), the relation

iv) $\deg\left[\sigma_{\varphi(y)}(\psi(x)) - \xi_n a^{p^{n+r}} b^{p^r} x^{p^{h+k+r}} y^{p^{h+k+n+r}} -\right.$
    $\left.\delta_n a^{p^r} b^{p^{n+r}} x^{p^{h+k+n+r}} y^{p^{h+k+r}}\right] = p^{h+k+u}(1+p^m)$.

If $m = n$, then, since $p^u(1+p^m) < p^r(1+p^n)$, we have that $u < r$ and, consequently, that $p^{h+k+u}(1+p^m) < p^{h+r}(1+p^{k+n})$. Furthermore if $m = 0$, then $u < r+n$ and $2p^{h+k+u} < p^{h+r}(1+p^{k+n})$. It follows by i), ii) and iv) that if we take k such that $p^{k+n}\neq p^i$ for any $i\in I_1$ and $p^{k+n}\neq p^j$ for any $j\in I_2$, for a suitable chosen value of c and d the polynomial

$$\sum_{i\in I_1} g_i(\eta_i(\varphi(x),\varphi(y))) + \sum_{j\in I_2} h_j(\eta_j(\psi(x),\psi(y))) + \sigma_{\varphi(y)}(\psi(x))$$

is a non-symmetrical factor system having degree equal to $(1+p^s)p^v$ with $s\neq n$, a contradiction.

With the same arguments we get that the condition $\deg\sigma_{y_0}(x_1) < p^r(1+p^n)$ leads to a contradiction. Hence the theorem is proved. $\quad\square$

The following corollary is a direct consequence of the above theorem and of Theorem 5.2.34.

6.4.8 Corollary. *Let G be a uni-minimal algebraic group defined over a field k of characteristic $p > 2$. If any two two-dimensional connected algebraic subgroups of G are isogenous, then G is a chain.* $\quad\square$

In Sections 5.4 and 5.5 we saw that for algebraic groups over fields of characteristic zero the property to be aligned or to be co-aligned is strong enough to obtain interesting theorems. In contrast to this, for algebraic groups over a field of characteristic $p > 0$ in Section 5.7 we have been forced to adjoin additional hypothesis to obtain treatable classes of groups. The next Theorem 6.4.9 shows that this procedure was necessary, since already in the class of three-dimensional unipotent algebraic groups there are many example of groups which are neither aligned nor co-aligned or which have just one of these two properties. But surprisingly there are also many three-dimensional unipotent algebraic groups which are aligned as well as co-aligned.

6.4.9 Theorem. *For a three-dimensional unipotent group G over a perfect field of characteristic greater than two the following holds:*

   i) *if G belongs to the class 1.a) of Theorem 6.2.1, then G is co-aligned but not aligned;*

   ii) *if G belongs to the class 1.b) of Theorem 6.2.1, then G is co-aligned but not aligned;*

   iii) *if G belongs to the class 1.c) of Theorem 6.2.1, then G is neither co-aligned nor aligned;*

   iv) *if G belongs to the class 1.d) of Theorem 6.2.1, then G is not co-aligned but aligned;*

   v) *if G belongs to the class 2.a) of Theorem 6.3.3, then G is aligned;*

   vi) *if G belongs to the class 2.b) of Theorem 6.3.3, then G is aligned as well as co-aligned.*

Proof. In the cases $i)$ and $ii)$ the group G is co-aligned, since in the case 1.$a)$ the connected component of the center is the only one-dimensional normal connected subgroup of G, whereas in the case 1.$b)$ the center of G is a two-dimensional chain and every one-dimensional connected normal subgroup of G is contained in the center. Since in both cases G is not uni-maximal and non-commutative, it follows from Theorem 6.4.7 that G is not aligned.

The same Theorem 6.4.7 yields that a group in the class 1.$c)$ is not aligned whereas a group in the class 1.$d)$ is aligned. Moreover, a group G in any of the two classes 1.$c)$ and 1.$d)$ is not co-aligned, since there exists a one-dimensional central subgroup of G different from G'.

Also the groups in the classes 2.$a)$ and 2.$b)$ are aligned (cf. Theorem 6.4.7). Moreover, if G belongs to the class 2.$b)$, then G is co-aligned, too, since the connected component of the center is the only one-dimensional connected normal subgroup of G. □

7

Normality of Subgroups

7.1 Quasi-Normal Subgroups

Let G be a connected algebraic group and P a connected algebraic subgroup of G. If for any closed subgroup H of G we have $PH = HP$, then P is a normal subgroup of G (see [87]). Observe that for algebraic subgroups P and H of G with $PH = HP$, the group PH is an algebraic subgroup, too (see [45], 7.4 Corollary, p. 54).

For affine connected algebraic groups we can sharpen Theorem 1 in [87].

7.1.1 Theorem. *A connected algebraic k-subgroup P of a connected affine algebraic group G defined over an infinite perfect field k such that $PH = HP$ for any k-closed subgroup H of G is normal in G.*

Proof. Let N be the normalizer of P in G. Every finite k-subgroup E is contained in N, because P is the connected component of the group $PE = EP$. For any given k-rational element $x \in G(\mathsf{k})$ we have $x_s, x_u \in G(\mathsf{k})$, because k is perfect (cf. [8], 4.2 Corollary 1 (3), p. 81). The unipotent part x_u of x is such that the intersection $\mathcal{A}(x_u)$ of all closed subgroups containing x_u is either a finite k-subgroup (if k has positive characteristic) or a one-dimensional vector k-subgroup V (if k has characteristic zero) (cf. [8], 2.1 (b), p. 57). In the former case we have that $x_u \in \mathcal{A}(x_u) \le N$, because $\mathcal{A}(x_u)$ is a finite k-subgroup. In the latter case, assume that $P^{x_u} \ne P$. Then $PV = PP^{x_u} = P^{x_u}P$ since P is maximal in the group $PV = VP$. As no group is a product of two distinct conjugate subgroups we obtain a contradiction. It follows that $x_u \in N$, as well as in the former case.

The semi-simple part x_s of x is such that the intersection $\mathcal{A}(x_s)$ of all closed subgroups containing x_s is a k-subgroup S of G (cf. [8], 2.1 (b), p. 57) and the group S is the direct product of its connected component, which is a k-torus T, with a finite subgroup F.

Every element of finite order m of $\mathcal{A}(x_s)$ lies in the k-closed finite kernel of the map $x \mapsto x^m$, hence it normalizes P. Since the torsion subgroup of T

is dense in T (cf. [8], 8.9 Corollary, p. 116), the group $\mathcal{A}(x_s)$ normalizes P (cf. [8], Proposition 1.7 (b), p. 52).

Summarizing we have found that $\mathcal{A}(x) \leq N$ for any given $x \in G(\mathsf{k})$. Since k is infinite and perfect, the subgroup of rational points $G(\mathsf{k})$ is dense in G and the assertion follows (cf. [8], Proposition 1.7 (b), p. 52). $\qquad\square$

We will see that the situation changes if we request for a connected algebraic subgroup permutability only with respect to *connected* closed subgroups. We call a connected algebraic subgroup P of a connected algebraic group G *quasi-normal*, if $PH = HP$ holds for every connected closed subgroups H of G.

7.1.2 Remark. The group $\mathfrak{J}_3$ is an example of an algebraic group, containing a connected algebraic subgroup P which permutes with any connected algebraic subgroup of $\mathfrak{J}_3$, but which is not normal. To see this we recall that the commutator subgroup $\mathfrak{J}'_3$ is the unique connected two-dimensional subgroup of $\mathfrak{J}_3$. Since $\mathfrak{J}'_3$ is a vector subgroup, containing the one-dimensional centre of $\mathfrak{J}_3$, it is enough now to take as P any one-dimensional connected subgroup of $\mathfrak{J}_3$ different from than $_3\mathfrak{J}_3$. $\qquad\square$

7.1.3 Proposition. *A connected algebraic subgroup P of a connected non-affine algebraic group G is quasi-normal in G if and only if the maximal affine connected subgroup $L(P)$ of P is quasi-normal in the maximal connected affine subgroup $L(G)$ of G.*

Proof. Let H be a connected algebraic subgroup of G. According to Theorem 1.3.2 we have $G = L(G)D(G)$ and $H = L(H)D(H)$. The factor group $HD(G)/D(G)$ is affine. Since $HD(G)/D(G)$ is isomorphic to $L(H)D(H)/(H \cap D(G))$ and the group $D(H)$ has no non-trivial epimorphic images, we have $D(H) \leq D(G)$. Hence $D(H)$ is central in G.

Let $P = L(P)D(P)$ be a quasi-normal subgroup of G and $H = L(H)D(H)$ a connected algebraic subgroup of G. We have $PH = L(P)L(H)D(P)D(H) = HP$ if and only if $L(P)L(H) = L(H)L(P)$. $\qquad\square$

This proposition shows that we can restrict our attention to quasi-normal subgroups in affine algebraic groups.

7.1.4 Proposition. *Let G be a connected algebraic group and let Q be a quasi-normal algebraic subgroup of G which is a maximal subgroup of G. Then Q is normal in G.*

Proof. For any $g \in G$ together with Q the group Q^g is quasi-normal. Then according to [45], 7.4 Corollary p. 54, the group QQ^g is a closed subgroup of G which is quasi-normal. Since no group is a product of two distinct conjugate subgroups, one has $Q = Q^g$. $\qquad\square$

If the ground field of the affine algebraic group G has characteristic zero, then G is generated by one-dimensional connected algebraic subgroups X and for any quasi-normal subgroup Q one has $QX = XQ$. In this group, Q is maximal, and by Proposition 7.1.3 and 7.1.4 we have the following

7.1.5 Corollary. *Let G be a connected algebraic group over a field of characteristic zero and let Q be a quasi-normal algebraic subgroup of G. Then Q is normal in G.* □

7.1.6 Lemma. *Let Q be a connected algebraic quasi-normal subgroup of a connected algebraic group G. Then the following assertions hold:*

(i) Q is normalized by any minimal closed connected subgroup H of G,
(ii) Q is subnormal in G.

Proof. (i) Assume that $H \neq Q$. Then because of the minimality of H the group Q is a maximal closed connected quasi-normal subgroup of the group QH. Now the assertion follows from Proposition 7.1.4.

(ii) Consider a maximal chain

$$Q = Q_0 < Q_1 < \cdots < Q_n = G$$

of closed connected quasi-normal subgroups of G. Since Q_i is normal in Q_{i+1} (Proposition 7.1.4), our claim follows. □

If G is a connected algebraic group and H is a connected algebraic subgroup of G, we denote by H_G the maximal connected normal subgroup of G contained in H.

7.1.7 Theorem. *Let G be a connected affine algebraic group and let Q be a closed connected quasi-normal subgroup of G which is not normal in G. Then Q/Q_G is unipotent.*

Proof. Let G be a group of minimal dimension containing closed connected quasi-normal subgroups P such that P is not normal and P/P_G is not unipotent. Among these subgroups choose Q to have maximal dimension. Since G/Q_G containing Q/Q_G as a quasi-normal subgroup, is a counter-example, too, one has $Q_G = 1$. The group G is generated by its unipotent radical R_u together with all of its tori, thus by R_u and all of its minimal closed connected subgroups. It follows from Lemma 7.1.6, (i) that $Q^{R_u} \neq R_u$. Because of the minimality of $\dim G$ we have $G = R_u Q$, and from Proposition 7.1.4 one concludes that $\dim R_u > 1$.

Assume that R_u is not uni-maximal and let M_1, M_2 be different maximal connected subgroups of R_u. Then again by the minimality of $\dim G$ we have $Q^{M_1} = Q$ and $Q^{M_2}Q$. But this contradicts the fact that $R_u = M_1 M_2$ and $Q^{R_u} \neq R_u$. Hence R_u has a unique maximal closed connected subgroup M and M is normal in G. Because of $\dim R_u > 1$ one has $M \neq 1$. So because of the minimality of $\dim G$ the equality $Q^M = Q$ holds and this has $[Q, M] \leq Q \cap M$ as a consequence. Since $G = R_u Q$, it follows that the group $Q \cap Z$ is normal in G, where Z denotes the connected component of the centre of R_u. Thus $Q \cap Z = 1$.

Since $\dim Q$ is maximal and QZ is a quasi-normal subgroup of G, it follows that group QZ is normal in G. Then $[Q, M] = [QZ, M]$ is normal in G, too. Thus because of $Q_G = 1$ and the normality of Q in QM we see that $[Q, MZ] = 1$. Again from the maximality of $\dim Q$ it follows that QM is normal in G. Hence $[Q, R_u] \leq [MQ, R_u] \leq MQ \cap R_u \leq M$, since R_u is uni-maximal and $QM \neq G$. Thus Q centralises the series $1 < M < R_u$ of normal subgroups. It follows that the group $Q/C_Q(R_u)$, where $C_Q(R_u)$ is the centraliser of R_u in Q, is unipotent. On the other hand one has $Q/C_Q(R_u) = Q/(C_G(R_u) \cap Q) \triangleleft R_u Q/(C_G(R_u) \cap Q) = G/(C_G(R_u) \cap Q)$ and we have reached the contradiction that Q is normal in G. $\qquad\square$

7.1.8 Corollary. *Let P be a connected quasi-normal subgroup of a non-commutative connected affine algebraic group G. If P does not contain a non-trivial normal subgroup of G, then P is contained in the unipotent radical of G.*

Proof. According to Proposition 7.1.4 any minimal connected subgroup of G normalizes P. Hence the normalizer of P in G contains the subgroup S generated by the maximal tori of G, because these are direct product of one-dimensional subtori.

By Theorem 7.1.7 the group P is a unipotent normal subgroup of PS. Hence it is contained in the unipotent radical of PS, which is contained in the unipotent radical of G. $\qquad\square$

The following corollary shows that non-normal quasi-normal subgroups of an algebraic group are phenomena which are essentially restricted to unipotent affine groups.

7.1.9 Corollary. *Let G be an arbitrary connected algebraic group and let Q be a closed connected quasi-normal subgroup of G which is not normal in G. Then Q/Q_G lies in the unipotent radical of $L(G/Q_G)$.*

Proof. Without loss of generality we may assume that the normal subgroup Q_G of G is trivial. Then Q lies in $L(G)$ since $D(G)$ is contained the centre of G, and the proposition follows from the previous corollary. $\qquad\square$

In analogy to the theory of abstract groups, we call an algebraic group *quasi-hamiltonian* if each connected closed subgroup is quasi-normal. Connected subgroups and epimorphic images of quasi-hamiltonian groups are obviously quasi-hamiltonian.

7.1.10 Proposition. *Let G be a connected algebraic group and E be a finite normal subgroup of G. The factor group G/E is quasi-hamiltonian if and only if G is quasi-hamiltonian.*

Proof. Let G/E be quasi-hamiltonian and let U, V be connected subgroups of G. Then $(UE/E)(VE/E) = (VE/E)(UE/E)$ which is equivalent to $UVE/E = VUE/E$. As a pre-image of an algebraic group, the set $UVE = VUE$ is an algebraic group. The connected component of the identity is $UV = VU$ and the assertion follows. If G is quasi-hamiltonian, then clearly G/E is quasi-hamiltonian.   □

7.1.11 Theorem. *A connected quasi-hamiltonian algebraic group is nilpotent.*

Proof. Since every connected algebraic group G is a central extension of the commutative group $D(G)$ by a connected affine group, we may assume that G itself is affine.

By Theorem 7.1.7 any maximal torus T is normal in the affine quasi-hamiltonian group G. Since the factor group G/T consists only of unipotent elements (see [45], 15.5, p. 112), it is unipotent. Therefore G is nilpotent.   □

7.1.12 Theorem. *Let G be a connected algebraic group. If $G/_3G$ is a chain, then G is quasi-hamiltonian.*

Proof. Let G be a counter-example of minimal dimension and let M be the maximal connected algebraic subgroup of G containing $_3°G$. Then $_3°G \leq _3°M$ and $M/_3M$ is a chain. Hence M fulfils the claim. Therefore, if P and Q are contained in M, then $PQ = QP$. If $P \not\leq M$, then $P_3°G = G$, because otherwise $P_3°G \leq M$ since M is the unique maximal connected subgroup containing $_3°G$. But in this case we have that P is normal in G, hence we have again $PQ = QP$, a contradiction.   □

7.1.13 Proposition. *Let G be a connected unipotent algebraic group having a unique maximal connected closed subgroup M. If M is quasi-hamiltonian, then G is quasi-hamiltonian, as well.*   □

7.1.14 Proposition. *Let G be a quasi-hamiltonian connected unipotent group. Then G contains a (unique) maximal connected vector subgroup, containing any one-dimensional connected subgroup.*

Proof. Let V be a vector subgroup, not contained in any other vector subgroup. Assume that a one-dimensional connected subgroup K exists such that K is not contained in V. As any element $x \in V$ is contained in a one-dimensional connected subgroup $P_x \leq V$, we can consider the subgroup $P_x K = K P_x$, which is a two-dimensional vector subgroup. Thus any element of V commutes with any element of K, forcing the subgroup $KV = VK$ to be a vector subgroup, which is a contradiction to the choice of V.   □

7.1.15 Remark. The above proposition does not yield a sufficient condition for G to be quasi-hamiltonian, even under the stronger assumption that a vector subgroup of co-dimension one exists in G, containing any one-dimensional connected algebraic subgroup. The following group gives in fact such an example:

Let G be the unipotent algebraic group defined on a 4-dimensional affine space by the multiplication

$$(x_0, x_1, x_2, x_3)(y_0, y_1, y_2, y_3) = (x_0 + y_0, x_1 + y_1 + x_0 y_0^p, x_2 + y_2, x_3 + y_3 + y_0 x_2).$$

We show that the vector subgroup $V = \{(x_0, x_1, x_2, x_3) \in G : x_0 = 0\}$ contains any one-dimensional connected algebraic subgroup K. In fact, the connected algebraic subgroup $N = \{(x_0, x_1, x_2, x_3) \in G : x_0, x_1 = 0\}$ is a normal subgroup of G and KN/N is (at most) a one-dimensional subgroup of the two-dimensional chain G/N, forcing $x_0 = 0$, for any element $(x_0, x_1, x_2, x_3) \in K$. Since

$$(x_0, x_1, 0, 0)(0, 0, x_2, 0) = (x_0, x_1, x_2, 0)$$

whilst

$$(0, 0, x_2, 0)(x_0, x_1, 0, 0) = (x_0, x_1, x_2, x_0 x_2),$$

the subgroups $P = \{(x_0, x_1, x_2, x_3) \in G : x_2, x_3 = 0\}$ and $Q = \{(x_0, x_1, x_2, x_3) \in G : x_0, x_1, x_3 = 0\}$ do not permute, thus G is not quasi-hamiltonian.

Now we want to show that every connected maximal subgroup $M \neq V$ of G is quasi-hamiltonian. Let $M \neq V$ be a maximal subgroup of G. By Proposition 7.2.5 it is enough to prove that the centre of M has co-dimension at most one in M. One sees that the centre of G is the two-dimensional connected subgroup

$$_3G = \{(x_0, x_1, x_2, x_3) : x_0 = 0, x_2 = 0\}.$$

Moreover, the commutator subgroup G' coincides with $_3G$. In fact an arbitrary commutator has the form

$$(0, x_0 y_0^p - x_0^p y_0, 0, y_0 x_2 - x_0 y_2)$$

and, for $x_0, y_0, x_2, y_2 \in k$, these elements cover a two-dimensional connected variety, as one sees computing the jacobian

$$\begin{pmatrix} 0 & y_0^p & 0 & -y_2 \\ 0 & -x_0^p & 0 & x_2 \\ 0 & 0 & 0 & y_0 \\ 0 & 0 & 0 & -x_0 \end{pmatrix}$$

which has rank two, for infinitely many $x_0, y_0, x_2, y_2 \in k$. □

In the following Proposition 7.1.16 we give a sufficient condition for a group having a vector subgroup of co-dimension one to be quasi-hamiltonian.

7.1.16 Proposition. *Let G be an n-dimensional connected unipotent algebraic group containing a vector group V as a maximal subgroup. Then:*

– if any maximal subgroup $M \neq V$ is uni-maximal, then G is quasi-hamiltonian,

– if G is quasi-hamiltonian, and $\dim G' = n-2$, then any maximal subgroup $M \neq V$ is uni-maximal.

Proof. Since any maximal connected algebraic subgroup of G is normal in G, in order to prove the first assertion we show that any connected algebraic subgroup H having co-dimension at least two in G is contained in V. In fact, if $M \neq V$ is a maximal subgroup of G containing H, then, as $M \cap V$ is the unique maximal subgroup of M, we have $H \leq M \cap V \leq V$.

Now assume that G is quasi-hamiltonian, that $\dim G' = n-2$, and that G contains a maximal subgroup $M \neq V$ which is not uni-maximal. Therefore, for a suitable maximal subgroup $H \neq G'$ of M and a one-dimensional subgroup P of V with $P \not\subseteq G'$, we have that $M = G'H$ and $V = G'P$. In particular, as $HP = PH$ is a maximal subgroup of G, it has even to contain G', and this leads to the following contradiction: $G = MP = G'HP = HP$. $\qquad\square$

In the second assertion of the above Proposition 7.1.16 it is indispensable that $\dim G' = n-2$. For example, the group $\mathfrak{J}_2 \times \mathbf{G}_a{}^{n-2}$ is quasi-hamiltonian even if it contains the maximal subgroup

$$M = \{((x_1, x_2), (y_1, \ldots, y_{n-2})) \in \mathfrak{J}_2 \times \mathbf{G}_a{}^{n-2} : y_{n-2} = 0\},$$

which is not uni-maximal.

7.1.17 Corollary. *Let G be a three-dimensional connected non-commutative unipotent algebraic group. Then G is quasi-hamiltonian if and only if it contains at most one two-dimensional vector subgroup.*

Proof. Assume that G is quasi-hamiltonian, and let V be a two-dimensional vector subgroup of G. If $G' = V$, then V is the unique maximal subgroup of G, otherwise $\dim G' = 1$ and by Proposition 7.1.16 any two-dimensional subgroup $M \neq V$ of G is a chain.

Conversely, if G contains exactly one two-dimensional vector subgroup V, then any maximal subgroup $M \neq V$ is uni-maximal and G is quasi-hamiltonian by Proposition 7.1.16. $\qquad\square$

7.1.18 Examples:

1) The uni-maximal group $\mathfrak{J}_3$ has a two-dimensional commutator subgroup, which is a vector group. Hence it is quasi-hamiltonian.

2) The three-dimensional unipotent group defined by the operation
$$(x_0, x_1, x_2)(y_0, y_1, y_2) = (x_0 + y_0, x_1 + y_1, x_2 + y_2 + \Phi_1(x_0, y_0) + y_0 x_1),$$
where Φ_1 is the first Witt cocycle ([18], V, § 1, no. 1), has a one-dimensional commutator subgroup. Moreover it has a unique two-dimensional vector subgroup, namely $V = \{(x_0, x_1, x_2) : x_0 = 0\}$, because $(x_0, x_1, x_2)^p = (0, 0, x_0^p)$.

3) Three-dimensional unipotent groups of type 1.c) and 1.d) of the Theorem 6.2.1 are quasi-hamiltonian since they have a two-dimensional vector group as their center.   □

7.1.19 Remark. As the following example shows, the direct product of quasi-hamiltonian groups in general is not a quasi-hamiltonian group.

Let G be the unipotent algebraic group defined on a three-dimensional affine space by the following product:

$$(x_0, x_1, x_2)(y_0, y_1, y_2) = (x_0 + y_0, x_1 + y_1, x_2 + y_2 + x_0 y_0^p + x_1^p y_1).$$

As we know the group G is isomorphic to the factor group $(H_1 \times H_2)/K$, where H_1 and H_2 are the two-dimensional unipotent groups defined by the following product

$$(x_0, x_1)(y_0, y_1) = (x_0 + y_0, x_1 + y_1 + x_0 y_0^p),$$

for H_1, respectively by the product

$$(x_0, x_1)(y_0, y_1) = (x_0 + y_0, x_1 + y_1 + x_0^p y_0),$$

for H_2, and K is the central subgroup of the direct product $H_1 \times H_2$ defined by $K = \{((x_0, x_1), (y_0, y_1)) \in H_1 \times H_2 : x_0 = y_0 = 0, x_1 + y_1 = 0\}$. Although the groups H_1 and H_2 are quasi-hamiltonian, the group G is not quasi-hamiltonian. To prove this it suffices to note, by Corollary 7.1.17, that G contains more than one two-dimensional vector group: Take for example the algebraic subgroups $T_1 = \{(x_0, x_1, x_2) \in G : x_0 = x_1\}$ and $T_2 = \{(x_0, x_1, x_2) \in G : x_0 + x_1 = 0\}$, which are two-dimensional commutative groups having exponent p.   □

As we noticed in Remark 7.1.2, the group $\mathfrak{J}_3$ has a connected one-dimensional algebraic subgroup which is not normal. Hence $\mathfrak{J}_n$ is hamiltonian if and only if $n \leq 2$. We cannot establish whether an upper bound for the nilpotency class of a quasi-hamiltonian algebraic group exists. For sure it is greater than 4, as the following remark shows.

7.1.20 Remark. The group $\mathfrak{J}_n$ is quasi-hamiltonian if and only if $n \leq 4$.

Proof. First we show that $\mathfrak{J}_4$ is quasi-hamiltonian. Since $\mathfrak{J}_4$ has a unique maximal connected subgroup, coinciding with its commutator subgroup $\mathfrak{J}_4' = \{(x_0, x_1, x_2, x_3) \in \mathfrak{J}_4 : x_0 = 0\}$, it suffices to prove that $\mathfrak{J}_4'$ is quasi-hamiltonian. But this follows immediately from Proposition 7.1.17, since every one-dimensional subgroup of $\mathfrak{J}_4$ lies, by Remark 3.1.7, in the two-dimensional vector subgroup $H_2 = \{(x_0, x_1, x_2, x_3) \in \mathfrak{J}_4 : x_0 = x_1 = 0\}$.

To prove that $\mathfrak{J}_5$ is not quasi-hamiltonian, we show that its subgroups S_1 and S_2, as defined in Section 3.1 after the Proposition 3.1.8, are not quasi-normal. Otherwise the two-dimensional algebraic subgroup S_1 should be a maximal connected subgroup of the product $S_1 S_2$, since $\dim S_1 S_2 = \dim S_1 + \dim S_2 - \dim (S_1 \cap S_2) = 3$. On the other hand, S_1 is not normalized by S_2, since for generic $x_1, x_3, y_2 \in k$ we have:

$$(0, 0, y_2, 0, 0)^{-1}(0, x_1, 0, x_3, 0)(0, 0, y_2, 0, 0) =$$

$$(0, 0, -y_2, 0, 0)(0, x_1, y_2, x_3, x_1 y_2^{p^2}) = (0, x_1, 0, x_3, x_1 y_2^{p^2} - x_1^{p^3} y_2),$$

which is not contained in S_1. $\square$

In abstract groups, quasi-normal subgroups (i.e. subgroups which are permutable with any other subgroup) have been a subject of interesting investigations. The most important paper in this direction is [93]. As a consequence of the results there one obtains that quasi-hamiltonian groups (i.e. groups in which any subgroup is quasi-normal) coincide with locally nilpotent groups having modular subgroup lattice (cf. [16], p. 739). The structure of such groups G is well-known (cf. [88], Chapter 2):

i) If G is periodic, then G is locally finite (cf. [16], Lemma 2.1, p. 741) and it is the direct product of modular p-groups the structure of which is determined in [47],

ii) if G is non-periodic and non-commutative, then the elements of finite order form a subgroup T and the factor group G/T is locally cyclic (cf. [48] or [16], p. 740).

7.2 Hamiltonian Groups

In the theory of groups, a group G is called hamiltonian if each subgroup of G is normal. A non-commutative hamiltonian group is the direct product of the quaternion group of order 8, an elementary abelian 2-group and a periodic commutative group in which every element has odd order (see [75], 5.3.7, p. 139).

In analogy to this, it might seem natural to call an algebraic group G hamiltonian if each closed subgroup of G is normal. The following claim shows that this condition would be too strong.

7.2.1 Theorem. *If G is a connected affine algebraic k-group and one of the following holds:*

1) every closed subgroup of G is normal,
2) every k-closed subgroup of G is normal and k is infinite and perfect,

then G is commutative.

Proof. Let T be a maximal k-subtorus of G (cf. [8], Theorem 18.2, p. 218). From the normality of T we have that G is, together with a Borel subgroup, nilpotent. So we can restrict attention to the unipotent radical U. Take $x \in U$ (respectively $x \in U(k)$) and consider the closure $X = \overline{\langle x \rangle}$ of the cyclic group generated by x. If char $k = p > 0$, then X is a finite cyclic normal subgroup (respectively k-subgroup), hence central. Thus U is commutative (respectively the subgroup $U(k)$, which is dense by [8], Corollary 18.3, p. 220, is central) and we are done. If char $k = 0$, the group X is a one-dimensional connected group. In this case X is central as well, since G centralises any one-dimensional normal connected subgroup. $\square$

According to this theorem, we feel justified to give the following

7.2.2 Definition. *A connected algebraic group G is called* hamiltonian *if each of its connected algebraic subgroups is normal.*

The following proposition shows that hamiltonian algebraic groups are stable under isogenies:

7.2.3 Proposition. *Let G be a connected algebraic group and E be a finite normal subgroup of G. The factor group G/E is hamiltonian if and only if G is hamiltonian.*

Proof. Let G/E be hamiltonian and U be a connected subgroup of G. Since UE/E is normal in G/E, the product UE is normal in G, hence $U = (UE)^\circ$ is normal in G. The converse is clear. $\square$

The following proposition shows that we can restrict our investigation to *affine* hamiltonian algebraic groups. Of course, Rosenlicht's Theorem 1.3.2 plays a fundamental rôle here, too.

7.2.4 Proposition. *A connected algebraic group G is hamiltonian if and only if its maximal connected affine subgroup $L(G)$ is hamiltonian.*

Proof. Let $L(G)$ be hamiltonian and let H be a connected algebraic subgroup of G. Moreover, let $H = L(H)D(H)$ be its decomposition and observe that $L(H) \leq L(G)$, so $L(H)$ is normal in $L(G)$, hence in G as well.
     As $HD(G)/D(G)$ is affine, one has $D(H) \leq D(G)$. Since $D(G)$ is central in G, the assertion follows. $\square$

In a connected hamiltonian affine algebraic group G any torus is normal, hence central. Thus G is nilpotent. Theorem 5.2.27 and Theorem 5.2.34 give necessary conditions for non-commutative algebraic groups to be hamiltonian. We give now another condition which makes a group hamiltonian.

7.2.5 Proposition. *If G is a connected algebraic group over a perfect field and* codim $_3°G = 1$, *then G is quasi-commutative and hamiltonian.*

Proof. Let U be a connected non-central subgroup of G. Since $_3°G$ is maximal connected, the product $U \cdot {_3°G}$ equals G. For any $x \in U$, and for any $g = uc \in G$ (where $u \in U, c \in {_3°G}$) we have $x^g = x^u \in U$. So U is normal. Furthermore, under the assumption that U is commutative, we have even $x^g = x^u = x$, i.e. U is central. $\square$

In Theorem 4.3.12 we classified affine algebraic groups G over perfect fields of positive characteristic such that the connected component of the centre is a maximal connected subgroup of G. These groups are examples for hamiltonian and quasi-commutative non-commutative algebraic groups. But the example 3.2.25 is a quasi-commutative algebraic group of dimension three such that the connected component of its centre, which coincides with the commutator subgroup of G, is the unique one-dimensional connected subgroup of G. This shows that there are hamiltonian and quasi-commutative algebraic groups in which the connected component of their centre is not a maximal connected subgroup of G.

7.2.6 Proposition. *Let* char $k > 2$ *and let G be a non-commutative hamiltonian unipotent algebraic group of dimension $d \leq 4$. Then G' is central in G.*

Proof. Let $d = 3$. If the centre of G had dimension one and G' had dimension two, then G' would be a chain. Hence G would be a chain as well, a contradiction to Corollary 4.2.10. Let $d = 4$ and let K be a connected one-dimensional subgroup of G. As K is normal in G, for the three-dimensional hamiltonian group G/K we have, as we have just now proved, $[G, G'K] \leq K$. Hence $K = [G, G']$ is the unique one-dimensional connected algebraic subgroup of G. Theorem 5.2.34 yields that $G' = K$ or $G' = C'$ where C is a chain of dimension $t \leq 4$. If $t = 4$, then $C = G$ and we are done by Corollary 4.2.10. If $t \leq 3$, then $G' = K$ by the same corollary. $\square$

Now we begin to study hamiltonian groups in detail. Connected subgroups and homomorphic images of hamiltonian groups are hamiltonian, as well, but direct products in general are not (see Propositions 7.2.12, 7.2.14).

7.2.7 Theorem. *Any connected hamiltonian algebraic group G over a field k is nilpotent and each of its one-dimensional subgroups is central. Moreover, if the characteristic of k is zero, then G is commutative.*

Proof. Since hamiltonian groups are quasi-hamiltonian, the nilpotency of G follows from Theorem 7.1.11. All one-dimensional closed connected subgroups of the maximal affine algebraic subgroup $L(G)$ of G are normal in G, hence even central in G. If k has characteristic zero, then it follows from Corollary 7.3.7, (4), that $L(G)$ is commutative, and the theorem is proved. $\square$

Since any one-dimensional normal subgroup of a unipotent algebraic group is central, from Theorem 7.2.7 we obtain the following

7.2.8 Remark. A three-dimensional non-commutative unipotent algebraic group G over a perfect field k is hamiltonian if and only if one of the following conditions hold:

i) G is uni-minimal;
ii) the center of G has dimension two. $\square$

As a consequence of this we get

7.2.9 Remark. Let G be a three-dimensional non-commutative unipotent algebraic group defined over a perfect field k of positive characteristic p.

If $\dim G' = 2$ and $p > 2$, then G is hamiltonian if and only if it is of type 2.a) in Theorem 6.3.3

If $\dim G' = 1$ and G is of type 1.b), 1.c) 1.d) in Theorem 6.2.1, then G is hamiltonian.

7.2.10 Theorem. *Any connected algebraic* k*-group G over a field* k *of characteristic zero such that any connected* k*-subgroup is normal is commutative.*

Proof. Clearly, we can assume that G is affine. Since a maximal k-subtorus is normal in G, together with a Borel subgroup the group G is nilpotent. Since k is perfect, the unipotent radical of G is defined over k, so we can directly assume that G is a unipotent counter-example of minimal dimension. Let L be a maximal connected k-subgroup of the k-split group G. By minimality on the dimension of G the subgroup L is commutative. Moreover L is a vector group and, together with any of its one-dimensional k-subgroups, the subgroup L is central. Therefore G, as a central extension of L with a one-dimensional subgroup, is commutative.   □

For hamiltonian uni-maximal algebraic groups there is even a stronger claim; groups with big centre arise (see 3.2).

7.2.11 Proposition. *Let G be a non-commutative hamiltonian algebraic group defined over a field k of characteristic greater than two. If G is uni-maximal, then $\mathfrak{z}°G$ is the maximal connected algebraic subgroup of G.*

Proof. Let G be hamiltonian and uni-maximal. From this and from Theorem 7.2.7 it follows that G is unipotent. Let M be the maximal connected algebraic subgroup of G. For any one-dimensional subgroup K of the hamiltonian group G, we can inductively assume that $[G/K, M/K] \leq K/K$, i.e. $[G, M] \leq K$. So we assume $[G, M] = K$. Then K is the unique one-dimensional connected algebraic subgroup of G, and we are done by Corollary 5.2.35.   □

Since the n-dimensional group $\mathfrak{J}_n$ defined in Section 3.1 has nilpotency class n, the group $\mathfrak{J}_n$ is uni-maximal and the connected component of the centre of $\mathfrak{J}_n$ is a minimal connected subgroup of $\mathfrak{J}_n$. Now the Proposition 7.2.11 shows that for $n \geq 3$ no group $\mathfrak{J}_n$ is hamiltonian.

Any two-dimensional unipotent group, any chain, as well as any algebraic group where the centre is maximal is an elementary example of a hamiltonian group (see Proposition 7.2.5). To exhibit further examples of hamiltonian groups, we now discuss some product constructions. As a consequence of the next proposition, we see that a direct product $G \times G$ is hamiltonian if and only if G is commutative. Actually we can prove a bit more. Recall that, for a group G and a central subgroup H of G, the *direct product of G by G with* *amalgamated central subgroup* H is the following epimorphic image of $G \times G$, the dimension of which is $d = 2 \dim G - \dim H$:

$$(G \curlyvee G)_H = (G \times G)/\Delta(H),$$

where $\Delta(H) = \{(h, h^{-1}) : h \in H\}$.

7.2.12 Proposition. *Let G be a hamiltonian group and H a central connected subgroup of G. Then $(G \curlyvee G)_H$ is hamiltonian if and only if G is commutative.*

Proof. Let $\Delta(G) = \{(g, g^{-1}) : g \in G\}$. Since $\Delta(G)/\Delta(H)$ is normal in $(G \curlyvee G)_H$, we see that $\Delta(G)$ is normal in $(G \times G)$. Because

$$(g, g^{-1})^{(x,1)} = (g^x, g^{-1})$$

for all $x, g \in G$, we see that G is commutative. $\square$

Now we ask under what conditions a group lying between G and $G \times G$ is hamiltonian. Recall that for a group G and a normal subgroup $N \trianglelefteq G$, the *direct product of G by G with amalgamated factor group* is the following connected algebraic subgroup of $G \times G$, the dimension of which is $d = \dim G + \dim N$:

$$(G \curlywedge G)_N = \{(g, g)(n_1, n_2) : g \in G, n_i \in N\} =$$

$$\{(g, g)(1, n) : g \in G, n \in N\}.$$

Note that for every element $x \in (G \curlywedge G)_N$ there is a unique $g \in G$ and a unique $n \in N$ such that $x = (g, g)(1, n)$.

7.2.13 Proposition. *Let G be a hamiltonian group and let N be a normal subgroup of G. Then $(G \curlywedge G)_N$ is hamiltonian if and only if N is central.*

Proof. Let $(G \curlywedge G)_N$ be hamiltonian and consider the subgroup $\Gamma(G) = \{(g, g) : g \in G\} \leq (G \curlywedge G)_N$. For every $x \in N$ we have $(g, g)^{(1,x)} = (g, g^x) \in \Gamma(G)$. Since $\Gamma(G)$ is normal in the hamiltonian group $(G \curlywedge G)_N$, the subgroup N is central.

Conversely, assume that N is central. For any subgroup T of $(G \curlywedge G)_N$, there exist a subgroup $H \leq G$ and a subgroup $S \leq N$ such that $T = (H \curlywedge H)_S$. Thus for any $g \in G, h \in H, z \in N, x \in S$ we have

$$(g, g)(1, z)(h, h)(1, x)(1, z^{-1})(g^{-1}, g^{-1}) = (h^g, h^g)(1, x).$$

It follows that T is normal in $(G \curlywedge G)_N$. $\square$

7.2.14 Proposition. *Let G_1, G_2 be a pair of hamiltonian algebraic groups. If for all connected algebraic subgroups $S \leq G_1$, for all algebraic normal subgroups $A \trianglelefteq G_2$, and for any (not necessarily algebraic) homomorphism $\gamma : S \longrightarrow G_2/A$, we have $[S, G_1] \leq \ker \gamma$ and $[S^\gamma, G_2/A] = 1$, then $G_1 \times G_2$ is hamiltonian.*

Proof. Let U be a connected algebraic subgroup of $G_1 \times G_2$. According to [88], Theorem 1.6.1, p. 35, there exists a connected algebraic subgroup S of G_1, a normal algebraic subgroup A of G_2 and an abstract group homomorphism $\gamma : S \longrightarrow G_2/A$ such that

$$U = \{(x, y) \in G_1 \times G_2 : x \in S \text{ and } x^\gamma = yA\}.$$

Let $(x, y) \in U$ and let $(g_1, g_2) \in G_1 \times G_2$, then $(x, y)^{(g_1, g_2)} = (x^{g_1}, y^{g_2})$. Since $x \in S$ we have $(x^{g_1})^\gamma = (x[x, g_1])^\gamma = x^\gamma$, whilst $y^{g_2} A = yA$, because $[S^\gamma, G_2/A] = 1$. Hence $(x^{g_1})^\gamma = x^\gamma = yA = y^{g_2} A$, forcing $(x, y)^{(g_1, g_2)} \in U$. $\square$

7.2.15 Corollary. *If G_1 is a vector group and G_2 is a chain, then $G_1 \times G_2$ is hamiltonian.*

Proof. Clearly we may assume that G_2 is non-commutative and defined over a perfect field of positive characteristic p. Hence G_2 is a unipotent chain. Let S be a connected algebraic subgroup of G_1 and let A be a normal subgroup of G_2. Since S is commutative of exponent p, for any homomorphism $\gamma : S \to G_2/A$ we have that S^γ has at most dimension one and the claim follows from the previous proposition. $\square$

In the following corollaries we make use of Proposition 3.2.12, finding that a chain with a one-dimensional commutator subgroup has a centre of co-dimension one.

7.2.16 Corollary. *If G_1 is a commutative algebraic group and G_2 is a chain with a one-dimensional commutator subgroup G_2', then $G_1 \times G_2$ is hamiltonian.* $\square$

7.2.17 Corollary. *Let G_1, G_2 be chains with one-dimensional commutator subgroups G_1', G_2' and let $\dim G_1 < \dim G_2$. Then $G_1 \times G_2$ is hamiltonian.* $\square$

Note that, as the direct product $G \times G$ is hamiltonian if and only if G is commutative, the direct product of two chains having the same dimension is not always hamiltonian. In particular, the group $G = \mathfrak{J}_2(\mathbf{F}^r) \times \mathfrak{J}_2(\mathbf{F}^r)$ is not hamiltonian, any diagonal subgroup being not normal. On the contrary, we show in the following example that, for $r \neq s$, the direct product $G = \mathfrak{J}_2(\mathbf{F}^r) \times \mathfrak{J}_2(\mathbf{F}^s)$ *is* a hamiltonian group.

7.2.18 Example. Recall that the group $\mathfrak{J}_2(\mathbf{F}^k)$ is defined on the affine plane (where $\operatorname{char} \mathsf{k} = p > 0$) by the multiplication

$$(x_0, x_1) \cdot (x_0', x_1') = (x_0 + x_0', x_1 + x_1' + x_0 x_0'^{p^k}).$$

Let U be a connected subgroup of $G = G_1 \times G_2$, with $G_1 = \mathfrak{J}_2(\mathbf{F}^r)$ and $G_2 = \mathfrak{J}_2(\mathbf{F}^s)$, $r \neq s$. If U is one-dimensional, then U is central, because G is quasi-commutative, being the direct product of two quasi-commutative groups. If U

is three-dimensional, then U is normal, being a maximal connected subgroup of G.

Let $\pi_i : U \longrightarrow G_i$ be the canonical projection, $i = 1, 2$. If $\dim \pi_i(U) \leq 1$ for $i = 1$ or $i = 2$, then $\pi_i(U)$ is central. As in the proof of Proposition 7.2.14, with $S = \pi_i(U)$, we find that U is normal. In fact, assume without loss of generality that $\dim \pi_1(U) = 1$ and write

$$U = \{(x, y) \in G_1 \times G_2 : x \in \pi_1(U) \text{ and } x^\gamma = yA\},$$

where $A = U \cap G_2$. Let $(x, y) \in U$ and let $(g_1, g_2) \in G_1 \times G_2$, then $(x, y)^{(g_1, g_2)} = (x^{g_1}, y^{g_2})$. Since $x \in S = \pi_1(U) \leq {}_3G_1$ we have $(x^{g_1})^\gamma = x^\gamma$, whilst $y^{g_2} A = yA$, because $[S^\gamma, G_2/A] = 1$. Hence $(x^{g_1})^\gamma = x^\gamma = yA = y^{g_2} A$, forcing $(x, y)^{(g_1, g_2)} \in U$.

Finally, we show that no two-dimensional connected subgroup U of G, different from G_i, is such that $\pi_i(U) = G_i$. This will prove that G is hamiltonian. Assume, arguing by contradiction, that U is a counter-example. Since $U {}_3G / {}_3G$ is one-dimensional, there exist two non-trivial p-polynomials g, f such that any element $((x_0, x_1); (y_0, y_1)) \in U$ fulfils $x_0 = g(t), y_0 = f(t)$ for some $t \in k$. The commutator subgroup U' is generated by the commutators

$$[((g(t), x_1); (f(t), y_1)), ((g(t'), x_1'); (f(t'), y_1'))] = ((0, u); (0, v)),$$

where $u = g(t)g(t')^{p^r} - g(t)^{p^r} g(t')$ and $v = f(t)f(t')^{p^s} - f(t)^{p^s} f(t')$. We show now that u and v are independent variables, in contradiction to the fact that $\dim U' = 1$. If there exists $t_0 \in k$ such that $g(t_0) \neq 0 = f(t_0)$, then we find, for any $t \in k$,

$$((0, g(t)g(t_0)^{p^r} - g(t)^{p^r} g(t_0)); (0, 0)) \in U',$$

forcing $U' = {}_3G_1$, for reasons of dimension and connectedness. Hence we have

$$f(t)f(t')^{p^s} - f(t)^{p^s} f(t') = 0,$$

for any $t, t' \in k$: a contradiction to $f \neq 0$. Without loss of generality we assume that the degree of g is less or equal to the degree of f.

A standard argument with the derivatives yields that a p-polynomial $h(x)$ has a multiple root if and only if

$$h(x) = h_0(x^p) = h_1(x)^p,$$

for a suitable p-polynomial h_1. Hence any p-polynomial is a p^i-power of a polynomial having simple roots only. Therefore $f(x) = g(x)^{p^k}$ for a suitable $k \geq 0$ since any root of g is a root of f. So any commutator of U' has the form

$$((0, x_0 x_0'^{p^r} - x_0^{p^r} x_0'); (0, (x_0 x_0'^{p^s} - x_0^{p^s} x_0')^{p^k})).$$

As $r \neq s$, the jacobian of the function

$$(x_0, x_0') \mapsto (x_0 x_0'^{p^r} - x_0^{p^r} x_0', x_0 x_0'^{p^s} - x_0^{p^s} x_0')$$

has the following determinant

$$\begin{vmatrix} x_0'^{p^r} & x_0'^{p^s} \\ -x_0^{p^r} & -x_0^{p^s} \end{vmatrix} = -x_0'^{p^r} x_0^{p^s} + x_0'^{p^s} x_0^{p^r},$$

which is non-trivial. Hence the two variables $u = (x_0 x_0'^{p^r} - x_0^{p^r} x_0')$ and $v = (x_0 x_0'^{p^s} - x_0^{p^s} x_0')$ in fact are independent.   □

For later use we finally consider the connected algebraic subgroup

$$H = \{((x_0, x_1); (y_0, y_1)) \in G : x_0 = y_0\}$$

of $\mathfrak{J}_2(\mathbf{F}^r) \times \mathfrak{J}_2(\mathbf{F}^s)$. By the same arguments as the previous remark one can see that the commutator subgroup H' of H, being generated by commutators

$$((0, x_0 x_0'^{p^r} - x_0^{p^r} x_0'); (0, (x_0 x_0'^{p^s} - x_0^{p^s} x_0')^{p^k})),$$

is two-dimensional.

Now we show that uni-minimal groups form a rich class of hamiltonian groups.

7.2.19 Theorem. *Any uni-minimal algebraic group over a field of characteristic greater than two is hamiltonian.*

Proof. By Theorem 5.2.34 the uni-minimal group G contains a chain C and two-dimensional connected normal algebraic subgroups $J_1, \cdots, J_r$ of exponent p such that $G = CJ_1 \cdots J_r$. Let K be the unique one-dimensional connected algebraic subgroup of G. From the fact that K is contained in C and in any J_k and that, for any $k = 1, \cdots, r$, we have $[G, J_k] \leq K$, we infer that G/K is isogenous to a vector group or to the direct product of a chain by a vector group. Hence G/K is hamiltonian by Corollary 7.2.15. Since any connected algebraic subgroup of G contains K, the group G is also hamiltonian.   □

7.2.20 Remark. Finally, we want to remark that some of the classes of groups we have introduced in these notes, namely quasi-commutative groups, groups where the centre has co-dimension one, chains and their products, hamiltonian groups, form distinct classes of groups.

The group given in 3.2.25 serves us as an example of a hamiltonian group, the centre of which is not maximal. As an example of a hamiltonian algebraic group which is not quasi-commutative, consider the group G defined on the three-dimensional affine space by the multiplication

$$(a_0, x_0, x_1)(b_0, y_0, y_1) = (a_0 + b_0, x_0 + y_0, x_1 + y_1 + \Phi_1(x_0, y_0) + a_0 b_0^p + a_0 y_0),$$

where Φ_1 is the first Witt cocycle ([18], V, § 1, no. 1). Since $\mathfrak{z}^\circ G = \{(a, x_0, x_1) \in G : a = x_0 = 0\}$ and G contains the commutative subgroup $W = \{(a, x_0, x_1) \in G : a = 0\}$, isomorphic to the Witt group $\mathfrak{W}_2$, the group G is not quasi-commutative.

Let K be a connected one-dimensional subgroup of G. Since we have $(a_0, x_0, x_1)^p = (0, 0, x_0^p)$, we find $K \leq J = \{(a_0, x_0, x_1) : x_0 = 0\}$. Since J is a chain, the group G has a unique one-dimensional subgroup. Thus G is hamiltonian, because of $\dim G = 3$.

The above group G is not a direct product of chains with amalgamated factor group, since otherwise it would be quasi-commutative, as a subgroup of the direct product of two chains.

Conversely, quasi-commutative groups exist which are not the direct product of chains, nor a direct product of chains with a amalgamated central subgroup. To see this, consider the connected algebraic three-dimensional subgroup H of the group $\mathfrak{J}_2(\mathbf{F}^r) \times \mathfrak{J}_2(\mathbf{F}^s)$ mentioned at the end of Example 7.2.18. As $\mathfrak{z}^\circ H$ has co-dimension one in H, it follows that H is quasi-commutative (see Proposition 7.2.5). If H were a direct product with a amalgamated central subgroup of two chains, then the commutator subgroup of the three-dimensional group H would have dimension one. But this is not the case as Example 7.2.18 shows.

Finally, a quasi-commutative group is not necessarily hamiltonian: in fact, for any quasi-commutative, non-commutative group G, the group $G \times G$ provides an example of a quasi-commutative algebraic group which is not hamiltonian, by Proposition 7.2.12.   □

7.3 Topological Quasi-Normality

In the previous sections we have seen that in the class of connected algebraic groups over a perfect field of positive characteristic there are many groups containing non-normal quasi-normal subgroups and many quasi-hamiltonian groups which are not hamiltonian, as well as many hamiltonian groups which are not commutative.

For algebraic groups over a field of characteristic zero and for Lie groups over the reals, the complex numbers or over a p-adic field the situation changes radically, even if one weakens the notion of quasi-normality. For a topological group, in particular a Lie group, one calls two closed subgroups A, B *topologically permutable* if the closures of the sets AB and BA coincide (see [52]). For a locally compact topological group G a closed subgroup is called *topologically quasi-normal* if it is topologically permutable with any closed subgroup of G. In this way one obtains the notion of a topologically quasi-hamiltonian group (see [52]). Generalizing the main result of [86], it was shown in [53] that in a connected locally compact topological group every topologically permutable subgroup is normal.

We will show that in a connected Lie group as well as in a connected algebraic group over a field of characteristic zero, a closed connected subgroup is normal if it is topologically permutable with every other closed connected subgroup. As a consequence, we obtain the result that every connected Lie group as well as every connected algebraic group over a field of characteristic zero is commutative if every two closed connected subgroups are topologically permutable. To prove this we choose the following method which also covers the case of p-adic Lie groups.

Let G be a group together with a topology on the set G given by a closure operator κ, such that H^κ is a subgroup of G for every subgroup of H of G and such that $(H^g)^\kappa = (H^\kappa)^g$ for all $g \in G$ and $H \leq G$. (Here we assume no separation axioms.)

Let $\mathcal{L}$ be a family of subgroups of G that are closed with respect to κ. We call $\mathcal{L}$ a *bush* if for every κ-closed subgroup H of G there is a unique greatest element $A \in \mathcal{L}$ such that $A \leq H$. Elements $A, B \in \mathcal{L}$ are called $\mathcal{L}$-*permutable* if $(AB)^\kappa = (BA)^\kappa$ holds. A bush $\mathcal{L}$ is called *neat* if for every pair A, B of $\mathcal{L}$-permutable elements one even has $AB = BA$ and $AB \in \mathcal{L}$.

7.3.1 Proposition. *Let G be a group and let $\mathcal{L}$ be a neat bush of subgroups of G. If Q is a maximal κ-closed subgroup of G which is $\mathcal{L}$-permutable with all κ-closed subgroups of G, then Q is normal in G.*

Proof. Since $\mathcal{L}$ is a neat bush, it follows that for any $g \in G$ the set QQ^g is a subgroup of G belonging to $\mathcal{L}$. If Q was not normal in G, the maximality of Q would imply $G = QQ^g$, a contradiction to the fact that no group is a product of two distinct conjugate subgroups. □

7.3.2 Proposition. *Let G be a group and let $\mathcal{L}$ be a neat bush of subgroups of G. If a subgroup Q of G is $\mathcal{L}$-permutable with all κ-closed subgroups of G, then Q is normalized by any minimal element of $\mathcal{L}$.*

Proof. Let H be a minimal element of $\mathcal{L}$ different from Q. Then the group Q is maximal among the elements of $\mathcal{L}$ lying in the subgroup $QH \in \mathcal{L}$. Now the assertion follows from Proposition 7.3.1. □

7.3.3 Remark. We give examples of neat bushes of subgroups in various categories of groups.

1) In any abstract group, the set of all subgroups forms a neat bush.
2) The set of all formal subgroups of a formal group over a field of characteristic zero is a neat bush (cf. [39], 14.2.3 Theorem, p. 81). Observe that the closure operator plays no rôle here.
3) In a topological Hausdorff group, the set of all compact subgroups is a neat bush.
4) The bush of all closed subgroups of a topological group does not have to be neat (cf. [52], p. 392).
5) In a connected real or complex Lie group G, the subgroups generated by the exponential images of the subalgebras of the Lie algebra form a neat bush (see [40], Theorem 3.2, p. 80 and [55], Theorem 1.15, p. 23).
6) Let G be a connected affine algebraic group. Then the set of all algebraic subgroups of G forms a neat bush, as well as the set of all *connected* algebraic subgroups (cf. [45], 7.4 Corollary, p. 54).

For a Lie group G it is usual to consider closed subgroups instead of analytic subgroups as in Remark 7.3.3, 5). In this case there are two natural possibilities to choose a bush. Either one takes the bush $\mathcal{C}(G)$ of all closed subgroups of G or the bush $\mathcal{C}_\circ(G)$ of all closed connected subgroups of G. Both bushes are not neat (see [52], p. 392). In [53], Theorem 0.3, p. 196, the following is shown: If G is a connected Lie group having as solvable radical N an exponential group (i.e. $N = \bigcup_\varphi \operatorname{im}(\varphi)$, where φ runs trough the set of continuous homomorphisms $\mathbb{R} \to G$), then every connected $\mathcal{C}(G)$-permutable subgroup of G is normal in G.

7.3.4 Definition. Let G be a group and let $\mathcal{L}$ be a bush of subgroups of G. If a subgroup $Q \in \mathcal{L}$ is $\mathcal{L}$-permutable with all subgroups in $\mathcal{L}$, then Q is called $\mathcal{L}$-*quasi-normal* in G.

7.3.5 Theorem. *Let G be group and let $\mathcal{L}$ be a bush of subgroups of G. A $\mathcal{L}$-quasi-normal subgroup Q is normal in G, if one of the following holds:*

(1) G is a formal group over a field of characteristic zero and $\mathcal{L}$ is the bush of all formal subgroups of G,

(2) G is a Lie group over $\mathbb{R}$ or $\mathbb{C}$ and $\mathcal{L}$ is the bush of all analytic subgroups of G,

(3) G is a Lie group over $\mathbb{R}$ or $\mathbb{C}$ and $\mathcal{L}$ is the bush of all closed connected subgroups of G,

(4) G is an algebraic group over a field of characteristic zero and $\mathcal{L}$ is the bush of all closed connected subgroups of G.

Proof. In the cases *(1)*, *(2)* and *(4)* there is a functorial correspondence between the groups and their Lie algebras. For formal groups see [39], Theorem 14.2.3, p. 81, for Lie groups and the bush of analytic subgroups compare [40], Theorem 3.2, p. 80 and [55], Theorem 1.15, p. 23, and for algebraic groups over a field of characteristic zero and closed connected subgroups this follows from [45], 13.1 Theorem, p. 87. Since in the cases *(1)*, *(2)* and *(4)* every group G is generated by minimal subgroups belonging to $\mathcal{L}$, the assertion follows from Proposition 7.3.2.

In the case *(3)* we also can apply the Proposition 7.3.2 to prove the theorem, since any real or complex Lie group is generated by its *closed* one-dimensional subgroups. □

The case *(3)* of the previous theorem 7.3.5 is a variant of the result in [86].

7.3.6 Definition. Let G be a group and let $\mathcal{L}$ be a bush of subgroups of G. If every κ-closed subgroup of G lying in $\mathcal{L}$ is $\mathcal{L}$-permutable with all subgroups in $\mathcal{L}$, then G is called $\mathcal{L}$-*quasi-hamiltonian*.

7.3.7 Corollary. *Let G be group and let $\mathcal{L}$ be a bush of subgroups of G such that G is $\mathcal{L}$-quasi-hamiltonian. Then G is commutative, if one of the following holds:*

(1) G is a formal group over a field of characteristic zero and $\mathcal{L}$ is the bush of all formal subgroups of G,

(2) G is a Lie group over $\mathbb{R}$ or $\mathbb{C}$ and $\mathcal{L}$ is the bush of all analytic subgroups of G,

(3) G is a Lie group over $\mathbb{R}$ or $\mathbb{C}$ and $\mathcal{L}$ is the bush of all closed connected subgroups of G,

(4) G is an algebraic group over a field of characteristic zero and $\mathcal{L}$ is the bush of all closed connected subgroups of G.

Proof. As we have seen in the proof Theorem 7.3.5, the Lie algebra of the group G is generated by one-dimensional subalgebras corresponding to κ-closed subgroups of G lying in $\mathcal{L}$. Denote by Ξ the family of these subgroups of G. Since every subgroup in Ξ is normal in G (Theorem 7.3.5), any two elements of Ξ centralise each other. The assertion follows because the family Ξ generates G. □

7.3.8 Proposition. *Let $\mathfrak{g}$ be a non-commutative Lie algebra over a field k such that for any two subalgebras $\mathfrak{a}$ and $\mathfrak{b}$ the subspace $\mathfrak{a} + \mathfrak{b}$ is a subalgebra. Then $\mathfrak{g}$ is almost-abelian, i.e. there exists a basis $\mathcal{B} = \{e_i\}_{i \in I}$ such that $[e_1, e_k] = e_k$ and $[e_h, e_k] = 0$ for any $h, k \neq 1$.*

Proof. We can assume that $\dim \mathfrak{g} > 2$. Since $\mathfrak{g}$ is not commutative, there are two vectors $e_1, e_2 \neq 0$ such that $[e_1, e_2] = e_2$ (see [49], p. 12). Now we take any two vectors $e_h, e_k \in \mathfrak{g}$ such that $\{e_1, e_2, e_h, e_k\}$ are linearly independent. If we put $[e_1, e_h] = ae_1 + be_h$, then $[e_1, e_2 + e_h] = e_2 + ae_1 + be_h$, forcing $b = 1$. Up to exchanging e_h with $ae_1 + e_h$, we can assume $[e_1, e_h] = e_h$. Put $[e_2, e_h] = ce_2 + de_h$, then $[e_2, e_1 + e_h] = (c - 1)e_2 + de_h$, which forces $d = 0$. Since $[e_h, e_1 + e_2] = -e_h - ce_2$, we obtain $c = 0$, that is $[e_2, e_h] = 0$.

Similarly we can assume that $[e_1, e_k] = e_k$ and $[e_2, e_k] = 0$. If we put $[e_k, e_h] = ae_h + be_k$ and compute $[e_1 + e_k, e_h] = e_h + ae_h + be_k$, we find $b = 0$. Since $[e_2 + e_h, e_k] = -ae_h$, we obtain $a = 0$, that is $[e_k, e_h] = 0$. □

7.3.9 Proposition. *Let k be a field, let C be a subgroup of k^* such that $|C| \geq 2$ and let*

$$H = \left\{ \begin{pmatrix} 1 & 0 \\ 0 & t \end{pmatrix} : t \in C \right\} \leq L_2(\mathsf{k}).$$

If $g = \begin{pmatrix} 1 & b \\ 0 & 1 \end{pmatrix} \in L_2(\mathsf{k})$, with $b \neq 0$, then $g^{-1}HgH = H^gH \neq HH^g$ holds in the subgroup generated by $\{H, H^g\}$.

Proof. Indeed we have

$$H^gH \left\{ \begin{pmatrix} 1 & by(1 - x) \\ 0 & xy \end{pmatrix} : x, y \in C \right\}$$

and

$$HH^g = \left\{ \begin{pmatrix} 1 & b - by \\ 0 & xy \end{pmatrix} : x, y \in C \right\}.$$

If $HH^g = H^g H$, then one would have $2y - y.x = 1$ for all $x, y \in C$. But this contradicts the assumption $|C| \geq 2$. □

Let $\mathfrak{g}$ be the Lie algebra of a Lie group G over a complete ultrametric field of characteristic zero. If $\mathfrak{a}$ is a subalgebra of $\mathfrak{g}$, we denote by $\mathcal{E}_{\mathfrak{a}}$ the set of Lie subgroups of G having $\mathfrak{a}$ as its Lie algebra.

7.3.10 Definition. For subalgebras $\mathfrak{a}, \mathfrak{b}$ of the Lie algebra of a Lie group G over a complete ultrametric field of characteristic zero, we call the families $\mathcal{E}_{\mathfrak{a}}$ and $\mathcal{E}_{\mathfrak{b}}$ *locally permutable* if there are subgroups $A \in \mathcal{E}_{\mathfrak{a}}$ and $B \in \mathcal{E}_{\mathfrak{b}}$ such that $AB = BA$ is an analytic subgroup of G. The family $\mathcal{E}_{\mathfrak{q}}$ is called *locally quasi-normal* if it is locally permutable with every family $\mathcal{E}_{\mathfrak{h}}$. If every family $\mathcal{E}_{\mathfrak{h}}$ is locally quasi-normal, then G is called *locally quasi-hamiltonian*.

If $\mathfrak{n}$ is an ideal of the Lie algebra of a Lie group G over a complete ultrametric field of characteristic zero, then the family $\mathcal{E}_{\mathfrak{n}}$ contains a subgroup N which is normal in an open subgroup of G (see [9], Chapter III, § 7, Proposition 2, p. 215). We call such a family *locally normal*.

7.3.11 Remark. If F is an infinite locally compact totally disconnected field of characteristic zero and if G is a finite-dimensional Lie group over F, then every $\mathcal{E}_{\mathfrak{h}}$ contains arbitrarily small compact subgroups. Assume that the families $\mathcal{E}_{\mathfrak{a}}$ and $\mathcal{E}_{\mathfrak{b}}$ are locally permutable. Using the exponential map (see [9], Chapter III, § 7, Proposition 6, pp. 221ff.) we always find permutable compact subgroup $A \in \mathcal{E}_{\mathfrak{a}}$ and $B \in \mathcal{E}_{\mathfrak{b}}$ such that $AB = BA$ is closed.

7.3.12 Proposition. *Let G be a Lie group over a complete ultrametric field of characteristic zero and let $\mathfrak{g}$ be the Lie algebra of G. If the family $\mathcal{E}_{\mathfrak{q}}$ is locally quasi-normal, then $\mathfrak{q} + \mathfrak{h}$ is a subalgebra for every subalgebra $\mathfrak{h}$ of $\mathfrak{g}$.*

Proof. We consider analytic subgroups $Q \in \mathcal{E}_{\mathfrak{q}}$ and $H \in \mathcal{E}_{\mathfrak{h}}$ such that $HQ = QH$ is an analytic subgroup of G. It follows from [9], Chapter III, § 4, Theorem 4, p. 171, that $\mathfrak{q} + \mathfrak{h}$ is the Lie algebra of QH. □

7.3.13 Theorem. *Let G be a Lie group over a locally compact totally disconnected field F of characteristic zero. If the family $\mathcal{E}_{\mathfrak{q}}$ is locally quasi-normal, then it is locally normal.*

Proof. First we remark that the field F is ultrametric (cf. [96], 27.1 Theorem, p. 268). Let $\mathfrak{g}$ be the Lie algebra of G. Then by Proposition 7.3.12 for every one-dimensional subalgebra $\mathfrak{e}$ of $\mathfrak{g}$ the subspace $\mathfrak{s} = \mathfrak{q} + \mathfrak{e}$ is a subalgebra of $\mathfrak{g}$. Assume that $\mathcal{E}_{\mathfrak{q}}$ is not locally normal in every group belonging to the family $\mathcal{E}_{\mathfrak{s}}$. Then there is an element $s \in \mathfrak{e}$ such that for all $X \in \mathcal{E}_{\mathfrak{q}}$ one has $X^{\exp s} \neq X$. The families $\mathcal{E}_{\mathfrak{q}}$ and $\mathcal{E}_{\mathfrak{q}(\mathrm{Ad}\,\exp s)}$ are locally permutable. By [9],

Chapter III, § 5, no. 4, p. 190-191, the equality $\mathcal{E}_{\mathfrak{q}(\text{Ad}\exp s)} = \mathcal{E}_{\mathfrak{q}}^{\exp s}$ holds. Hence there are subgroups Q_1 and Q_2 having $\mathfrak{q}$ as Lie algebra such that the group $Q_1 Q_2^{\exp s} = Q_2^{\exp s} Q_1$ belongs to the family $\mathcal{E}_{\mathfrak{s}}$. This contradicts the fact that the product of two conjugate subgroups never is a group.   □

7.3.14 Theorem. *A locally quasi-hamiltonian Lie group over a complete ultrametric field* F *of characteristic zero contains an open commutative subgroup.*

Proof. By Proposition 7.3.12 the sum of any pair of subalgebras of the Lie algebra $\mathfrak{g}$ of G is a subalgebra of $\mathfrak{g}$. It follows from Proposition 7.3.8 that $\mathfrak{g}$ is almost-commutative. If $\mathfrak{g}$ were not commutative, then $\mathfrak{g}$ would contain a subalgebra $\mathfrak{l}$ isomorphic to the two-dimensional non-commutative algebra $\mathfrak{l}_2$. Thus the group G would contain a non-commutative subgroup L isomorphic to an open subgroup of the linear group $L_2(\mathsf{F})$ defined before Proposition 5.3.4. But every open subgroup of L has the property that no non-normal subgroup C with $|C| \geq 2$ is permutable with its conjugates (Proposition 7.3.9), a contradiction. Hence the Lie algebra $\mathfrak{g}$ is commutative and the assertion follows from [9], Chapter III, § 4, no. 1, Corollaire 3, p. 168.   □

7.4 Super-Hamiltonian Groups

If we consider a group G in which every subgroup is characteristic, then this group is commutative (see [75], 5.3.7, p. 139). If G is a torsion group in which every subgroup is characteristic, then the primary components of G are Prüfer groups or finite cyclic groups (see [6], Theorem 5.1, p. 109 and Sec. 6, Corollary, p. 117). Commutative torsion-free groups, however, have with respect to this condition already a more complicated structure. Namely, there are infinitely many torsion-free commutative groups of rank one in which every subgroup is characteristic (cf. [28], Sec. 116, p. 274). Hence it is not surprising that the situation does not simplify if we restrict, for algebraic groups or Lie groups, the condition to be a characteristic subgroup only to connected closed subgroups.

Super-Hamiltonian Algebraic Groups

We call a connected algebraic group G *super-hamiltonian* if every closed connected subgroup of G is invariant under all algebraic automorphisms on G. As a hamiltonian group, by Theorem 7.2.7 any super-hamiltonian algebraic group G is nilpotent, and, if the characteristic of the ground field is zero, then G is commutative. The algebraic group G is super-hamiltonian if its maximal connected affine subgroup $L(G)$ and its minimal connected subgroup $D(G)$ having no non-trivial affine epimorphic image are super-hamiltonian. Conversely, under the further assumption that $L(G) \cap D(G)$ is connected, we have

7.4.1 Proposition. *Let G be a super-hamiltonian algebraic group such that $L(G) \cap D(G)$ is connected. Then G is the direct product of the super-hamiltonian group $D(G)$ with an affine super-hamiltonian group $H \leq L(G)$ having a maximal torus of dimension at most one.*   □

Proof. By Theorem 7.2.7 the group G is nilpotent. Thus $L(G) = U \times T$, where U is the unipotent radical and T is a torus. Since $L(G) \cap D(G)$ is connected, $T \cap D(G)$ and $U \cap D(G)$ are also connected. It follows directly that $T = (T \cap D(G)) \times S$ for a suitable subtorus S of T. For the unipotent group U we also find a decomposition $U = (U \cap D(G)) \times W$: This follows by Theorem 1.3.2 if the characteristic of the ground field is positive, whereas in characteristic zero the group G is commutative by Theorem 7.2.7 and the vector group U is the direct product $U = (U \cap D(G)) \times W$ for a suitable vector subgroup W. Putting $H = S \times W$ we have $G = H \times D(G)$, and any automorphism of H, as well as any automorphism of $D(G)$ lifts to an automorphism of G; and the assertion follows.   □

The proof of the above proposition yields for affine algebraic groups the following

7.4.2 Corollary. *Let G be a connected affine algebraic group defined over a field of characteristic zero. The group G is super-hamiltonian if and only if G is a commutative group having the unipotent radical as well as the maximal torus of dimension at most one.*   □

7.4.3 Corollary. *Let G be a connected affine algebraic group defined over a field of positive characteristic. The group G is super-hamiltonian if and only if G is the direct product of a connected super-hamiltonian unipotent group with a torus of dimension at most one.*   □

7.4.4 Proposition. *i) A direct product G of Witt groups is super-hamiltonian if and only if G is a Witt group.*

ii) A direct product G of simple abelian varieties is super-hamiltonian if and only if no two of the simple abelian subvarieties of G are isogenous.

Proof. Let $G = A \times B \times C$ where A and B are isogenous to Witt groups. Let $\alpha : A \longrightarrow B$ be a homomorphism having as kernel the unique maximal connected algebraic subgroup of A. Then the mapping $A \times B \longrightarrow A \times B$ defined by $(x, y) \mapsto (x, y + \alpha(x))$ is an automorphism of $A \times B$ such that $\{(x, 0) : x \in A\}^{\alpha} = \{(x, \alpha(x)) : x \in A\}$. Since this automorphism extends to an algebraic automorphism of G, the claim i) is proved.

Let $G = A \times B \times C$ be a product of abelian varieties such that A and B are simple isogenous varieties. Let α be an isogeny of A into B. Then the automorphism $A \times B \longrightarrow A \times B$ defined by $(x, y) \mapsto (x, y + \alpha(x))$ extends to an algebraic automorphism of G. Conversely, if no two of simple abelian subvarieties of G are isogenous, then G is super-hamiltonian.   □

7.4.5 Corollary. *i) Any connected algebraic subgroup of a commutative affine connected algebraic group G over a field of positive characteristic is invariant under all isogenies of G onto G if and only if G is a connected subgroup of the direct product of a one-dimensional torus and a Witt group.*

ii) Any connected subgroup of an abelian variety G is invariant under all isogenies of G onto G if and only if no two of the simple abelian subvarieties of G are isogenous.

Proof. A commutative affine connected algebraic group G is a direct product of a torus T and a commutative unipotent algebraic group U which is isogenous to a direct product W of Witt groups. For any isogeny α of G onto G we have $T^\alpha = T$ and $U^\alpha = U$. Since any automorphism of T can be extended to an automorphism of G, the torus T has dimension at most one.

Let β be an isogeny from U onto W and γ be an isogeny from W onto U. It follows from [70], Hilfsatz 1, that $\gamma\beta$ gives the identity on the lattice of all connected algebraic subgroups.

If G is isogenous to a direct product V of simple abelian varieties, then there exist isogenies $\beta : G \longrightarrow V$ and $\gamma : V \longrightarrow G$ and also in this case it follows from [70], Hilfsatz 1, that $\gamma\beta$ gives the identity on the lattice of all connected subgroups of the abelian variety V.

Now the assertion follows from Proposition 7.4.4. □

To determine super-hamiltonian algebraic groups over fields of positive characteristic seems to be a very difficult task. This is shown by the following discussion for three-dimensional connected unipotent groups G over perfect fields of positive characteristic, where we can decide whether a given group G is super-hamiltonian or not. A three-dimensional super-hamiltonian affine algebraic group is either unipotent or the direct product of a one-dimensional torus with a two-dimensional unipotent chain.

For our next purposes we recall that for any factor system of the form $f(\Phi_1(x, y)) + \sum_i g_i(xy^{p^i})$ the following property holds

$$f(\Phi_1(x,y)) + \sum_i g_i(xy^{p^i}) \in \mathrm{B}^2(\mathbf{G}_a, \mathbf{G}_a) \Longleftrightarrow f = g_i = 0, \text{ for any } i. \quad (7.1)$$

7.4.6 Proposition. *Let G be a three-dimensional connected unipotent algebraic group such that $\dim G' = 1$ and both $\mathfrak{z}^\circ G$ and G/G' are isomorphic to a two-dimensional vector group. Then G is hamiltonian, but not super-hamiltonian.*

Proof. By Proposition 7.2.5 the group G is hamiltonian. By Proposition 6.2.2 the group G is isomorphic to the group defined by the following product:

$$(x_0, x_1, x_2) \cdot (y_0, y_1, y_2) =$$

$$(x_0 + y_0, x_1 + y_1, x_2 + y_2 + \sum_i \alpha_i \Phi_1^{p^i}(x_0, y_0) + \sum_{h,k} \lambda_{h,k} x_0^{p^h} y_0^{p^{h+k}}),$$

that is: G is the direct product of a one-dimensional connected algebraic subgroup with a two dimensional non-commutative chain. As a simple computation shows, the map $\Upsilon : (x_0, x_1, x_2) \mapsto (x_0, x_1, x_2 + x_1)$ is an automorphism of G. If K is the one-dimensional subgroup $\{(x_0, x_1, x_2) : x_0 = x_2 = 0\}$, we have $\Upsilon(K) = \{(x_0, x_1, x_2) : x_0 = 0, x_1 = x_2\} \neq K$, which proves that G is not super-hamiltonian. □

By Proposition 7.2.6, if $\dim G' = 2$ and $\dim {}_3 G = 1$, then the three-dimensional unipotent group G is not hamiltonian. To decide whether a three-dimensional unipotent group which is not a chain is super-hamiltonian, it remains to consider the following cases:

a) G/G' is a two-dimensional chain and ${}_3°G$ is a two-dimensional vector group;
b) ${}_3°G$ is a two-dimensional chain and G/G' is a two-dimensional vector group;
c) ${}_3°G$ is one-dimensional and G/G' is a two-dimensional vector group;
d) ${}_3°G = G'$ is a two-dimensional vector group.

Let $\mathcal{A}$ be the class of three-dimensional connected unipotent algebraic groups G such that $\dim G' = 1$ and G/G' is a chain. Every group $G \in \mathcal{A}$ is uni-maximal and the unique maximal connected algebraic subgroup of G is central. Thus G is hamiltonian by Proposition 7.2.5.

7.4.7 Proposition. *The class $\mathcal{A}$ contains super-hamiltonian groups as well as non-super-hamiltonian groups.*

Proof. By Proposition 6.2.3, if the group $G \in \mathcal{A}$ is not a chain, then it is isomorphic to the group defined by the following product:

$$(x_0, x_1, x_2) \cdot (y_0, y_1, y_2) =$$

$$(x_0 + y_0, x_1 + y_1 + \sum_i \alpha_i \Phi_1^{p^i}(x_0, y_0), x_2 + y_2 + \sum_j \beta_j \, \Phi_1^{p^j}(x_0, y_0) + \sum_{h,k} \lambda_{h,k} \, x_0^{p^h} y_0^{p^{h+k}}).$$

Since any algebraic automorphism of the group $\mathbf{G}_a$ is of the shape $x \mapsto ax$ and the connected component ${}_3° G = \{(x_0, x_1, x_2) : x_0 = 0\}$ of the centre of G, as well as the commutator subgroup $G' = \{(x_0, x_1, x_2) : x_0 = x_1 = 0\}$, is characteristic in G, any automorphism Υ of G has the form

$$\Upsilon(x_0, x_1, x_2) = \Upsilon(x_0, 0, 0) + \Upsilon(0, x_1, 0) + \Upsilon(0, 0, x_2) =$$

$$(ax_0, \varphi(x_0), \psi(x_0)) + (0, bx_1, \rho(x_1)) + (0, 0, cx_2) =$$

$$(ax_0, bx_1 + \varphi(x_0), cx_2 + \psi(x_0) + \rho(x_1)),$$

where $a, b, c \neq 0$, $\rho \in \mathsf{End}(\mathbf{G}_a)$ and $\varphi, \psi : \mathbf{G}_a \longrightarrow \mathbf{G}_a$ are algebraic morphisms. From the fact that Υ is a homomorphism and that $\Phi_1(ax, ay) = a^p \Phi_1(x, y)$, it follows that

i) $\sum_i \alpha_i \left(b - a^{p^{i+1}}\right) \Phi_1^{p^i}(x, y) = \varphi(x) + \varphi(y) - \varphi(x + y),$

ii) $\sum_j \beta_j \left(c - a^{p^{j+1}}\right) \Phi_1^{p^j}(x, y) + \rho(\sum_i \alpha_i \Phi_1^{p^i}(x, y)) +$

$\sum_{h,k} \lambda_{h,k} \left(c - a^{p^h + p^{h+k}}\right) x^{p^h} y^{p^{h+k}} = \psi(x) + \psi(y) - \psi(x + y)$

for any $x, y \in \mathbf{G}_a$.

As recalled in (7.1), the condition i) yields that $b = a^{p^{i+1}}$ for any i and that $\varphi \in \mathsf{End}(\mathbf{G}_a)$. Analogously, if $i > j$ for any i, j corresponding to non-zero summands of $\sum_i \alpha_i \Phi_1^{p^i}, \sum_j \beta_j \Phi_1^{p^j}$, or $\beta_j = 0$ for any j, then the condition ii) yields that $\rho = 0$, $\psi \in \mathsf{End}(\mathbf{G}_a)$ and

$c = a^{p^h + p^{h+k}}$ for any h, k, if $\beta_j = 0$ for any j

$c = a^{p^{j+1}} = a^{p^h + p^{h+k}}$ for any j, h, k, otherwise.

From these results it follows that the group $G_1 \in \mathcal{A}$ defined by the product

$$(x_0, x_1, x_2) \cdot (y_0, y_1, y_2) = (x_0 + y_0, x_1 + y_1 + \Phi_1^p(x_0, y_0), x_2 + y_2 + \Phi_1(x_0, y_0) + x_0 y_0^p)$$

is super-hamiltonian, whilst the group $G_2 \in \mathcal{A}$ defined by the product

$$(x_0, x_1, x_2) \cdot (y_0, y_1, y_2) = (x_0 + y_0, x_1 + y_1 + \Phi_1(x_0, y_0) + \Phi_1^p(x_0, y_0), x_2 + y_2 + x_0 y_0^p)$$

is not super-hamiltonian. In fact, any automorphism of G_1 has the form

$$\Upsilon(x_0, x_1, x_2) = (a x_0, b x_1 + \varphi(x_0), c x_2 + \psi(x_0)),$$

where $b = a^{p^2}$ and $c = a^p = a^{p+1}$. From the latter we get that any automorphism of G_1 fixes $\mathfrak{z}° G_1$ element-wise. As $\mathfrak{z}° G_1$ is the unique two-dimensional subgroup of G_1, it follows that G_1 is super-hamiltonian.

In contrast, any automorphism of G_2 has the form

$$\Upsilon(x_0, x_1, x_2) = (a x_0, b x_1 + \varphi(x_0), c x_2 + \psi(x_0)),$$

where $b = a^p = a^{p^2}$ and $c = a^{p+1}$. The set $K = \{(x_0, x_1, x_2) : x_0 = 0, x_1 = x_2\}$ is a one-dimensional connected algebraic subgroup of G_2. Therefore, taking $a \notin \{0, 1\}$ satisfying the condition $a^p = a^{p^2}$ and putting $\varphi = \psi = 0$, we have that $\Upsilon(K) = \{(x_0, x_1, x_2) : x_0 = 0, x_2 = a x_1\} \neq K$. It follows that G_2 is not super-hamiltonian. □

The groups in the class $\mathcal{B}$ of the following proposition have two-dimensional centre. Hence they are hamiltonian.

7.4.8 Proposition. *Let $\mathcal{B}$ be the class of all three-dimensional connected unipotent algebraic groups G such that $\dim G' = 1$ and the connected component $\mathfrak{z}° G$ of the centre of G is a two-dimensional chain. The class $\mathcal{B}$ contains super-hamiltonian as well as non-super-hamiltonian groups.*

Proof. We consider the groups G in $\mathcal{B}$ which are not chains. By Theorem 5.1.1 we can therefore assume that G/G' is a two-dimensional vector group. By Proposition 6.2.2, any group $G \in \mathcal{B}$ is isomorphic to a group defined by the following product:

$$(x_0, x_1, x_2) \cdot (y_0, y_1, y_2) =$$

$$(x_0 + y_0, x_1 + y_1, x_2 + y_2 + \sum_i \alpha_i \Phi_1^{p^i}(x_0, y_0) + \sum_{h,k} \lambda_{h,k} \, x_0^{p^h} y_0^{p^{h+k}} + \sum_j \beta_j \, \Phi_1^{p^j}(x_1, y_1)).$$

Writing any element (x_0, x_1, x_2) as the product $(x_0, 0, 0)(0, x_1, 0)(0, 0, x_2)$, by similar arguments as in the proof of the previous Proposition 7.4.7 we see that any automorphism Υ of G has the form

$$\Upsilon(x_0, x_1, x_2)(ax_0, bx_1 + \varphi(x_0), cx_2 + \psi(x_0) + \rho(x_1)),$$

where $a, b, c \neq 0$ and for the functions ψ, ρ and the p-polynomial φ one has

$$\sum_i \alpha_i \, (a^{p^{i+1}} - c) \Phi_1^{p^i}(x_0, y_0) + \sum_{h,k} \lambda_{h,k} \, (a^{p^h + p^{h+k}} - c) \, x_0^{p^h} y_0^{p^{h+k}} +$$

$$\sum_j \beta_j \, \Phi_1^{p^j}(bx_1 + \varphi(x_0), by_1 + \varphi(y_0)) - c \sum_j \beta_j \, \Phi_1^{p^j}(x_1, y_1) =$$

$$\psi(x_0 + y_0) - \psi(x_0) - \psi(y_0) + \rho(x_1 + y_1) - \rho(x_1) - \rho(y_1)$$

for any $x_0, x_1, y_0, y_1 \in \mathbf{G}_a$.

Putting $x_0, y_0 = 0$ in the above equation we get, for any $x_1, y_1 \in \mathbf{G}_a$, the relation

$$\sum_j \beta_j \, (b^{p^{j+1}} - c) \, \Phi_1^{p^j}(x_1, y_1) = \rho(x_1 + y_1) - \rho(x_1) - \rho(y_1),$$

from which it follows, as recalled in (7.1), that $c = b^{p^{j+1}}$ for any j and that $\rho \in \mathsf{End}(\mathbf{G}_a)$. Analogously, putting $x_1 = y_1 = 0$ we obtain for any $x_0, y_0 \in \mathbf{G}_a$ the relation

$$\sum_i \alpha_i \, (a^{p^{i+1}} - c) \Phi_1^{p^i}(x_0, y_0) + \sum_j \beta_j \, \Phi_1^{p^j}(\varphi(x_0), \varphi(y_0)) +$$

$$\sum_{h,k} \lambda_{h,k} \, (a^{p^h + p^{h+k}} - c) \, x_0^{p^h} y_0^{p^{h+k}} = \psi(x_0 + y_0) - \psi(x_0) - \psi(y_0).$$

If $j > i$ for any i, j corresponding to non-zero summands of $\sum_i \alpha_i \Phi_1^{p^i}, \sum_j \beta_j \Phi_1^{p^j}$, or $\alpha_i = 0$ for any i, then the above relation yields that $\varphi = 0$, $\psi \in \mathsf{End}(\mathbf{G}_a)$ and

$$c = a^{p^h + p^{h+k}} \text{ for any } h, k, \text{ if } \alpha_i = 0 \text{ for any } i,$$
$$c = a^{p^{i+1}} = a^{p^h + p^{h+k}} \text{ for any } i, h, k, \text{ otherwise.}$$

From these results it follows that the group $G_1 \in \mathcal{B}$ defined by the product

$$(x_0, x_1, x_2) \cdot (y_0, y_1, y_2) = (x_0 + y_0, x_1 + y_1, x_2 + y_2 + \Phi_1(x_0, y_0) + x_0 y_0^p + \Phi_1^p(x_1, y_1))$$

is super-hamiltonian, whilst the group $G_2 \in \mathcal{B}$ defined by the product

$$(x_0, x_1, x_2) \cdot (y_0, y_1, y_2) = (x_0 + y_0, x_1 + y_1, x_2 + y_2 + \Phi_1(x_1, y_1) + \Phi_1^p(x_1, y_1) + x_0 y_0^p)$$

is not super-hamiltonian. In fact, any automorphism of G_1 has the form

$$\Upsilon(x_0, x_1, x_2) = (ax_0, bx_1, cx_2 + \psi(x_0) + \rho(x_1)),$$

where $c = b^{p^2} = a^p = a^{p+1}$. It follows that $a = 1$ and we get that any auto-morphism of G_1 fixes G_1/G_1' element-wise. In particular, as G_1 is uni-minimal (because $_3° G_1$ is a two-dimensional chain and G_1 is not commutative), any connected two-dimensional algebraic subgroup H of G_1 contains G_1' and this implies, by the above results, that the subgroup H is invariant.

In contrast, any automorphism of G_2 has the form

$$\Upsilon(x_0, x_1, x_2) = (ax_0, bx_1, cx_2 + \psi(x_0) + \rho(x_1)),$$

where $c = b^p = b^{p^2} = a^{p+1}$. The set $H = \{(x_0, x_1, x_2) : x_0 = x_1\}$ is a two-dimensional connected algebraic subgroup of G_2. Therefore, putting $b = 1$, $\psi = \rho = 0$ and choosing $a \neq 1$ satisfying the condition $a^{p+1} = 1$, we have that $\Upsilon(H) = \{(x_0, x_1, x_2) : x_0 = ax_1\} \neq H$. It follows that G_2 is not super-hamiltonian.   □

The class $\mathcal{C}$ of the following proposition is the only one which contains non-hamiltonian unipotent groups.

7.4.9 Proposition. *Let $\mathcal{C}$ be the class of all three-dimensional connected unipotent algebraic groups G such that $G' = _3° G$ has dimension one. The class $\mathcal{C}$ contains super-hamiltonian as well as non-hamiltonian groups.*

Proof. In order to exhibit a group in $\mathcal{C}$ which is super-hamiltonian, let us consider the group G defined by the following product:

$$(x_0, x_1, x_2) \cdot (y_0, y_1, y_2) =$$

$$(x_0 + y_0, x_1 + y_1, x_2 + y_2 + \Phi_1(x_0, y_0) + x_0 y_0^p + \Phi_1^p(x_1, y_1) + x_1 y_1^{p^2}).$$

Let Υ be an algebraic automorphism of G. Since $G' = \{(x_0, x_1, x_2) : x_0 = x_1 = 0\}$ is a characteristic subgroup of G, there exist $a \neq 0$, a morphism $F : \mathbf{G}_a \times \mathbf{G}_a \longrightarrow \mathbf{G}_a$ and an invertible matrix $\begin{pmatrix} \varphi_1 & \psi_1 \\ \varphi_2 & \psi_2 \end{pmatrix} \in M_2(\mathrm{End}(\mathbf{G}_a))$ such that

$$\Upsilon(x_0, x_1, x_2) = (\varphi_1(x_0) + \varphi_2(x_1), \psi_1(x_0) + \psi_2(x_1), ax_2 + F(x_0, x_1))$$

for any $(x_0, x_1, x_2) \in G$. Using the fact that Υ is a homomorphism and putting $x_1 = y_1 = 0$ we get for any x_0, y_0 the following relation:

$$\varphi_1(x_0)\varphi_1^p(y_0) - ax_0 y_0^p + \Phi_1(\varphi_1(x_0), \varphi_1(y_0)) - a\Phi_1(x_0, y_0) +$$

$$\psi_1(x_0)\psi_1^{p^2}(y_0) + \Phi_1^p(\psi_1(x_0), \psi_1(y_0)) = F(x_0, 0) + F(y_0, 0) - F(x_0 + y_0, 0).$$

It follows that the polynomial $\varphi_1(x_0)\varphi_1^p(y_0) - ax_0 y_0^p + \psi_1(x_0)\psi_1^{p^2}(y_0)$ has to be symmetric. This yields $\psi_1 = 0$, $\varphi_1(t) = bt$ with $b^{p+1} = a$, and the relation

$$\Phi_1(bx_0, by_0) - a\Phi_1(x_0, y_0) = F(x_0, 0) + F(y_0, 0) - F(x_0 + y_0, 0)$$

for any x_0, y_0. From this identity and being $b^{p+1} = a$ we obtain $b^p = b^{p+1} = a$. This forces $a = b = 1$.

Using again the fact that Υ is a homomorphism and putting $x_1 = y_1 = 0$ we get, by similar arguments, $\varphi_2 = 0$ and $\psi_2(t) = t$. Therefore any automorphism of G fixes the factor group G/G' element-wise. Since any two-dimensional connected algebraic subgroup H of G contains G', the group G is super-hamiltonian.

The three-dimensional group of all 3×3 unipotent matrices is an example of a group in the class $\mathcal{C}$ which is not hamiltonian. $\qquad\square$

The groups in the class $\mathcal{D}$ of the following proposition are all hamiltonian, having centre of dimension two.

7.4.10 Proposition. *Let $\mathcal{D}$ be the class of all three-dimensional connected unipotent algebraic groups G such that $G' = {}_3°\,G$ is isomorphic to a two-dimensional vector group. The class $\mathcal{D}$ contains super-hamiltonian as well as non-super-hamiltonian groups.*

Proof. By Theorem 6.3.3 any group $G \in \mathcal{D}$ is isomorphic to a group defined by the following product:

$$(x_0, x_1, x_2) \cdot (y_0, y_1, y_2) =$$

$$\left(x_0 + y_0, x_1 + y_1 + \sum_i \alpha_i \Phi_1^{p^i}(x_0, y_0) + \sum_{h,k} \lambda_{h,k}\, x_0^{p^h} y_0^{p^{h+k}}, \right.$$

$$\left. x_2 + y_2 + \sum_j \beta_j\, \Phi_1^{p^j}(x_0, y_0) + \sum_{r,s} \mu_{r,s}\, x_0^{p^r} y_0^{p^{r+s}} \right).$$

Since the commutator subgroup G' is characteristic in G and any automorphism of a two-dimensional vector group is of the shape $(x, y) \mapsto (\varphi_1(x) + \varphi_2(y), \psi_1(x) + \psi_2(y))$, for a given invertible matrix $\begin{pmatrix} \varphi_1 & \psi_1 \\ \varphi_2 & \psi_2 \end{pmatrix} \in M_2(\mathsf{End}(\mathbf{G}_a))$, any automorphism Υ of G has the form

$$\Upsilon(x_0, x_1, x_2) = (ax_0, \varphi_1(x_1) + \varphi_2(x_2) + \rho(x_0), \psi_1(x_1) + \psi_2(x_2) + \delta(x_0))$$

for suitable $a \neq 0$ and polynomial maps $\rho, \delta : \mathbf{G}_a \longrightarrow \mathbf{G}_a$. In particular, since Υ is a homomorphism, for any x, y we have

$i)$ $\sum_i \alpha_i\, a^{p^{i+1}}\, \Phi_1^{p^i}(x, y) - \varphi_1(\sum_i \alpha_i \Phi_1^{p^i}(x, y)) +$

$\sum_{h,k} \lambda_{h,k}\, a^{p^h + p^{h+k}}\, x^{p^h} y^{p^{h+k}} - \varphi_1(\sum_{h,k} \lambda_{h,k}\, x^{p^h} y^{p^{h+k}}) -$

$\varphi_2(\sum_j \beta_j\, \Phi_1^{p^j}(x, y)) - \varphi_2(\sum_{r,s} \tau_{r,s}\, x^{p^r} y^{p^{r+s}}) = \rho(x + y) - \rho(x) - \rho(y),$

ii) $\sum_j \beta_j\, a^{p^{j+1}}\, \Phi_1^{p^j}(x,y) - \psi_2(\sum_j \beta_j \Phi_1^{p^j}(x,y)) +$

$\quad \sum_{r,s} \tau_{r,s}\, a^{p^r + p^{r+s}}\, x^{p^r} y^{p^{r+s}} - \psi_2(\sum_{r,s} \tau_{r,s}\, x^{p^r} y^{p^{r+s}}) -$

$\quad \psi_1(\sum_i \alpha_i \Phi_1^{p^i}(x,y)) - \psi_1(\sum_{h,k} \lambda_{h,k}\, x^{p^h} y^{p^{h+k}}) = \delta(x+y) - \delta(x) - \delta(y).$

From these results it follows that any algebraic automorphism of the group $G_1 \in \mathcal{D}$ defined by the product

$$(x_0, x_1, x_2) \cdot (y_0, y_1, y_2) = (x_0 + y_0, x_1 + y_1 + \Phi_1(x_0, y_0) + x_0 y_0^p, x_2 + y_2 + x_0 y_0^{p^2})$$

has the form

$$\Upsilon(x_0, x_1, x_2) = (x_0, x_1 + \rho(x_0), x_2 + \delta(x_0)),$$

where ρ, δ turn out in this case to be p-polynomials. As any automorphism of G_1 fixes $\mathfrak{z}^\circ\, G_1$ element-wise and $\mathfrak{z}^\circ\, G_1$ is the unique two-dimensional subgroup of G_1, it follows that G_1 is super-hamiltonian.

Again from *i)* and *ii)*, it follows that any algebraic automorphism of the group $G_2 \in \mathcal{D}$ defined by the product

$$(x_0, x_1, x_2) \cdot (y_0, y_1, y_2) = (x_0 + y_0, x_1 + y_1 + x_0 y_0^p, x_2 + y_2 + x_0 y_0^{p^2})$$

has the form

$$\Upsilon(x_0, x_1, x_2) = (ax_0, a^{p+1} x_1 + \rho(x_0), a^{p^2+1} x_2 + \delta(x_0)),$$

where $a \neq 0$ and ρ, δ are p-polynomials. Therefore, taking $a \neq 0$ such that $a^{p+1} \neq a^{p^2+1}$, for the one-dimensional connected subgroup $H = \{(x_0, x_1, x_2) : x_0 = 0, x_1 = x_2\}$ of G_2, we have that $\Upsilon(H) = \{(x_0, x_1, x_2) : x_0 = 0, x_1 = a^{p^2-p} x_2\} \neq H$. Therefore G_2 is not super-hamiltonian. $\qquad \square$

Super-Hamiltonian Lie Groups

Analogously to the case of algebraic groups we call a connected real or complex Lie group G *super-hamiltonian*, if every connected closed subgroup of G is characteristic in G.

Since any super-hamiltonian real or complex Lie group is hamiltonian, it follows from Corollary 7.3.7 that any super-hamiltonian real Lie group is a connected subgroup of the group $\mathbb{R} \times SO_2(\mathbb{R})$, and that any complex super-hamiltonian Lie group is a connected closed subgroup of the direct product having as factors a one-dimensional vector group, a one-dimensional linear torus and a non-trivial toroidal group X which is super-hamiltonian.

If X is a direct product $X = X_1 \times \cdots \times X_r$ of Shafarevich extensions X_i of a simple complex torus by a simple complex torus (see Section 2.2), then X is super-hamiltonian if and only if for any two Shafarevich extensions X_i and X_j there is no non-trivial homomorphism form X_i to X_j. This can be proved with the same arguments as Proposition 7.4.4 *ii*). In all other cases the decision whether X is super-hamiltonian seems to be difficult.

In the next proposition we show that there exist super-hamiltonian toroidal groups of maximal rank n having real rank $n+1$.

7.4.11 Proposition. *Let $X = \mathbb{C}^n/\Lambda$ be a toroidal group of maximal rank n, having real rank $n+1$, and let $P = (I_n \ G)$ be a period matrix of X. If $G = -\overline{G}$, that is if the column G has real part equal to zero, then for any endomorphism f of X there exists $d \in \mathbb{Z}$ such that $\rho_a(f)(\mathbf{v}) = d \cdot \mathbf{v}$ and $\rho_r(f)(\mathbf{w}) = d \cdot \mathbf{w}$, for any $\mathbf{v} \in \mathbb{C}^n$, $\mathbf{w} \in \Lambda$.*

Proof. Applying the Hurwitz relations we find

$$C(I_n \ G) = (I_n \ G) \begin{pmatrix} A & B \\ E & d \end{pmatrix}$$

where $A \in M_{n \times n}(\mathbb{Z})$, $B \in M_{n \times 1}(\mathbb{Z})$, $E \in M_{1 \times n}(\mathbb{Z})$ and $d \in \mathbb{Z}$. It follows that $C = A + GE$ and $AG + GEG = B + Gd$. Since GEG and B are real matrices whereas AG and Gd are purely imaginary matrices, we have $GEG = B$ and $(A - dI_n)G = 0$. By the latter and by the irrationality conditions (2.7) we find $A = dI_n$.

Let $0 \neq B' \in M_{1 \times n}(\mathbb{Z})$ be such that $B'B = 0$. From $GEG = B$ we get now $B'GEG = 0$, forcing one of the complex numbers $B'G$ and EG to be zero. By the irrationality conditions we have $E = 0$, since B' is non-zero. It follows immediately that $C = A$ and $B = GEG = 0$. □

According to Proposition 2.2.1, we see that the automorphism group of any toroidal group X of the above proposition has order two. Moreover, any of subgroup of X is invariant. The situation changes radically if we just assume that $q > 1$, as the following example shows.

7.4.12 Example. It is possible to give examples of toroidal groups $X = \mathbb{C}^n/\Lambda$ of maximal rank n, having real rank $n + q$ with $q > 1$, and period matrix $P = (I_n \ G)$ such that $E_1 P = P E_2$ for suitable permutation matrices $E_1 \in M_{n \times n}(\mathbb{Z})$, $E_2 \in M_{(n+q) \times (n+q)}(\mathbb{Z})$. For instance, let X be the three-dimensional toroidal group having

$$P = \begin{pmatrix} 1 & 0 & 0 & i & i \\ 0 & 1 & 0 & i\sqrt{2} & 0 \\ 0 & 0 & 1 & 0 & i\sqrt{2} \end{pmatrix}$$

as a period matrix. One sees that for

$$E_1 = \begin{pmatrix} 1 & 0 & 0 \\ 0 & 0 & 1 \\ 0 & 1 & 0 \end{pmatrix}, \quad E_2 = \begin{pmatrix} 1 & 0 & 0 & 0 & 0 \\ 0 & 0 & 1 & 0 & 0 \\ 0 & 1 & 0 & 0 & 0 \\ 0 & 0 & 0 & 0 & 1 \\ 0 & 0 & 0 & 1 & 0 \end{pmatrix}$$

we have $E_1 P = P E_2$. □

7.4.13 Proposition. *A two dimensional complex torus is super-hamiltonian if and only if it is indecomposable or isogenous to the direct product of two non-isogenous tori.*

Proof. Clearly we only have to show that a non-simple two-dimensional complex torus G isogenous to the direct product of two isomorphic one-dimensional tori has an automorphism which moves one of its one-dimensional subtori. To see this, we assume that a period matrix of G is

$$P = \begin{pmatrix} 1 & z & \alpha & \beta \\ 0 & 0 & 1 & z \end{pmatrix},$$

where the matrix $\Sigma = (\alpha, \beta)$ is such that $\Sigma = (1, z)M - a(1, z)$ with $a \in \mathbb{C}$ and $M \in M_{2,2}(\mathbb{Z})$. Let $\epsilon = \pm 1$ and

$$C = \begin{pmatrix} \epsilon & 1 \\ 0 & \epsilon \end{pmatrix}, \qquad Q = \begin{pmatrix} \epsilon & 0 & 1 & 0 \\ 0 & \epsilon & 0 & 1 \\ 0 & 0 & \epsilon & 0 \\ 0 & 0 & 0 & \epsilon \end{pmatrix}.$$

Since $CP = PQ$, the linear transformation induced on $\mathbb{C}^2$ by the matrix C gives an automorphism of G which has infinite order and leaves only one sub-torus invariant. $\square$

References

1. Y. ABE, K. KOPFERMANN, *Toroidal groups. Line bundles, cohomology and quasi-Abelian varieties*, Lecture Notes in Mathematics 1759. Springer (2001).
2. J. M. ANCOCHEA-BERMUDEZ, R. CAMPOAMOR, Completable filiform Lie algebras, Linear algebra and its applications 367, 185-191 (2003).
3. J. M. ANCOCHEA-BERMUDEZ, M. GOZE, Classification des algèbres de Lie filiformes de dimension 8, Arch. Math. 50, 511-525 (1988).
4. S. ARIMA, Commutative group varieties, J. Math. Soc. Japan 12, 227-237 (1960).
5. R. BAER, Erweiterungen von Gruppen und ihren Isomorphismen, Math. Zeit. 38, 375-416 (1934).
6. R. BAER, Primary abelian groups and their automorphisms, Amer. J. Math. 59, 99-116 (1937).
7. C. BIRKENHAKE, H. LANGE, *Complex tori*, Progress in Mathematics 177, Birkhäuser 1999.
8. A. BOREL, *Linear Algebraic Groups*, Springer 1991.
9. N. BOURBAKI, *Éléments de mathématique. Fasc. XXXVII. Groupes et algèbres de Lie. Chapitre II: Algèbres de Lie libres. Chapitre III: Groupes de Lie*, Actualités Scientifiques et Industrielles, No. 1349. Hermann, Paris, 1972.
10. P. J. CAMERON, Permutation groups, p. 611-645 in *Handbook of combinatorics*, Graham, R. L. (ed.) et al., Vol. 1-2, Elsevier, Amsterdam (1995).
11. C. CHEVALLEY, *Theory of Lie groups I*, Princeton University Press 1946.
12. C. CHEVALLEY, *Théories des groupes de Lie II. Groupes Algebriques*, Hermann 1951.
13. C. CHEVALLEY, *Théorèmes généraux sur les algèbres de Lie*, Hermann 1955.
14. M. COWLING, A. H. DOOLEY, A. KORANYI, F. RICCI, H-Type Groups, Adv. Math. 87, 1-41 (1991).
15. D. VAN DANTZIG, Über topologisch homogene Kontinua, Fund. Math. 15, 102-125 (1930).
16. F. DE GIOVANNI, C. MUSELLA, Y. P. SYSAK, Groups with almost modular subgroup lattice, J. Algebra 243, 738-764 (2001).
17. I. DÉCHÈNE, Arithmetic of Generalized Jacobians, Algorithmic Number Theory Symposium - ANTS VII, volume 4076, Lecture Notes in Computer Science, pp. 421-435, Springer 2006.
18. M. DEMAZURE, P. GABRIEL, *Groupes Algébriques*, Masson & Cie 1991.

19. J. DIEUDONNÉ, Lie groups and Lie hyperalgebras over a field of characteristic $p > 0$ (VIII), Amer. J. Math. 80, 740-772 (1958).

20. J. DIXMIER, L'application exponentielle dans les groupes de Lie résolubles, Bull. Soc. Math. Fr. 85, 113-121 (1957).

21. A. ELDUQUE, A note on noncentral simple minimal nonabelian Lie algebras, Commun. Algebra 15, 1313-1318 (1987).

22. G. FALCONE, Frobenius Algebraic Groups, Forum Math. 12, 513-544 (2000).

23. G. FALCONE, M. A. VACCARO, Kronecker modules and reductions of a pair of bilinear forms, Acta Univ. Palacki. Olomuc., Fac. Rerum Nat., Math. 43, 55-60 (2004).

24. G. FALCONE, P. PLAUMANN, K. STRAMBACH, Monothetic Algebraic Groups, J. Austr. Math. Soc. 82, 315-324 (2007).

25. R. FARNSTEINER, On ad-semisimple Lie algebras, J. Algebra 83, 510-519 (1983).

26. A. FAUNTLEROY, Defining normal subgroups of unipotent algebraic groups, Proc. Amer. Math. Soc. 50, 14-20 (1975).

27. H. FREUDENTHAL, DE VRIES, *Linear Lie Groups*, Pure and Applied Mathematics Vol. 35, Academic Press 1969.

28. L. FUCHS, *Infinite Abelian Groups II*, Academic Press 1973.

29. M. A. GAUGER, On the classification of metabelian Lie algebras, Trans. Amer. Math. Soc. 179, 293-328 (1973).

30. A. G. GEJN, Minimal nonabelian Lie algebras of prime characteristic, C. R. Acad. Bulg. Sci. 37, 1291-1293 (1984).

31. A. G. GEJN, Minimal noncommutative and minimal nonabelian algebras, Commun. Algebra 13, 305-328 (1985).

32. A. G. GEJN, S. V. KUZNETSOV, YU. N. MUKHIN, On minimal nonabelian Lie algebras (Russian), Mat. Zap. Sverdl. 8, No.3, 18-27 (1973).

33. J. R. GÓMEZ, A. JIMENEZ-MERCHÁN, Y. KHAKIMJANOV, Low-dimensional filiform Lie algebras, J. Pure Appl. Algebra 130, 133-158 (1998).

34. M. GOZE, J. M. ANCOCHEA, On the classification of rigid Lie algebras, J. Algebra 245, No.1, 68-91 (2001).

35. M. GOZE, Y. KHAKIMJANOV, Sur le algebres de Lie nilpotent admettant un tore de derivations, Manuscr. Math., No.2, 115-124 (1994).

36. M. GOZE, YU. KHAKIMJANOV, *Nilpotent Lie algebras*, Kluwer Academic Publishers (1996).

37. M. GOZE, YU. KHAKIMJANOV, Nilpotent and solvable Lie algebras, in *Handbook of Algebra*, vol. 2 (Ed. M.Hazewinkel), 615-663, Elsevier (2000).

38. G. HAVAS, J. S. RICHARDSON, Groups of exponent five and class four, Comm. Algebra 11, No.3, 287-304 (1983).

39. M. HAZEWINKEL, *Formal groups and applications*, Academic Press 1978.

40. G. HOCHSCHILD, *The structure of Lie groups*, Holden-Day 1965.

41. K. H. HOFMANN, Lie algebras with subalgebras of co-dimension one, Illinois J. Math. 9, 636-643 (1965).

42. K. H. HOFMANN, Über das Nilradikal lokal kompakter Gruppen, Math. Z. 91, 206-215 (1966).

43. K. H. HOFMANN, A. MORRIS, *The structure of compact groups. A primer for the student—a handbook for the expert*, Walter de Gruyter, Berlin 1998.

44. K. H. HOFMANN, P. S. MOSTERT, Splitting in Topological Groups, Mem. Am. Math. Soc. 43 (1963).

45. J. E. HUMPHREYS, *Linear Algebraic Groups*, 2nd ed., Springer 1981.

46. B. HUPPERT, *Endliche Gruppen I*, Springer 1983.

47. K. IWASAWA, Über die endlichen Gruppen und die Verbände ihrer Untergruppen, J. Fac. Sci. Imp. Univ. Tokyo. Sect. I. **4**, 171–199 (1941).

48. K. IWASAWA, On the structure of infinite M-groups, Jap. J. Math. **18**, 709–728 (1943).

49. N. JACOBSON, *Lie Algebras*, Interscience Publishers 1962.

50. N. JACOBSON, *Basic Algebra II*, W. H. Freeman and Company 1980.

51. T. KAMBAYASHI, M. MIYANISHI, M. TAKEUCHI, *Unipotent algebraic groups*, Lecture Notes in Mathematics, Vol. 414. Springer-Verlag, Berlin-New York 1974.

52. F. KÜMMICH, Topologisch quasihamiltonsche Gruppen, Arch. Math 29, 392-397 (1977).

53. F. KÜMMICH, H. SCHEERER, Sottogruppi topologicamente quasi-normali dei gruppi localmente compatti, Rend. Sem. Mat. Univ. Padova 69, 195-210 (1983).

54. T. Y. LAM, *Lectures on Modules and Rings*. Springer 1991.

55. D. H. LEE, *The structure of complex Lie groups*, CRC Research Notes in Mathematics 429, Chapman & Hill 2002.

56. F. LEVSTEIN, A. TIRABOSCHI, Classes of 2-step nilpotent Lie algebras. Comm. Algebra, 27, 2425-2440 (1999).

57. G. A. MILLER, H. C. MORENO, Non-abelian groups in which every subgroup is abelian, Trans. Amer. Math. Soc., 4, 389-404 (1993).

58. D. MONTGOMERY, L. ZIPPIN, *Topological transformation groups* (Interscience Tracts in Pure and Applied Mathematics) Interscience Publishers 1955.

59. A. MORIMOTO, Non-compact Complex Lie groups without Non-costant Holomorphic Functions, Proceedings of the Conference of Complex Analysis Minneapolis, 256-272, Springer 1965.

60. A. MORIMOTO, On the classification of noncompact complex abelian Lie groups, Trans. Amer. Math. Soc., No.123, 200-228 (1966).

61. V. V. MOROZOV, Classification of Nilpotent Lie Algebras up to Dimension 6, Izv. Vyssh. Uchebn. Zaved., Mat. 1958, No.4(5), 161-171 (1958).

62. G. M. MUBARAKZJANOV, On solvable Lie algebra (Russian), Izv. Vyssh. Uchebn. Zaved., Mat. 1963, No.1(32), 114-123 (1963).

63. P. MÜLLER, On simple semiabelian p-adic Lie algebras, Comm. Algebra, 20, 1041-1049 (1992).

64. D. MUMFORD, *Abelian Varieties*, Oxford University Press 1974.

65. J. NEUKIRCH, A. SCHMIDT, K. WINGBERG, *Cohomology of Number Fields*, Springer 2000.

66. E. A. O'BRIEN, M. VAUGHAN-LEE, The 2-generator restricted Burnside group of exponent 7, Internat. J. Algebra Comput. 12, No.4, 575–592 (2002).

67. J. OESTERLÉ, Nombres de Tamagawa et groupes unipotents en caractéristique p, Invent. Math. 78, 13-88 (1984).

68. A. L. ONISHCHIK, E. B. VINBERG, V. V. GORBATSEVICH, *Lie groups and Lie algebras III. Structure of Lie groups and Lie algebras*, Encyclopaedia of Mathematical Sciences 41, Springer 1994.

69. F. OORT, YU. G. ZARHIN, Complex tori. Indag. Math. (N.S.) 7, no. 4, 473–487 (1996).

70. P. PLAUMANN, K. STRAMBACH, G. ZACHER, Der Verband der zusammenhängenden Untergruppen einer kommutativen algebraischen Gruppe, Arch. Math. 85, No.1, 37-48 (2005).

71. P. PLAUMANN, K. STRAMBACH, G. ZACHER, Complements of connected subgroups in algebraic groups, Rend. Sem. Mat. Univ. Padova, 115, 253-264 (2006).
72. G. PRASAD, A. S. RAPINCHUK, Irreducible Tori in Semisimple Groups, Int. Math. Res. Not. 23, 1229-1242 (2001).
73. L. RÉDEI, Das schiefe Produkt in der Gruppentheorie, Comment. Math. Helvet. 20, 225-267 (1947).
74. L. RÉDEI, Algebra, Erster Teil, Akademische Verlaggesellschaft Geest & Portig, Leipzig 1959.
75. D. J. S. ROBINSON, A Course in the Theory of Groups, Springer 1980.
76. M. ROSENLICHT, Generalized Jacobian varieties, Ann. of Math. (2) 59, 505-530 (1954).
77. M. ROSENLICHT, Some basic theorems on algebraic groups, Amer. J. Math. LXXVIII, 2, 401-443 (1956).
78. M. ROSENLICHT, Commutative algebraic group varieties, Princeton Math. Ser. 12, 151-156 (1957).
79. M. ROSENLICHT, Extensions of vector groups by Abelian varieties, Amer. J. Math. LXXX, 3, 685-714 (1958).
80. M. ROSENLICHT, Toroidal algebraic groups, Proc. Amer. Math. Soc. 12, 984-988 (1961).
81. M. ROSENLICHT, The definition of field of definition, Bol. Soc. Mat. Mex., II. Ser. 7, 39-46 (1962).
82. M. ROSENLICHT, Questions of rationality for solvable algebraic groups over nonperfect fields, Annali di Mat. IV, 61, 97-120 (1962).
83. M. ROSENLICHT, Another proof of a theorem on rational cross section, Pac. J. Math., 20 N. 1, 129-133 (1967).
84. H. SALZMANN, Zur Klassifikation topologischer Ebenen III. Abh. Math. Semin. Univ. Hamb. 28, 250-261 (1965).
85. R. SCHARLAU, Paare alternierender Formen, Math. Z. 147, 13-19 (1976).
86. C. SCHEIDERER, Topologisch quasinormale Untergruppen zusammenhängender lokalkompakter Gruppen, Monatsh. Math. 98, 75-81 (1984).
87. C. SCHEIDERER, Quasi-normal subgroups of algebraic groups, Arch. Math. 45, 8-11 (1985).
88. R. SCHMIDT, Subgroup Lattices of Groups, de Gruyter 1994.
89. J. P. SERRE, Groupes algébriques et corps de classes, Actualités scientifiques et industrielles 1264. Publ. de l'Institut de Mathématique de l'Université de Nancago, Hermann 1959.
90. G. SHIMURA, On analytic families of polarized abelian varieties and automorphic functions, Ann. of Math. 78, 149-193 (1963).
91. T. A. SPRINGER, Linear algebraic groups, 2nd ed., Progress in Mathematics 9, Birkhäuser 1998.
92. E. L. STITZINGER, Minimal nonnilpotent solvable Lie algebras, Proc. Am. Math. Soc. 28, 47-49 (1971).
93. S. E. STONEHEWER, Permutable subgroups of infinite groups, Math. Z. 125, 1–16 (1972).
94. J. TITS, Tabellen zu den einfachen Lie Gruppen und ihren Darstellungen, Springer 1967.
95. V. S. VARADAJAN, Lie groups, Lie algebras and their representations, Prentice-Hall 1984.
96. S. WARNER, Topological rings, North-Holland Mathematics Studies 178, North-Holland 1993.

97. B. Wehrfritz, *Infinite Linear Groups. An account of the group-theoretic properties of infinite groups of matrices*, Ergebnisse der Mathematik und ihrer Grenzgebiete Band 76, Springer 1973.

98. A. Weil, On algebraic groups of transformations, Am. J. Math. 77, 355-391 (1955).

99. A. Weil, On algebraic groups and homogeneous spaces, Am. J. Math. 77, 493-512 (1955).

100. A. Weil, *Foundations of algebraic geometry*, AMS Colloquium Publications Volume XXIX, 1962.

101. A. Weil, *Courbes algébriques et variétés abéliennes*, Hermann, Paris 1971.

102. E. Witt, Zyklische Körper und Algebren der Charakteristik p vom Grade p^n. Struktur diskret bewerteter perfekter Körper mit vollkommenem Restklassenkörper der Charakteristik p, J. Reine Angew. Math. 176, 126-140 (1936).

Notation

k	a field	
k_s	separable closure of the field k	
$k^{p^{-\infty}}$	perfect closure of the field k	
G°	connected component of the algebraic or topological group G containing the identity	
$\mathfrak{z}^\circ G$	connected component of the centre of the group G	
$G(\mathsf{k})$	group of the k-rational elements of the algebraic k-group G	
$\mathbf{G}_a$	one-dimensional additive group	
$\mathbf{G}_m$	one-dimensional multiplicative group	
$\mathfrak{W}$	ring of Witt vectors	p. 34
$\mathfrak{W}_n$	n-dimensional Witt group	p. 11
$\mathfrak{I}_n(\alpha)$	n-dimensional unipotent group of maximal nilpotency class	p. 30
$\mathfrak{C}_n(\mathcal{S})$	n-dimensional unipotent chain	p. 52
$M_{n,m}$	set of $n \times m$ matrices	
GL_n	general linear group	
SL_n, PSL_n	special, projective special, linear group	
SO_n, SU_n	special orthogonal, special unitary group	
$\mathbf{F}$	Frobenius homomorphism $x \mapsto x^p$	
$\mathsf{k}[\mathbf{F}]$	ring of p-polynomials	p. 7
$\langle M \rangle_{\mathsf{k}[\mathbf{F}]}$	k[$\mathbf{F}$]-module generated by the set M of polynomials with coefficients in k	

$x^y = y^{-1}xy$	for elements x, y in a group G	
$[x, y] = x^{-1}y^{-1}xy$	the commutator of the elements x, y in a group G	
$[H_1, H_2]$	the subgroup of G generated by the commutators $[h_1, h_2]$, $h_i \in H_i \le G$, $i = 1, 2$	
$G_1 \curlyvee G_2$	direct product with amalgamated central subgroup	p. 11
$G_1 \curlywedge G_2$	direct product with amalgamated factor group	p. 12
δ^1	coboundary operator	p. 13
δ^2	cocycle operator	p. 12
Φ_n	factor system of the extension $\mathfrak{W}_{n+1}$ of $\mathbf{G}_a$ by $\mathfrak{W}_n$	p. 52

Index

Lecture Notes in Mathematics

For information about earlier volumes
please contact your bookseller or Springer
LNM Online archive: springerlink.com

Vol. 1803: G. Dolzmann, Variational Methods for Crystalline Microstructure – Analysis and Computation (2003)

Vol. 1804: I. Cherednik, Ya. Markov, R. Howe, G. Lusztig, Iwahori-Hecke Algebras and their Representation Theory. Martina Franca, Italy 1999. Editors: V. Baldoni, D. Barbasch (2003)

Vol. 1805: F. Cao, Geometric Curve Evolution and Image Processing (2003)

Vol. 1806: H. Broer, I. Hoveijn. G. Lunther, G. Vegter, Bifurcations in Hamiltonian Systems. Computing Singularities by Gröbner Bases (2003)

Vol. 1807: V. D. Milman, G. Schechtman (Eds.), Geometric Aspects of Functional Analysis. Israel Seminar 2000-2002 (2003)

Vol. 1808: W. Schindler, Measures with Symmetry Properties (2003)

Vol. 1809: O. Steinbach, Stability Estimates for Hybrid Coupled Domain Decomposition Methods (2003)

Vol. 1810: J. Wengenroth, Derived Functors in Functional Analysis (2003)

Vol. 1811: J. Stevens, Deformations of Singularities (2003)

Vol. 1812: L. Ambrosio, K. Deckelnick, G. Dziuk, M. Mimura, V. A. Solonnikov, H. M. Soner, Mathematical Aspects of Evolving Interfaces. Madeira, Funchal, Portugal 2000. Editors: P. Colli, J. F. Rodrigues (2003)

Vol. 1813: L. Ambrosio, L. A. Caffarelli, Y. Brenier, G. Buttazzo, C. Villani, Optimal Transportation and its Applications. Martina Franca, Italy 2001. Editors: L. A. Caffarelli, S. Salsa (2003)

Vol. 1814: P. Bank, F. Baudoin, H. Föllmer, L.C.G. Rogers, M. Soner, N. Touzi, Paris-Princeton Lectures on Mathematical Finance 2002 (2003)

Vol. 1815: A. M. Vershik (Ed.), Asymptotic Combinatorics with Applications to Mathematical Physics. St. Petersburg, Russia 2001 (2003)

Vol. 1816: S. Albeverio, W. Schachermayer, M. Talagrand, Lectures on Probability Theory and Statistics. Ecole d'Eté de Probabilités de Saint-Flour XXX-2000. Editor: P. Bernard (2003)

Vol. 1817: E. Koelink, W. Van Assche (Eds.), Orthogonal Polynomials and Special Functions. Leuven 2002 (2003)

Vol. 1818: M. Bildhauer, Convex Variational Problems with Linear, nearly Linear and/or Anisotropic Growth Conditions (2003)

Vol. 1819: D. Masser, Yu. V. Nesterenko, H. P. Schlickewei, W. M. Schmidt, M. Waldschmidt, Diophantine Approximation. Cetraro, Italy 2000. Editors: F. Amoroso, U. Zannier (2003)

Vol. 1820: F. Hiai, H. Kosaki, Means of Hilbert Space Operators (2003)

Vol. 1821: S. Teufel, Adiabatic Perturbation Theory in Quantum Dynamics (2003)

Vol. 1822: S.-N. Chow, R. Conti, R. Johnson, J. Mallet-Paret, R. Nussbaum, Dynamical Systems. Cetraro, Italy 2000. Editors: J. W. Macki, P. Zecca (2003)

Vol. 1823: A. M. Anile, W. Allegretto, C. Ringhofer, Mathematical Problems in Semiconductor Physics. Cetraro, Italy 1998. Editor: A. M. Anile (2003)

Vol. 1824: J. A. Navarro González, J. B. Sancho de Salas, $\mathscr{C}^{\infty}$ – Differentiable Spaces (2003)

Vol. 1825: J. H. Bramble, A. Cohen, W. Dahmen, Multiscale Problems and Methods in Numerical Simulations, Martina Franca, Italy 2001. Editor: C. Canuto (2003)

Vol. 1826: K. Dohmen, Improved Bonferroni Inequalities via Abstract Tubes. Inequalities and Identities of Inclusion-Exclusion Type. VIII, 113 p, 2003.

Vol. 1827: K. M. Pilgrim, Combinations of Complex Dynamical Systems. IX, 118 p, 2003.

Vol. 1828: D. J. Green, Gröbner Bases and the Computation of Group Cohomology. XII, 138 p, 2003.

Vol. 1829: E. Altman, B. Gaujal, A. Hordijk, Discrete-Event Control of Stochastic Networks: Multimodularity and Regularity. XIV, 313 p, 2003.

Vol. 1830: M. I. Gil', Operator Functions and Localization of Spectra. XIV, 256 p, 2003.

Vol. 1831: A. Connes, J. Cuntz, E. Guentner, N. Higson, J. E. Kaminker, Noncommutative Geometry, Martina Franca, Italy 2002. Editors: S. Doplicher, L. Longo (2004)

Vol. 1832: J. Azéma, M. Émery, M. Ledoux, M. Yor (Eds.), Séminaire de Probabilités XXXVII (2003)

Vol. 1833: D.-Q. Jiang, M. Qian, M.-P. Qian, Mathematical Theory of Nonequilibrium Steady States. On the Frontier of Probability and Dynamical Systems. IX, 280 p, 2004.

Vol. 1834: Yo. Yomdin, G. Comte, Tame Geometry with Application in Smooth Analysis. VIII, 186 p, 2004.

Vol. 1835: O.T. Izhboldin, B. Kahn, N.A. Karpenko, A. Vishik, Geometric Methods in the Algebraic Theory of Quadratic Forms. Summer School, Lens, 2000. Editor: J.-P. Tignol (2004)

Vol. 1836: C. Năstăsescu, F. Van Oystaeyen, Methods of Graded Rings. XIII, 304 p, 2004.

Vol. 1837: S. Tavaré, O. Zeitouni, Lectures on Probability Theory and Statistics. Ecole d'Eté de Probabilités de Saint-Flour XXXI-2001. Editor: J. Picard (2004)

Vol. 1838: A.J. Ganesh, N.W. O'Connell, D.J. Wischik, Big Queues. XII, 254 p, 2004.

Vol. 1839: R. Gohm, Noncommutative Stationary Processes. VIII, 170 p, 2004.

Vol. 1840: B. Tsirelson, W. Werner, Lectures on Probability Theory and Statistics. Ecole d'Eté de Probabilités de Saint-Flour XXXII-2002. Editor: J. Picard (2004)

Vol. 1841: W. Reichel, Uniqueness Theorems for Variational Problems by the Method of Transformation Groups (2004)

Vol. 1842: T. Johnsen, A. L. Knutsen, K₃ Projective Models in Scrolls (2004)

Vol. 1843: B. Jefferies, Spectral Properties of Noncommuting Operators (2004)

Vol. 1844: K.F. Siburg, The Principle of Least Action in Geometry and Dynamics (2004)

Vol. 1845: Min Ho Lee, Mixed Automorphic Forms, Torus Bundles, and Jacobi Forms (2004)

Vol. 1846: H. Ammari, H. Kang, Reconstruction of Small Inhomogeneities from Boundary Measurements (2004)

Vol. 1847: T.R. Bielecki, T. Björk, M. Jeanblanc, M. Rutkowski, J.A. Scheinkman, W. Xiong, Paris-Princeton Lectures on Mathematical Finance 2003 (2004)

Vol. 1848: M. Abate, J. E. Fornaess, X. Huang, J. P. Rosay, A. Tumanov, Real Methods in Complex and CR Geometry, Martina Franca, Italy 2002. Editors: D. Zaitsev, G. Zampieri (2004)

Vol. 1849: Martin L. Brown, Heegner Modules and Elliptic Curves (2004)

Vol. 1850: V. D. Milman, G. Schechtman (Eds.), Geometric Aspects of Functional Analysis. Israel Seminar 2002-2003 (2004)

Vol. 1851: O. Catoni, Statistical Learning Theory and Stochastic Optimization (2004)

Vol. 1852: A.S. Kechris, B.D. Miller, Topics in Orbit Equivalence (2004)

Vol. 1853: Ch. Favre, M. Jonsson, The Valuative Tree (2004)

Vol. 1854: O. Saeki, Topology of Singular Fibers of Differential Maps (2004)

Vol. 1855: G. Da Prato, P.C. Kunstmann, I. Lasiecka, A. Lunardi, R. Schnaubelt, L. Weis, Functional Analytic Methods for Evolution Equations. Editors: M. Iannelli, R. Nagel, S. Piazzera (2004)

Vol. 1856: K. Back, T.R. Bielecki, C. Hipp, S. Peng, W. Schachermayer, Stochastic Methods in Finance, Bressanone/Brixen, Italy, 2003. Editors: M. Fritelli, W. Runggaldier (2004)

Vol. 1857: M. Émery, M. Ledoux, M. Yor (Eds.), Séminaire de Probabilités XXXVIII (2005)

Vol. 1858: A.S. Cherny, H.-J. Engelbert, Singular Stochastic Differential Equations (2005)

Vol. 1859: E. Letellier, Fourier Transforms of Invariant Functions on Finite Reductive Lie Algebras (2005)

Vol. 1860: A. Borisyuk, G.B. Ermentrout, A. Friedman, D. Terman, Tutorials in Mathematical Biosciences I. Mathematical Neurosciences (2005)

Vol. 1861: G. Benettin, J. Henrard, S. Kuksin, Hamiltonian Dynamics – Theory and Applications, Cetraro, Italy, 1999. Editor: A. Giorgilli (2005)

Vol. 1862: B. Helffer, F. Nier, Hypoelliptic Estimates and Spectral Theory for Fokker-Planck Operators and Witten Laplacians (2005)

Vol. 1863: H. Führ, Abstract Harmonic Analysis of Continuous Wavelet Transforms (2005)

Vol. 1864: K. Efstathiou, Metamorphoses of Hamiltonian Systems with Symmetries (2005)

Vol. 1865: D. Applebaum, B.V. R. Bhat, J. Kustermans, J. M. Lindsay, Quantum Independent Increment Processes I. From Classical Probability to Quantum Stochastic Calculus. Editors: M. Schürmann, U. Franz (2005)

Vol. 1866: O.E. Barndorff-Nielsen, U. Franz, R. Gohm, B. Kümmerer, S. Thorbjønsen, Quantum Independent Increment Processes II. Structure of Quantum Lévy Processes, Classical Probability, and Physics. Editors: M. Schürmann, U. Franz, (2005)

Vol. 1867: J. Sneyd (Ed.), Tutorials in Mathematical Biosciences II. Mathematical Modeling of Calcium Dynamics and Signal Transduction. (2005)

Vol. 1868: J. Jorgenson, S. Lang, Pos$_n$(R) and Eisenstein Series. (2005)

Vol. 1869: A. Dembo, T. Funaki, Lectures on Probability Theory and Statistics. Ecole d'Eté de Probabilités de Saint-Flour XXXIII-2003. Editor: J. Picard (2005)

Vol. 1870: V.I. Gurariy, W. Lusky, Geometry of Müntz Spaces and Related Questions. (2005)

Vol. 1871: P. Constantin, G. Gallavotti, A.V. Kazhikhov, Y. Meyer, S. Ukai, Mathematical Foundation of Turbulent Viscous Flows, Martina Franca, Italy, 2003. Editors: M. Cannone, T. Miyakawa (2006)

Vol. 1872: A. Friedman (Ed.), Tutorials in Mathematical Biosciences III. Cell Cycle, Proliferation, and Cancer (2006)

Vol. 1873: R. Mansuy, M. Yor, Random Times and Enlargements of Filtrations in a Brownian Setting (2006)

Vol. 1874: M. Yor, M. Émery (Eds.), In Memoriam Paul-André Meyer - Séminaire de Probabilités XXXIX (2006)

Vol. 1875: J. Pitman, Combinatorial Stochastic Processes. Ecole d'Eté de Probabilités de Saint-Flour XXXII-2002. Editor: J. Picard (2006)

Vol. 1876: H. Herrlich, Axiom of Choice (2006)

Vol. 1877: J. Steuding, Value Distributions of L-Functions (2007)

Vol. 1878: R. Cerf, The Wulff Crystal in Ising and Percolation Models, Ecole d'Eté de Probabilités de Saint-Flour XXXIV-2004. Editor: Jean Picard (2006)

Vol. 1879: G. Slade, The Lace Expansion and its Applications, Ecole d'Eté de Probabilités de Saint-Flour XXXIV-2004. Editor: Jean Picard (2006)

Vol. 1880: S. Attal, A. Joye, C.-A. Pillet, Open Quantum Systems I, The Hamiltonian Approach (2006)

Vol. 1881: S. Attal, A. Joye, C.-A. Pillet, Open Quantum Systems II, The Markovian Approach (2006)

Vol. 1882: S. Attal, A. Joye, C.-A. Pillet, Open Quantum Systems III, Recent Developments (2006)

Vol. 1883: W. Van Assche, F. Marcellàn (Eds.), Orthogonal Polynomials and Special Functions, Computation and Application (2006)

Vol. 1884: N. Hayashi, E.I. Kaikina, P.I. Naumkin, I.A. Shishmarev, Asymptotics for Dissipative Nonlinear Equations (2006)

Vol. 1885: A. Telcs, The Art of Random Walks (2006)

Vol. 1886: S. Takamura, Splitting Deformations of Degenerations of Complex Curves (2006)

Vol. 1887: K. Habermann, L. Habermann, Introduction to Symplectic Dirac Operators (2006)

Vol. 1888: J. van der Hoeven, Transseries and Real Differential Algebra (2006)

Vol. 1889: G. Osipenko, Dynamical Systems, Graphs, and Algorithms (2006)

Vol. 1890: M. Bunge, J. Funk, Singular Coverings of Toposes (2006)

Vol. 1891: J.B. Friedlander, D.R. Heath-Brown, H. Iwaniec, J. Kaczorowski, Analytic Number Theory, Cetraro, Italy, 2002. Editors: A. Perelli, C. Viola (2006)

Vol. 1892: A. Baddeley, I. Bárány, R. Schneider, W. Weil, Stochastic Geometry, Martina Franca, Italy, 2004. Editor: W. Weil (2007)

Vol. 1893: H. Hanßmann, Local and Semi-Local Bifurcations in Hamiltonian Dynamical Systems, Results and Examples (2007)

Vol. 1894: C.W. Groetsch, Stable Approximate Evaluation of Unbounded Operators (2007)

Vol. 1895: L. Molnár, Selected Preserver Problems on Algebraic Structures of Linear Operators and on Function Spaces (2007)

Vol. 1896: P. Massart, Concentration Inequalities and Model Selection, Ecole d'Été de Probabilités de Saint-Flour XXXIII-2003. Editor: J. Picard (2007)

Vol. 1897: R. Doney, Fluctuation Theory for Lévy Processes, Ecole d'Été de Probabilités de Saint-Flour XXXV-2005. Editor: J. Picard (2007)

Vol. 1898: H.R. Beyer, Beyond Partial Differential Equations, On linear and Quasi-Linear Abstract Hyperbolic Evolution Equations (2007)

Vol. 1899: Séminaire de Probabilités XL. Editors: C. Donati-Martin, M. Émery, A. Rouault, C. Stricker (2007)

Vol. 1900: E. Bolthausen, A. Bovier (Eds.), Spin Glasses (2007)

Vol. 1901: O. Wittenberg, Intersections de deux quadriques et pinceaux de courbes de genre 1, Intersections of Two Quadrics and Pencils of Curves of Genus 1 (2007)

Vol. 1902: A. Isaev, Lectures on the Automorphism Groups of Kobayashi-Hyperbolic Manifolds (2007)

Vol. 1903: G. Kresin, V. Maz'ya, Sharp Real-Part Theorems (2007)

Vol. 1904: P. Giesl, Construction of Global Lyapunov Functions Using Radial Basis Functions (2007)

Recent Reprints and New Editions